Generation of cDNA Libraries

METHODS IN MOLECULAR BIOLOGY™

John M. Walker, SERIES EDITOR

236. **Plant Functional Genomics**: *Methods and Protocols*, edited by *Erich Grotewold, 2003*

235. ***E. coli* Plasmid Vectors**: *Methods and Applications*, edited by *Nicola Casali and Andrew Preston, 2003*

234. **p53 Protocols**, edited by *Sumitra Deb and Swati Palit Deb, 2003*

233. **Protein Kinase C Protocols**, edited by *Alexandra C. Newton, 2003*

232. **Protein Misfolding and Disease**: *Principles and Protocols*, edited by *Peter Bross and Niels Gregersen, 2003*

231. **Directed Evolution Library Creation**: *Methods and Protocols*, edited by *Frances H. Arnold and George Georgiou, 2003*

230. **Directed Enzyme Evolution**: *Screening and Selection Methods*, edited by *Frances H. Arnold and George Georgiou, 2003*

229. **Lentivirus Gene Engineering Protocols**, edited by *Maurizio Federico, 2003*

228. **Membrane Protein Protocols**: *Expression, Purification, and Characterization*, edited by *Barry S. Selinsky, 2003*

227. **Membrane Transporters**: *Methods and Protocols*, edited by *Qing Yan, 2003*

226. **PCR Protocols, Second Edition,** edited by *John M. S. Bartlett and David Stirling, 2003*

225. **Inflammation Protocols,** edited by *Paul G. Winyard and Derek A. Willoughby, 2003*

224. **Functional Genomics**: *Methods and Protocols,* edited by *Michael J. Brownstein and Arkady B. Khodursky, 2003*

223. **Tumor Suppressor Genes**: *Volume 2: Regulation, Function, and Medicinal Applications,* edited by *Wafik S. El-Deiry, 2003*

222. **Tumor Suppressor Genes**: *Volume 1: Pathways and Isolation Strategies,* edited by *Wafik S. El-Deiry, 2003*

221. **Generation of cDNA Libraries**: *Methods and Protocols,* edited by *Shao-Yao Ying, 2003*

220. **Cancer Cytogenetics**: *Methods and Protocols,* edited by *John Swansbury, 2003*

219. **Cardiac Cell and Gene Transfer**: *Principles, Protocols, and Applications,* edited by *Joseph M. Metzger, 2003*

218. **Cancer Cell Signaling**: *Methods and Protocols,* edited by *David M. Terrian, 2003*

217. **Neurogenetics**: *Methods and Protocols,* edited by *Nicholas T. Potter, 2003*

216. **PCR Detection of Microbial Pathogens**: *Methods and Protocols,* edited by *Konrad Sachse and Joachim Frey, 2003*

215. **Cytokines and Colony Stimulating Factors**: *Methods and Protocols,* edited by *Dieter Körholz and Wieland Kiess, 2003*

214. **Superantigen Protocols,** edited by *Teresa Krakauer, 2003*

213. **Capillary Electrophoresis of Carbohydrates,** edited by *Pierre Thibault and Susumu Honda, 2003*

212. **Single Nucleotide Polymorphisms**: *Methods and Protocols,* edited by *Pui-Yan Kwok, 2003*

211. **Protein Sequencing Protocols, Second Edition,** edited by *Bryan John Smith, 2003*

210. **MHC Protocols,** edited by *Stephen H. Powis and Robert W. Vaughan, 2003*

209. **Transgenic Mouse Methods and Protocols,** edited by *Marten Hofker and Jan van Deursen, 2003*

208. **Peptide Nucleic Acids**: *Methods and Protocols,* edited by *Peter E. Nielsen, 2002*

207. **Recombinant Antibodies for Cancer Therapy**: *Methods and Protocols,* edited by *Martin Welschof and Jürgen Krauss, 2002*

206. **Endothelin Protocols,** edited by *Janet J. Maguire and Anthony P. Davenport, 2002*

205. ***E. coli* Gene Expression Protocols,** edited by *Peter E. Vaillancourt, 2002*

204. **Molecular Cytogenetics**: *Protocols and Applications,* edited by *Yao-Shan Fan, 2002*

203. **In Situ Detection of DNA Damage**: *Methods and Protocols,* edited by *Vladimir V. Didenko, 2002*

202. **Thyroid Hormone Receptors**: *Methods and Protocols,* edited by *Aria Baniahmad, 2002*

201. **Combinatorial Library Methods and Protocols,** edited by *Lisa B. English, 2002*

200. **DNA Methylation Protocols,** edited by *Ken I. Mills and Bernie H, Ramsahoye, 2002*

199. **Liposome Methods and Protocols,** edited by *Subhash C. Basu and Manju Basu, 2002*

198. **Neural Stem Cells**: *Methods and Protocols,* edited by *Tanja Zigova, Juan R. Sanchez-Ramos, and Paul R. Sanberg, 2002*

197. **Mitochondrial DNA**: *Methods and Protocols,* edited by *William C. Copeland, 2002*

196. **Oxidants and Antioxidants**: *Ultrastructure and Molecular Biology Protocols,* edited by *Donald Armstrong, 2002*

195. **Quantitative Trait Loci**: *Methods and Protocols,* edited by *Nicola J. Camp and Angela Cox, 2002*

194. **Posttranslational Modifications of Proteins**: *Tools for Functional Proteomics,* edited by *Christoph Kannicht, 2002*

193. **RT-PCR Protocols,** edited by *Joe O'Connell, 2002*

192. **PCR Cloning Protocols, Second Edition,** edited by *Bing-Yuan Chen and Harry W. Janes, 2002*

191. **Telomeres and Telomerase**: *Methods and Protocols,* edited by *John A. Double and Michael J. Thompson, 2002*

190. **High Throughput Screening**: *Methods and Protocols,* edited by *William P. Janzen, 2002*

189. **GTPase Protocols**: *The RAS Superfamily,* edited by *Edward J. Manser and Thomas Leung, 2002*

188. **Epithelial Cell Culture Protocols,** edited by *Clare Wise, 2002*

187. **PCR Mutation Detection Protocols,** edited by *Bimal D. M. Theophilus and Ralph Rapley, 2002*

186. **Oxidative Stress Biomarkers and Antioxidant Protocols,** edited by *Donald Armstrong, 2002*

185. **Embryonic Stem Cells**: *Methods and Protocols,* edited by *Kursad Turksen, 2002*

184. **Biostatistical Methods,** edited by *Stephen W. Looney, 2002*

183. **Green Fluorescent Protein**: *Applications and Protocols,* edited by *Barry W. Hicks, 2002*

182. **In Vitro Mutagenesis Protocols, Second Edition,** edited by *Jeff Braman, 2002*

181. **Genomic Imprinting**: *Methods and Protocols,* edited by *Andrew Ward, 2002*

180. **Transgenesis Techniques, Second Edition**: *Principles and Protocols,* edited by *Alan R. Clarke, 2002*

METHODS IN MOLECULAR BIOLOGY™

Generation of cDNA Libraries

Methods and Protocols

Edited by

Shao-Yao Ying

Keck School of Medicine, University of Southern California, Los Angeles, CA

Humana Press ✳ Totowa, New Jersey

Production Editor: Jessica Jannicelli.

Cover design by Patricia F. Cleary.

Cover Illustration: Figure 3 from Chapter 10, "Amplification of Representative cDNA Pools from Microscopic Amounts of Animal Tissue," by M. V. Matz.

For additional copies, pricing for bulk purchases, and/or information about other Humana titles, contact Humana at the above address or at any of the following numbers: Tel.: 973-256-1699; Fax: 973-256-8341; E-mail: humana@humanapr.com; or visit our Website: www.humanapress.com

Printed in the United States of America. 10 9 8 7 6 5 4 3 2 1

Library of Congress Cataloging in Publication Data

Generation of cDNA libraries : methods and protocols / edited by Shao-Yao Ying.
    p. cm. -- (Methods in molecular biology ; 221)
  Includes bibliographical references and index.
  ISBN 1-58829-066-2 (alk. paper) eISBN: 1-59259-359-3
  1. Antisense DNA--Laboratory manuals. I. Series.

  QP624.5.A57G46 2003
  611'.0186--dc21

                                                                2002192170

Preface

Since its invention and subsequent development nearly 20 years ago, polymerase chain reaction (PCR) has been extensively utilized to identify numerous gene probes in vitro and in vivo. However, attempts to generate complete and full-length complementary cDNA libraries were, for the most part, fruitless and remained elusive until the last decade, when simple and rapid methods were developed. With current decoding and potential application of human genome information to genechips, there are urgent needs for identification of functional significance of these decoded gene sequences. Inherent in bringing these applications to fruition is the need to generate a complete and full-length cDNA library for potential functional assays of specific gene sequences.

Generation of cDNA Libraries: Methods and Protocols serves as a laboratory manual on the evolution of generation of cDNA libraries, covering both background information and step-by-step practical laboratory recipes for which protocols, reagents, operational tips, instrumentation, and other requirements are detailed. The first chapter of the book is an overview of the basics of generating cDNA libraries, which include the following: (a) the definition of a cDNA library, (b) different kinds of cDNA libraries, (c) differences between methods for cDNA library generation using conventional approaches and novel strategies, including reverse generation of RNA repertoires from cDNA libraries, and (d) the quality of cDNA libraries. In subsequent chapters, various methods are presented to provide the reader with a wide range of methodologies for enhancing the generation of complete and full-length libraries. Again, each method of cDNA library generation contains a balanced presentation of both background information and practical procedures. The remainder of this book explains how to confirm the quality of the cDNAs generated and some of the applications, including (a) electrophoresis, (b) Northern blotting, (c) microarray analysis, (d) subtractive hybridization, (e) subtractive cloning, (f) gene cloning, and (g) peptide library generation.

The final chapter of the book outlines the future use of full-length cDNA libraries in biomedical research, diagnostic utilization, drug development, and clinical therapy.

The authors contributing the various chapters are all experts in their fields, and they have either developed and/or routinely performed the methodologies described herein. It is anticipated that the subject matter covered in *Generation of cDNA Libraries: Methods and Protocols* will be particularly useful for biologists, biochemists, molecular biologists, and clinicians, which would furnish them several ready-to-use methodologies for attacking the problems in their specific areas of interests.

Shao-Yao Ying

Contents

Preface v

Contributors xi

1 Complementary DNA Libraries: *An Overview*
   Shao-Yao Ying *1*

2 Rapid Amplification of cDNA Ends
   Yue Zhang *13*

3 cDNA Generation on Paramagnetic Beads
   Zhaohui Wang and Michael G. K. Jones *25*

4 Construction of a Normalized cDNA Library by mRNA–cDNA
   Hybridization and Subtraction
   Ye-Guang Chen *33*

5 Amplification of cDNA Ends Using PCR Suppression Effect
   and Step-Out PCR
   Mikhail V. Matz, Naila O. Alieva, Alex Chenchik,
   and Sergey Lukyanov *41*

6 Use of Inverse PCR to Clone cDNA Ends
   Sheng-He Huang, Steven H. M. Chen,
   and Ambrose Y. Jong *51*

7 Construction of Size-Fractionated cDNA Library Assisted by
   an In Vitro Recombination Reaction
   Osamu Ohara *59*

8 Construction of a Full-Length Enriched and 5'-End Enriched cDNA
   Library Using the Oligo-Capping Method
   Yutaka Suzuki and Sumio Sugano *73*

9 cDNA Library Construction Using In Vitro Transcriptional
   Amplification
   Shi-Lung Lin and Henry Ji *93*

10 Amplification of Representative cDNA Pools from Microscopic
   Amounts of Animal Tissue
   Mikhail V. Matz *103*

vii

11 Single-Cell cDNA Library Construction Using Cycling aRNA
      Amplification
      Shi-Lung Lin *117*

12 mRNA/cDNA Library Construction Using RNA–Polymerase Cycling
      Reaction
      Shi-Lung Lin and Shao-Yao Ying *129*

13 Quality Assessment of cDNA Libraries
      Hans-Jürgen Fülle *145*

14 Assessment of the Quality of mRNA Libraries by Agarose Gel
      Electrophoresis
      Tsen-Yin Lin and Shao-Yao Ying *155*

15 PACS RT-PCR: *A Method for the Generation and Measurement
      of any Poly(A)-Containing mRNA Not Affected by Contaminating
      Genomic DNA*
      Igor Nepluev and Rodney J. Folz *161*

16 Single-Cell mRNA Library Analysis by Northern Blot Hybridization
      Shi-Lung Lin *169*

17 Generation of cDNA Libraries for Profiling Gene Expression
      of Given Tissues or Cells
      Xin Zhang, Qiu-Hua Huang, and Ze-Guang Han *179*

18 Screening Poly [dA/dT(−)] cDNA for Gene Identification
      **San Ming Wang, Scott C. Fears, Lin Zhang, Jian-Jun Chen,
      and Janet D. Rowley** *197*

19 Generation of Longer cDNA Fragments from SAGE Tags for Gene
      Identification
      **Jian-Jun Chen, Sanggyu Lee, Guolin Zhou, Janet D. Rowley,
      and San Ming Wang** *207*

20 Generation of Full-Length cDNA Libraries Enriched for Differentially
      Expressed Genes
      **Bakhyt Zhumabayeva, Cynthia Chang, Joseph McKinley,
      Luda Diatchenko, and Paul D. Siebert** *223*

21 Subtractive Hybridization for the Identification of Differentially
      Expressed Genes Using Uracil–DNA Glycosylase
      and Mung-Bean Nuclease
      Tsen-Yin Lin and Shao-Yao Ying *239*

22 Subtractive Cloning of Differential Genes Using RNA-PCR
      Shao-Yao Ying and Shi-Lung Lin *253*

23 Strategy for Construction of a cDNA Encoding a Repetitive Amino
    Acid Sequence
    Masahiro Asada and Toru Imamura *261*
24 Preparing Lambda Libraries for Expression of Proteins
    in Prokaryotes or Eukaryotes
    Rebecca L. Mullinax and Joseph A. Sorge *271*
25 Peptide Library Construction from RNA-PCR-Derived RNAs
    Shi-Lung Lin *289*
26 Identifying Interacting Proteins in an *Escherichia coli*-Based
    Two-Hybrid System
    Bonnie Wu, Rebecca L. Mullinax, and Joseph A. Sorge *295*
27 Future Perspectives
    Shao-Yao Ying *311*
Index *331*

Contributors

NAILA O. ALIEVA • *Whitney Laboratory, University of Florida, St. Augustine, FL*

MASAHIRO ASADA • *Age Dimension Research Center, National Institute of Advanced Industrial Science and Technology, Ibaraki, Japan*

CYNTHIA CHANG • *BD Biosciences Clontech, Palo Alto, CA*

JIAN-JUN CHEN • *Section of Hematology and Oncology, University of Chicago Medical Center, Chicago, IL*

STEVEN H. M. CHEN • *Division of Hematology, Department of Pediatrics, University of Southern California, Childrens Hospital Los Angeles, Los Angeles, CA*

YE-GUANG CHEN • *Division of Biomedical Sciences, University of California, Riverside, CA*

ALEX CHENCHIK • *BD Biosciences Clontech, Palo Alto, CA*

LUDA DIATCHENKO • *BD Biosciences Clontech, Palo Alto, CA*

SCOTT C. FEARS • *Section of Hematology and Oncology, University of Chicago Medical Center, Chicago, IL*

RODNEY J. FOLZ • *Department of Medicine and Cell Biology, Division of Pulmonary and Critical Care Medicine, Duke University Medical Center, Durham, NC*

HANS-JÜRGEN FÜLLE • *Department of Cell and Neurobiology, Keck School of Medicine, University of Southern California, Los Angeles, CA*

ZE-GUANG HAN • *Chinese National Human Genome Center at Shanghai, Shanghai, China*

QIU-HUA HUANG • *Chinese National Human Genome Center at Shanghai, Shanghai, China*

SHENG-HE HUANG • *Division of Infectious Diseases, Department of Pediatrics, University of Southern California, Childrens Hospital Los Angeles, Los Angeles, CA*

TORU IMAMURA • *Age Dimension Research Center, National Institute of Advanced Industrial Science and Technology, Ibaraki, Japan*

HENRY JI • *Epiclone Inc., San Diego, CA*

MICHAEL G. K. JONES • *Western Australia State Agricultural Biotechnology Center, Murdoch University, Perth, WA, Australia*

AMBROSE Y. JONG • *Division of Hematology, Department of Pediatrics, University of Southern California, Childrens Hospital Los Angeles, Los Angeles, CA*

SANGGYU LEE • *Section of Hematology and Oncology, University of Chicago Medical Center, Chicago, IL*

SHI-LUNG LIN • *Epiclone Inc., San Diego, CA*

TSEN-YIN LIN • *Department of Cell and Neurobiology, Keck School of Medicine, University of Southern California, Los Angeles, CA*

SERGEY LUKYANOV • *Evrogen GSC, Shemyakin-Ovchinnikov Institute of Bioorganic Chemistry, Russian Academy of Sciences, Moscow, Russia*

MIKHAIL V. MATZ • *Whitney Laboratory, University of Florida, St. Augustine, FL*

JOSEPH MCKINLEY • *BD Biosciences Clontech, Palo Alto, CA*

REBECCA L. MULLINAX • *Stratagene, La Jolla, CA*

IGOR NEPLUEV • *Duke University Medical Center, Durham, NC*

OSAMU OHARA • *Department of Human Gene Research, Kazusa DNA Research Institute, Chiba, Japan; Immunogenomics Research Team, Research Center for Allergy and Immunology, The Institute of Physical and Chemical Research (RIKEN), Yokohama, Japan*

JANET D. ROWLEY • *Section of Hematology and Oncology, University of Chicago Medical Center, Chicago, IL*

PAUL D. SIEBERT • *BD Biosciences Clontech, Palo Alto, CA*

JOSEPH A. SORGE • *Stratagene, La Jolla, CA*

SUMIO SUGANO • *Laboratory of Genome Structure Analysis, HGC, Institute of Medical Science, University of Tokyo, Tokyo, Japan*

YUTAKA SUZUKI • *Laboratory of Genome Structure Analysis, HGC, Institute of Medical Science, University of Tokyo, Tokyo, Japan*

SAN MING WANG • *Section of Hematology and Oncology, University of Chicago Medical Center, Chicago, IL*

ZHAOHUI WANG • *Western Australia State Agricultural Biotechnology Center, Murdoch University, Perth, WA, Australia*

BONNIE WU • *Stratagene, La Jolla, CA*

SHAO-YAO YING • *Department of Cell and Neurobiology, Keck School of Medicine, University of Southern California, Los Angeles, CA*

LIN ZHANG • *Oncology Center, Johns Hopkins University School of Medicine, Baltimore, MD*

XIN ZHANG • *Chinese National Human Genome Center at Shanghai, Shanghai, China*

YUE ZHANG • *Department of Molecular Genetics and Microbiology, Center for Infectious Diseases/CMM, State University of New York at Stony Brook, Stony Brook, NY*

GUOLIN ZHOU • *Section of Hematology and Oncology, University of Chicago Medical Center, Chicago, IL*

BAKHYT ZHUMABAYEVA • *BD Biosciences Clontech, Palo Alto, CA*

1

Complementary DNA Libraries

An Overview

Shao-Yao Ying

1. Introduction

Complementary DNA libraries reflect gene expression at certain times for specific cells, whereas genomic DNA libraries represent all genetic information in somatic cells. The complexity of cellular organization reflects a genetic program that encodes a collection of genes and the means to use them by manufacturing proteins for cellular structures, functional activities, and reproduction of cells themselves. The essential aspect of this process is protein synthesis based on the information stored in the sequence of nucleotides that make up a gene (a transcribable segment of a DNA molecule) as the blueprint. The information is transcribed as a complementary sequence of the nucleotides (mRNA or the transcript) that carries the genetic information from the nucleus to the protein-synthesizing machinery in the cytoplasm. Then, mRNA is translated into the sequence of amino acids that make up a protein. The basis of the widely used novel strategies for the generation of cDNA libraries are base pair complementarities, reverse transcription, and polymerase chain reactions. This chapter presents some general information on the principles of, biology behind, basic protocols of, and reagents used in the generation of cDNA libraries. Hopefully, this information will help researchers overcome problems encountered in actual construction of cDNA libraries.

1.1. Base Pair Complementarities

Nucleic acids exhibit base pair complementarities that faithfully convert one strand of RNA/DNA to a complementary one. Although all genetic information

From: *Methods in Molecular Biology, vol. 221: Generation of cDNA Libraries: Methods and Protocols*
Edited by: S.-Y. Ying © Humana Press Inc., Totowa, NJ

in the somatic cells of a specific organism can be expressed as a transcript, many DNA sequences are not transcribed. These segments of DNA are the coding exons and the noncoding introns. Basically, the genetic information is stored as a strand of a DNA molecule consisting of four bases: adenine, thymine, guanine, and cytosine. A second complementary strand of DNA can be formed by DNA polymerase. Polymerases, enzymes that function in DNA replication and RNA transcription, synthesize a nucleic acid from the genetic information encoded by the template strand. The polymerases are unique because they take direction from another nucleic acid template, which is either DNA or RNA. During the formation of a second strand of DNA, bases are generated according to the Watson–Crick base-pairing pattern. That is to say, every cytosine is replaced by a guanine, every guanine by a cytosine, every adenine by a thymine, and every thymine by an adenine. In this way, information in DNA is correctly transcribed into RNA.

1.2. Probe Hybridization

Another unique feature of the base pair complementarity is probe hybridization. The findings of Gillespie and Spiegelman *(1)* that viral genomic DNA and RNA in infected cells showed a base pair complementarity opened an avenue for specific hybridization between a gene and its transcript as a DNA–RNA hybrid. Subsequently, the DNA–DNA or DNA–RNA hybrids have been employed in a large number of powerful techniques for the identification and manipulation of the geneitic information stored in DNA and used by the cell via RNA. Usually, a labeled-probe nucleic acid is hybridized with a target nucleic acid. After removal of any unreacted probe, the remaining labeled probe is identified and the intensity of the labeling of the hybrid duplex is determined. As a result, the regions of complementarity between the probe and the target nucleotides are detected *(2)*. Frequently, the number of targets is quite low, perhaps only a few copies. In such cases, amplification techniques are performed to produce large numbers of copies of the target, thus increasing the amount of hybrid duplex and the observed signal. In addition, immobilization of the target on a surface, such as a nitrocellulose or nylon filter and many other solid-phase materials, is used to solve the competitive equilibrium problem. Thus, nucleic acid sequences can be quantified by molecular hybridization using complementary nucleic acids as probes, with complementarity as the essential feature for hybridization.

1.3. Polymerases Are Essential for DNA Synthesis

Polymerases that use RNA as a template to form a complementary DNA are RNA-direct DNA polymerases *(3,4)*. One of these enzymes is reverse transcriptase, usually observed as a part of the viral particle, during the life cycle of retroviruses and other retrotransposable elements. Purified reverse transcriptase is used to generate complementary DNA from polyadenylated mRNAs; therefore, double-stranded DNA molecules can be formed from the single-stranded RNA templates. The synthesis of DNA on an RNA template mediated by the enzyme reverse transcriptase is known as reverse transcription *(5)*.

1.4. A Primer is Required for Reverse Transcription

Although polymerases copy genetic information from one nucleotide into another, including copying a mRNA to generate a complementary DNA strand in the presence of reverse transcriptase, they do need a "start signal" to tell them where to begin making the complementary copy. The short piece of DNA that is annealed to the template and serves as a signal to initiate the copying process is the primer *(6)*. The primer is annealed to the template by basepairing so that its 3′-terminus possesses a free 3′-OH group and chain growth is exclusively from 5′ end to the 3′ end for polymerization. Wherever such as primer–template pair is found, DNA polymerase will begin adding bases to the primer to create a complementary copy of the template.

1.5. Formation of cDNA

Generally, the cDNA of cells can be formed according to the following steps:

1. Isolation of the mRNA template: The source mRNAs can be enriched by increasing the abundance of specific classes of rare mRNAs via one of the following approaches: (1) antibody precipitation of the protein of interest that is synthesized in cell lines, (2) increasing the concentrations of relevant RNAs by drug-induced overexpression of genes of interest, and (3) inhibition of protein synthesis by inhibitors, resulting in extended transcription of the early genes of mammalian DNA virus.

   The integrity of the mRNA is essential for the quality of cDNA generation. The size of mRNAs isolated should range from 500 bp to 8.0 kb, and the sequence should retain the capability of synthesizing the polypeptide of interest in vitro, such as in cell-free reticulocytes. When fractionated by electrophoresis and stained with ethidium bromide, a good preparation of mRNA should appear as a smear from 500 bp to 8 kb.

2. A short oligo(dT) primer is bound to the poly(A) of each mRNA at the 3′ end.

3. The mRNA is transcribed by reverse transcriptase (the primer is needed to initiate DNA synthesis) to form the first strand of DNA, usually in the presence of a reagent to denature any regions of the secondary structure. RNAse is used to prevent RNA degradation.

4. DNA–RNA hybids are formed.

5. The RNA is nicked by treatment with RNAse H to generate the free 3′-OH groups.

6. DNA polymerase I is added to digest the RNA, using the RNA fragments as primers, and replace the RNA with DNA. In some cases, a primer–adapter method is carried out as follows: (1) terminal transferase is added to the first strand cDNA [add (dC) to provide free 3′ hydroxyl groups]; (2) the tail of hybrided cDNA with oligo(dG) serves as the primer.
7. Double-stranded cDNA is formed.

1.6. PCR

Another important development in generating DNA from mRNA is the enzymatic amplification of DNA by a technique known as polymerase chain reaction (PCR). The technique was originally reported by Saike et al. *(7)*, who employed a heat-stable DNA polymerase–*Taq* polymerase with two primers that are complementary to DNA sequences at the 3′ ends of the region of the DNA to be amplified. The oligonucleotides serve as primers to which nucleotides are added during the subsequent replication steps. Because a DNA strand can only add nucleotides at the 3′ hydroxyl terminus of an existing strand, a strand of DNA that provides the necessary 3′-OH terminus, in this case, is also called a primer. All DNA polymerases require a template and a primer.

The PCR is well established as the default method for DNA and RNA analysis. More robust formats have been introduced, improved thermal cyclers developed, and new labeling and detection methods developed. Because gene expression profiling relies on mRNA extraction from defined types and numbers of cells, in some cases the use of small number of cells or even a few cells is necessary. In this situation, the PCR technique has been used to allow synthesis of cDNAs from a small amount of mRNA *(8,9)*. Other techniques of amplifying mRNA have been developed *(10)*. For instance, the cDNA can be generated by mRNA extracted and amplified by poly(A) reverse transverse transcription and PCR.

2. Definitions
2.1. Complementary DNA

If a chromosome is defined as a supercoiled, linear DNA molecule consisting of numerous transcribable segments as genes (specific segments of DNA that code for a specific protein), the complementary DNA (cDNA) can be defined as the transcriptionally active segment of a DNA molecule that shows the base pair complementarity between the gene and its transcribed and processed mRNA molecules—the transcript. To define it differently, cDNAs are complementary DNA copies of mRNA that are generated by the enzyme—reverse transcriptase. In contrast to genomic DNA, the extra, nontranscribed DNAs in a genome are removed by this process because DNA polymerase activity depends on the presence of an RNA template. As a result, the cDNA represents only the 3% of the genomic DNA in human cells that are transcriptionally active genes. Consequently, the generation of cDNA is a powerful tool for examining cell- and tissue-specific gene expression. Not only are cDNAs the expressed genes of a cell at a specific time with a specific function, they are also the faithful and stable double-strand DNA copies of transcribable portions of mRNA. This occurs because they are prepared from a population of RNA in which any intervening sequences (i.e., introns) have been previously removed. Therefore, cDNAs commonly contain an uninterrupted sequence encoding the gene product. For this reason, cDNA reflects both expressible RNA and gene products (polypeptide or proteins).

2.2. Complementary DNA Libraries

A molecular library is defined as a collection of various molecules that can be screened for individual species that show specific properties. Different libraries are developed for different purposes. For example, genomic libraries (raw DNA sequences harvested from an organism's chromosomes) represent the entire genomic DNA sequence of an organism. This type of library is typically not expressed. Complementary DNA libraries are composed of processed nucleic acid sequences harvested from the RNA pools of cells or tissues and represent all of the cDNA sequence prepared at a certain time for genes expressed in certain cells or tissues. This type of library is derived from DNA copies of messenger RNA (mRNA) (generated by reverse transcriptase), which are interspersed throughout a gene and are arranged contiguously within DNA. Messenger RNA libraries represent the transcript expressed at a certain time of certain cells or tissues. With recombinant DNA technologies *(11,12)*, genetic sequences of interest can be recombined with a replication-competent DNA vector, such as plasmid or bacteriophage, or built in a form of primer-binding sites. The libraries can be amplified by PCR, thereby generating combinatorial libraries. Other methods of amplification of DNA libraries have also been developed *(13–15)*. Analogously, polypeptide or protein libraries are collections of gene products of cells or tissues.

2.3. Why cDNA Libraries?

Complementary DNA libraries are preferable to mRNA libraries for the following reasons:

1. cDNA can represent the gene that is expressed as mRNA in a specific tissue or specific cells at a specific time; therefore, the mRNAs in two different types of cell or the same type of cell with different treatments may vary because the expression of genes varies.

2. cDNA libraries usually provide reading frames encoded within the DNA insert after the noncoding intervening sequences are removed; therefore, cDNA reflects both a mRNA transcript and a protein translation product. cDNAs can be used as probes for screening the mRNA transcript as well as in the rapid identification of amino acid sequences of polypeptides or proteins. Because there are no introns in a cDNA molecule, they are frequently used in protein synthesis in vitro.

3. The protein-encoding mRNA may not be present in all cells showing the specific protein because the mRNA is easily degraded and the protein formed in the cell could be present as a stable form from an earlier expression of the mRNA.

4. Because different numbers of copies of different mRNAs are present in a cell (low, middle, and high abundance), a desirable characteristic of cDNA libraries is that they increase the number of the less abundant species and reduce the relative number of high and middle abundant species. By manipulating the rate of strand reannealing in a denatured cDNA preparation, the high and middle abundance species of mRNA can be removed. The resulting cDNA generated is representative of the rarer species. Other modifications can be used to achieve the enrichment of cell-, tissue-, or stage-specific mRNA species in the preparation of cDNA libraries.

5. Messenger RNA are difficult to maintain, clone, and amplify; therefore, they are converted to more stable cDNA, which is less susceptible than mRNA to degradation by contaminating molecules.

For the above-mentioned reasons, cDNA libraries are preferred over mRNA libraries for genetic manipulations.

3. Conventional vs Novel Strategies for cDNA Generation

The conventional method of the generation of cDNA is based on the isolation of clones after transformation of bacteria or bacteriophages with an enriched but impure population of cDNA molecules ligated to a vector *(16–18)*. This method is good for abundant mRNA such as globin, immunoglobin, and ovalbumin. About 30% of mRNAs in cells, which are present at less than 14 copies per cell, cannot be identified with this method. After transcription of RNA into cDNA, the cDNA is digested by restriction endonucleases at specific sequence sites to form fragments of different size. Same-length DNA fragments from any cDNA species that contains at least two restriction sites are produced. Then, a second specific cleavage with a restriction endonuclease capable of cleaving the desired sequence at an internal site is performed. After separation from the contaminants, the subfragments of the desired sequence may be joined using DNA ligase to reconstitute the original sequence. The purified fragment can then be recombined with a cloning vector and transformed into a suitable host strain.

Polymerase chain reaction is commonly used in recent approaches to generating cDNA libraries, which are randomly primed and amplified from a small amount of DNA *(12,19)*. As a result, the use of PCR simplifies and improves the method of cDNA generation. To facilitate the formation of cDNAs from rare mRNAs, modifications of 3′ and 5′ ends of the DNA strand with a primer were adapted *(8,9)*. To avoid multiple purification or precipitation steps in the conventional method of cDNA library preparation, paramagenetic beads or other types of immobilization methods were developed *(20)*.

Subsequently, strategies that included a means of reducing the number of clones in a cDNA library in order to detect rare transcripts, a process known as normalization, were introduced *(21)*. Because the quality of the cDNA library generated is dependent on the quality of the mRNA, efforts were made to maintain the integrity or to amplify the copies of mRNAs to provide pure, undegraded, enriched mRNAs for generation of cDNA libraries. Another recently developed method of increasing mRNA copies is the use of amplified antisense RNA (aRNA) *(22)*. For the purpose of cloning and screening libraries efficiently, numerous vectors that are compatible with cDNA synthesis have been developed *(23)*. Another goal is the generation of full-length cDNA libraries. The method of amplification of DNA end regions has been effective *(24)*. In this approach, a small stretch of a known DNA sequence, a gene-specific primer at one end, and a universal primer at the other end, is used to form the flanking unknown sequence region *(25)*. Inverse PCR, a method that amplifies the flanking unknown sequence by using two gene-specific primers to reduce nonspecific amplification, generates full-length cDNA libraries *(26)*. Recently, a method coupling the prevention of mRNA degradation and thermocycling amplification was developed to generate full-length cDNA libraries *(27)*.

4. Different Methods in Generating cDNA Libraries

Generation of cDNAs has been previously reported, using the method described by Sambrook et al. *(28)*. This method involves the tedious procedures of reverse transcription, restriction, adaptor ligation, and vector cloning. The resulting cDNA libraries usually are incomplete because although the method is good for highly abundant mRNAs, rare species of mRNAs cannot be transcribed, particularly when the starting material is limited. Subsequently, a random priming polymerase chain reaction reverse transcription PCR (RT-PCR) was introduced to construct normalized cDNA libraries *(21)*. Although complete cDNA libraries can be fully amplified with this method, the use of random-primer amplification greatly reduces the integrity of the cDNA sequence because the normalized cDNA library usually loses part of the end sequences during cloning into a vector; this kind of low integrity may introduce significant difficulty in sequence analysis. Furthermore, the random amplification procedure also increases nonspecific contamination of primer dimers, resulting in false-positive sequences in the cDNA library.

Subsequently, the generation of aRNA was developed to increase transcriptional copies of specific mRNAs from limited amounts of cDNAs. In this method, an oligo(dT) primer is coupled to a T7 RNA polymerase promoter sequence [oligo(dT)-promoter] during reverse transcription (RT), and the single copy mRNA can be amplified up to 2000-fold by aRNA amplification *(22)*. This method was used for the characterization of the expression pattern of certain gene transcripts in cells *(29)*. Using this method, 50–75% of total intracellular mRNA was recovered from a single neuron *(22,29)*, suggesting that the prevention of mRNA degradation is necessary for the generation of complete full-length libraries. However, using this method for identification of rare mRNAs from a single cell still results in low completeness of the cDNA library *(29)*.

Recently, a novel technology has been developed to clone complete cDNA libraries from as few as 20 cells, called single-cell cDNA library amplification (*see* the flowchart in Chapter 9). In this method, a fast, simple, and specific means for generating a complete full-length cDNA library from single cells is provided. This approach combines the amplification of aRNAs from single cells and in-cell RT-PCR from mRNA *(22,29,30)*. First, during the initial reverse transcription of intracellular mRNAs, an oligo(dT)-promoter primer is introduced as a recognition site for subsequent transcription of newly reverse-transcribed cDNAs. These cDNAs are further tailed with a polynucleotide; now, the polynucleotide and the promoter primer of these cDNAs form binding templates for specific PCR amplification. After one round of reverse transcription, transcription, and PCR, a single copy of mRNA can be multiplied 2×10^9-fold. Coupling this method with a cell fixation and permeabilization step, the complete full-length cDNA library can be directly generated from a few single cells, avoiding mRNA degradation. Therefore, cell-specific full-length cDNA libraries are prepared.

In addition, preparations of single cells from histological slides for gene analysis were recently reported. In this method, single cells of a tissue specimen can be obtained from histological tissue sections that were routinely formalin-fixed and paraffin-embedded *(31)*. Briefly, the prepared tissue is shielded with a transparent film, and stained cells are identified and microdissected with a laser microbeam. In this way, a clear-cut gap is formed around the selected area and the dissected cells are adhered to the film; then, the specimen is directly delivered to a common microfuge tube containing the extraction buffer. Subsequently, studies of gene analysis or identification of expressed genes of a small number of specific cells can be performed by RT-PCR. This method has been used for the isolation of a single cell from archival colon adenocarcinoma, with subsequent detection of point mutations within codon 12 of c-Ki-ras2 mRNA after RT-PCR *(32)*. This method is highly precise, avoids contamination, and is easy to apply. To take advantage of the above-described features, complete full-length cDNA libraries from epithelial cells of three prostate cancer patients were generated *(27)*. The libraries so generated showed a gene expression pattern similar to that observed in human prostate cancer cell lines. This technique provides better resolution than most other methods for the analysis of cell-specific gene expression and its relation to the disease.

5. Quality of cDNA Libraries

The quantity of mRNA usually is assessed by the final product, the cDNA library generated. Because a large amount of mRNA libraries can be generated by the RNA-PCR method, a few tests to ascertain the quality of mRNAs can be performed. First and foremost, the mRNA libraries are fractionated by electrophoresis in a 1% formaldehyde–agarose gel with ethidium bromide. A uniform smearing pattern of the product, viewed under ultraviolet (UV) light, indicates that good quality of mRNAs is achieved. In most cases, the size of RNAs should range from 500 bp to 8 kp (*see* Chapter 11).

Subsequently, Northern blot analysis is performed to ascertain certain genes of interest that are eluted at the right position. A variety of internal standards can be used. Routinely, we use probes for GAPDH and β-actin, Rb, and p16 or p21 to identify the housekeeping, abundant, and rare species of mRNAs, respectively (*see* Chapter 16). In situations in which the gene of interest is larger than 8 kb, the probe of a cytoplasmic protein such as PTPL1, a widely distributed cytoplasmic protein tyrosine phosphatase with a size of 9.4 kb *(33)*, can be used. In some cases, a ribosomal RNA marker, as a negative control, is added to ensure that no contamination with rRNAs occurs in the mRNA library preparation. The quality of cDNA libraries can be assessed by Northern blot analysis (*see* Chapter 16), polymerase chain reaction coupled reverse transcription (RT-PCR) *(34)*, differential display *(35)*, subtractive hybridization (*see* Chapter 21), subtractive cloning (*see* Chapter 22), RNA microarray, and cDNA cloning (*see* Chapter 13).

To assay the quality of mRNA generated, a pretest or control test array of the selected Genechips can be used. These tests arrays are designed to optimize the labeling and hybridization conditions and determine the linear dynamic range of gene expression levels, but, most of all, to also assess differential gene expression of known abundant and rare genes. Therefore, the quality of the mRNA preparations can be determined.

6. Potential Applications

The most common application of mRNA/cDNA libraries is the identification of genes of interest. They are also used for other mRNA/cDNA manipulations to determine differentially expressed gene levels associated with structural and functional changes that are of high relevance to disease controls or pathways of specific molecule modulations. In the past, one rate-limiting step in this type of study is the lack of high-quality human mRNA generated from limited and heterogenous pathological specimens. The RNA-PCR method and other methods for generating high-quality mRNAs will solve this problem. Coupled with microarray technologies and microdissected single cells, mRNA/cDNA libraries so generated can be used to monitor a large number of genes and provide a powerful tool for assessing differential mRNA expression levels for the identification of disease-associated genes. With the antisense knockout techniques, double-stranded mRNA silencing of posttranscriptional gene expression, and a newly developed cDNA–mRNA hybrid interference of gene expression, the function of an overexpressed genes can be examined.

After generation of stage-specific cDNA libraries, one may examine other genes of interests and determine whether these genes are differentially expressed. Altered gene expression of certain molecules and their related receptors may shed light on the developmental, physiological, and pathological significance of these molecules. In addition, one can examine differential gene expression of other genes in the presence and absence of the gene of interest by modulating the levels of each gene of interest. In this manner, each marker, growth factor and/or its receptors, or genes associated with a physiological or pathological phenomenon could be thoroughly monitored for altered levels of expression. For instance, aberrations of gene expression may be found to be crucial to the development of an organ or a certain protein or its associated isoforms. Such proteins may be essential for cell invasion, migration, and angiogenesis. Obviously, these molecules will be important potential targets for the development of new therapies. This approach may ultimately contribute to specific drug design or therapy for regulation of the expression of those genes responsible for prostatic cancer. Another potential approach is to couple the generation of full-length mRNA/cDNA libraries with transcriptional inference of a specific gene for identification of functional significance of an overexpressed gene. The remaining challenge is to develop a system to overexpress the downregulated genes as well as to determine their functional significance in an in vivo system.

References

1. Gillespie, D. and Spiegelman, S. (1965) A quantitative assay for DNA–RNA hybrids with DNA immobilized on a memrane. *J. Mol. Biol.* **12,** 829–842.
2. Kessler, C. (1992) Non-radioactive analysis of biomolecules, in *Nonisotopic DNA Probe Techniques* (Kricka, L., ed.), Academic, San Diego, CA, pp. 29–92.
3. Temin, H. M. and Mizutani, S. (1970) RNA-dependent DNA polymerase in virions of Rous sacroma virus. *Nature* **226,** 1211–1213.

4. Baltimore D. (1970) RNA-dependent DNA polymerase in virions of RNA tumour viruses. *Nature* **226,** 1209–1211.
5. Kohlstaedt, L. A., Wang, J., Friedman, J. M., Rice, P.A., and Steitz, T.A. (1992) Crystal structure at 3.5 A resolution of HIV-1 reverse transcriptase complexed with an inhibitor. *Science* **256,** 1783–1790.
6. Kornbert, A. and Baker, T. A. *DNA Replication*, 2nd ed., WH Freeman, New York, pp. 106–109.
7. Saiki, R. K., Bugawan, T. L., Horn, G. T., Mullis, K. B., and Erlich, H. A. (1986) Analysis of enzymatically amplified beta-globin and HLA-DQ alpha DNA with allele-specific oligonucleotide probes. *Nature* **324,** 163–166.
8. Forhman, M. A., Dush, M. K., and Martin, G. R. (1988) Rapid production of full-length cDNAs from rare transcripts by amplification using a single gene-specific oligonucleotide primer. *Proc. Natl. Acad. Sci. USA* **85,** 8998–9002.
9. Ohara, O., Dorit, R. L., and Gilbert, W. (1989) One-sided polymerase chain reaction: the amplification of cDNA. *Proc. Natl. Acad. Sci. USA* **86,** 5673–5677.
10. Bashirdes, S. and Lovett, M. (2001) cDNA detection and analysis. *Curr. Opin. Chem. Biol.* **5,** 15–20.
11. Lawn, R. M., Fritsch, E. F., Parker, R. C., Blake, G., and Maniatis, T. (1978) The isolation and characterization of linked delta- and beta-globin genes from a cloned library of human DNA. *Cell* **15,** 1157–1174.
12. Mullis, K., Faloona, F., Scharf, S., Saiki, R., Horn, G., and Erlich, H. (1986) Specific enzymatic amplification of DNA in vitro: the polymerase chain reaction. *Cold Spring Harb. Symp. Quant. Biol.* **51,** 263–273.
13. Barany, F. (1991) Genetic disease detection and DNA amplification using cloned thermostable ligase. *Proc. Natl. Acad. Sci. USA* **88,** 189–193.
14. Breaker, R. R. and Joyce, G. F. (1994) Emergence of a replicating species from an in vitro RNA evolution reaction. *Proc. Natl. Acad. Sci. USA* **91,** 6093–6097.
15. Walker, G. T., Fraiser, M. S., Schram, J. L., Little, M. C., Nadeau, J. G., and Malinowski, D. P. (1992) Strand displacement amplification—an isothermal, in vitro DNA amplification technique. *Nucleic Acids Res.* **20,** 1691–1696.
16. Ullrich, A., Shine, J., Chirgwin, J., Pictet, R., Tischer, E., Rutter, W. J., et al. (1977) Rat insulin genes: construction of plasmids containing the coding sequences. *Science* **196,** 1313–1319.
17. Seeburg, P. H., Shine, J., Martial, J. A., Baxter, J. D., and Goodman, H. M. (1977) Nucleotide sequence and amplification in bacteria of structural gene for rat growth hormone. *Nature* **270,** 486–494.
18. Goodman, H. M., Seeburg, P. H., Shine, J., Martial, J. A., and Baxter, J. D. (1979) *Specific Eukaryotic Genes: Structure, Organization, Function, Proc. Alfred Benzon Symp.*, p. 179.
19. Saiki, R. K., Gelfand, D. H., Stofgfel, S., Scharf, S. J., Hinguchi, R., Horn, G. T., et al. (1988) Primed-directed enzymatic acmplification of DNA with a thermostable DNA polymerase. *Science* **239,** 487–491.

20. Lambert, K. N. and Williamson, V. M. (1993) cDNA library construction from small amounts of RNA using paramagenetoc beads and PCR. *Nucleic Acids Res.* **21,** 775–776.

21. Patanjali, S. R., Parimoo, S., and Weissman, S. M. (1911) Construction of a uniform-abundance (normalized) cDNA library. *Proc. Natl. Acad. Sci. USA* **88,** 1943–1947.

22. Eberwine, J., Yeh, H., Miyashiro, K., Cao, Y., Nair, S., Finnell, R., et al. (1992) Analysis of gene expression in single live neurons. *Proc. Natl. Acad. Sci. USA* **89,** 3010–3014.

23. Short, J. M., Fernandez, J. M., Sorge, J. A., and Huse, W. D. (1988). Lambda bacteriophage lambda expression vector with in vivo excision properties. *Nucleic Acids Res.* **16,** 7583–7600.

24. Huang S. H., Jong, A. Y., Yang, W., and Holcenberg, J. (1993) Amplification of gene ends from gene libraries by PCR with single-sided specificity. *Methods Mol. Biol.* **15,** 357–363.

25. Huang, S. H., Hu, Y. Y., Wu, C. H., and Holcenbert, J. (1990) A simple method for idrect cloning cDNA sequence that flanks a region of known sequence from total RNA by applying the inverse polymerase chain reaction. *Nucleic Acids Res.* **18,** 1922.

26. Ochman, H., Gerber, A. S., and Hartl, D. J. (1988) Genetic applications of an inverse polymerase chain reaction. *Genetics* **120,** 621–625.

27. Lin, S. L., Chuong, C. M., Widelitz, R. B., and Ying, S. Y. (1999) In vivo analysis of cancerous gene expression by RNA-polymerase chain reaction. *Nucleic Acids Res.* **27,** 4585–4589.

28. Sambrook, J., Fritsch, E. F., and Maniatis, T. (1989) *Molecular Cloning*, 2nd ed., Cold Spring Harbor Laboratory, Cold Spring Harbor, NY, pp. 8.11–8.35.

29. Crino, P. B., Trajanowski, J. Q., Dichter, M. A., and Eberwine, J. (1996) Embryonic neuronal markers in tuberous sclerosis: single-cell molecular pathology. *Proc. Natl. Acad. Sci. USA* **93,** 14,152–14,157.

30. O'Dell, D. M., Raghupathi, P., Crino, P. B., Morrison, B., Eberwine, J. H., and McIntosh, T. K. (1998) Amplification of mRNAs from single, fixed, TUNEL positive cells. *BioTechniques* **25,** 566–570.

31. Becker, I., Becker, K. F., Rohrl, M. H., and Hofler, H. (1997) Leser-assisted preparation of single cells from stained histological slides fro gene analysis. *Histochem. Cell. Biol.* **108,** 447–451.

32. Schutze, K. and Lahr, G. (1998) Identification of expressed genes by laser-mediated manipulation of single cells. *Nat. Biotechnol.* **16,** 737–742.

33. Saras, J., Claesson-Welsh, L., Heldin, C. H., and Gonez, L. J. (1994) Cloning and characterization of PTPL1, a protein tyrosine phosphatase with similarities to cytoskeletal-associated proteins. *J. Biol. Chem.* **269,** 24,082–24,089.

34. Ghosh, S., Gifford, A. M., Riviere, L. R., Tempst, P., Nolan, G. P., and Baltimore, D. (1990). Cloning of the p50 DNA binding subunit of NF-kappa B: homology to rel and dorsal. *Cell* **62,** 1019–1929.

35. Liang, P. and Pardee, A. B. (1992). Differential display of eukaryotic messenger RNA by means of the polymerase chain reaction. *Science* **259,** 967–997.

2

Rapid Amplification of cDNA Ends

Yue Zhang

1. Introduction

Rapid amplification of complementary DNA (cDNA) ends (RACE) is a powerful technique for obtaining the ends of cDNAs when only partial sequences are available. In essence, an adaptor with a defined sequence is attached to one end of the cDNA; then, the region between the adaptor and the known sequences is amplified by polymerase chain reaction (PCR). Since the initial publication in 1988 (1), RACE has greatly facilitated the cloning of new genes. Currently, RACE remains the most effective method of cloning cDNAs ends. It is especially useful in the studies of temporal and spatial regulation of transcription initiation and differential splicing of mRNA. The methods described in this chapter are quite simple and efficient. A linker at the 3′ end and an adaptor at the 5′ end are added to the first strand of cDNA during reverse transcription; amplification of virtually any transcript to either end can then make use of this same pool of cDNAs. In addition to being simple, the efficiency of 5′-RACE is dramatically increased because the adaptor is added only to full-length cDNAs.

Since the initial description of RACE, many labs have developed significant improvements on the basic approach. The methods described here were developed from more recent reports from Frohman's and Roeder's groups. Among these, adaptor addition accompanying reverse transcription was developed from the CapFinding (2–4) technique of Clonetech (Palo Alto, CA): Moloney murine leukemia virus reverse transcriptase (MMLV RT) adds an extra two to four cytosines to the 3′ ends of newly synthesized cDNA strands upon reaching the cap structure at the 5′ end of mRNA templates. When an oligonucleotide with multiple G's at its 3′-most end is present in the reaction mixture, its terminal G nucleotides base pair with the C's of the

From: *Methods in Molecular Biology, vol. 221: Generation of cDNA Libraries: Methods and Protocols*
Edited by: S.-Y. Ying © Humana Press Inc., Totowa, NJ

newly synthesized cDNA. Through a so-called "template switch" process, this oligonucleotide will serve as a continuing template for the RT. Thus, the reverse complement adaptor sequence can be easily incorporated into the 3' end of the newly synthesized first strand of cDNA, which is at the beginning of the new cDNA (*see* **Fig. 1**). CapFinding obligates the addition of the adaptor in a Cap-dependent manner, resulting in adaptor attachment to full-length cDNA clones only. Therefore, because there is no additional enzymatic modification of the cDNAs after reverse transcription, this results in a simplified method with improved overall efficiency.

This protocol also utilizes biotin–streptavidin interactions to facilitate the elimination of excessive adaptors before carrying out PCR (*see* **Fig. 1**). The importance of adaptor elimination has been documented since the initial description of RACE *(1)*. The presence of extra adaptors is detrimental to the following PCR reaction because their sequence or complement sequence is present in ALL cDNAs in the reaction mixture, resulting in heavy background amplification and failure to amplify the specific product if not removed.

2. Materials

1. 5X Reverse transcription buffer: 250 mM Tris-HCl pH 8.3 (at 45°C), 30 mM MgCl$_2$, 10 mM MnCl$_2$, 50 mM dithiothreitol, 1 mg/mL bovine serum albumin (BSA).
2. Biotin-labeled primer P$_{total}$ (biotin-labeled primers can be ordered from Invitrogen). The sequences of P$_{total}$ and the following primers are listed in **Fig. 2**.
3. CapFinder adaptor.
4. RNasin (Promega Biotech).
5. dNTPs: 10 mM solutions (PL-Biochemicals/Pharmacia or Roche).
6. SuperScript II RNase H$^-$ Reverse Transcriptase (Invitrogen/Life Technologies).
7. Streptavidin MagneSphere Particles and MagneSphere Magnetic Separation Stand (components of PolyATract mRNA Isolation System from Promega).
8. TE: 10 mM Tris-HCl (pH 7.5), 1 mM EDTA.
9. PCR cocktail: hot start polymerase systems are recommended (e.g., Stratagene Hercules, or Roche Expand™ High Fidelity). Assemble the PCR cocktail according to the manufacturer's instructions. One should also use the extension temperature specified. For simplicity, 72°C is used in the following methods.
10. Gene specific primer 1 (GSP1), GSP2, and P$_o$ and P$_i$ primers for 3'-RACE or reverse gene specific primer 1 (RGSP1), RGSP2, and U$_o$ and U$_i$ primers for 5'-RACE.

3. Methods

3.1. Reverse Transcription to Generate cDNA Templates (see Notes 1–7)

1. Assemble reverse transcription components on ice: 4 µL of 5X reverse transcription buffer, 2 µL of dNTPs, 1 µL of CapFinder adaptor (10 µM), and 0.25 µL (10 U) of RNasin.

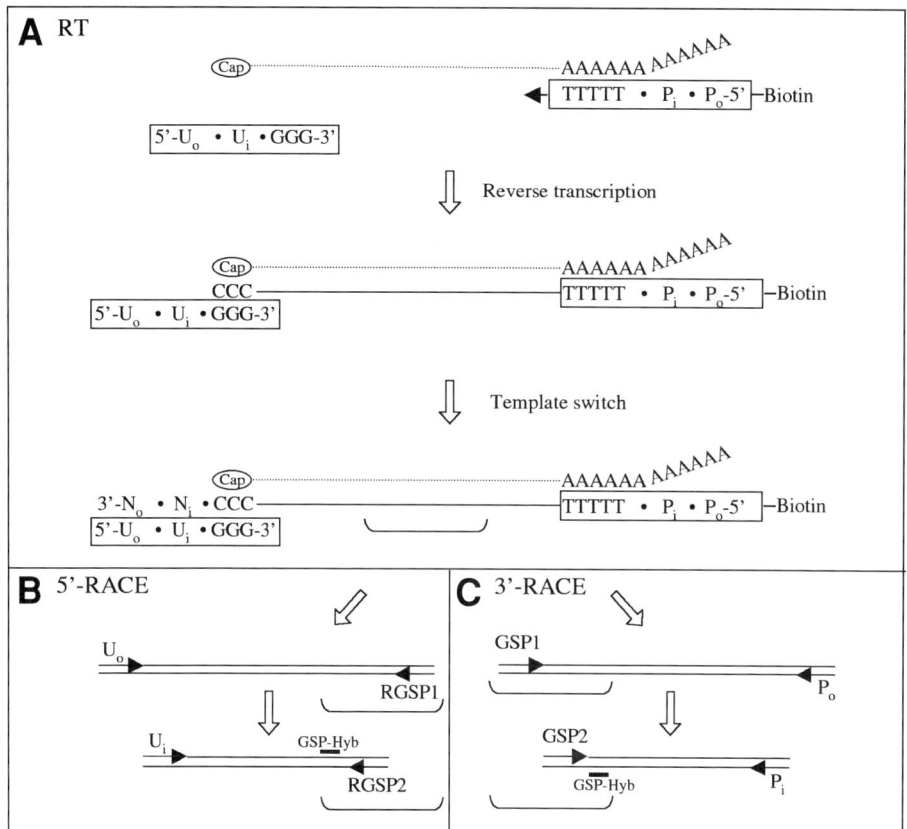

Fig. 1. Schematic representation of RACE. (**A**) Reverse transcription, template switch, and incorporation of adaptor sequences at the 3′ end of the first strand of cDNA. Biotin-labeled primer P_{total} is used to initiate reverse transcription through hybridization of the poly(dT) tract with the mRNA polyA tail. After reaching the 5′ end of the mRNA, oligo(dC) is added by the reverse transcriptase in a Cap-dependent manner. Then, through template switching via base pairing between the oligo(dC) and the oligo(dG) at the end of CapFinder Adaptor, the reverse complementary sequence of the CapFinder oligo is incorporated to the first strand of the cDNA. The dotted line indicates mRNA, the solid line indicates cDNA, and the rectangle indicates the primer. The brace indicates the known region. (**B**) 5′-RACE. The first round of PCR uses primer U_o and RGSP1 (reverse gene-specific primer 1); the second round uses U_i and RGSP2. GSP-Hyb is also within the known region; it can be used to confirm the authenticity of the RACE product. (**C**) 3′-RACE. Similar to 5′-RACE, but note that GSP1 and GSP2 are in the same sequences as the gene, whereas the P_o and P_i are the reverse complement.

```
CapFinder: 5'-TGGTTGCCATAAGCGGATCATCGGGAGGAGAAACGGG-3'
U_o:        5'-TGGTTGCCATAAGCGGATC-3'
U_i:                            5'-TCATCGGGAGGAGAAACGG-3':
```

$$P_{total}: 3'-G/A/C(T)_{17}CTATCGCTCCGCTCGCAAGGTTTGGGTCAGGTTGGTTT-5'$$
$$P_i: 3'-CTATCGCTCCGCTCGCAAG-5'$$
$$P_o: 3'-GGTTTGGGTCAGGTTGGTTT-5$$

For convenience: The 5'->3' sequences of the P primers:

$$P_{total}: 5'-TTTGGTTGGACTGGGTTTGGAACGCTCGCCTCGCTATC(T)_{17}C/A/G-3'$$
$$P_o: 5'-TTTGGTTGGACTGGGTTTGG-3'$$
$$P_i: 5'-GGAACGCTCGCCTCGCTATC-3'$$

Fig. 2. Primer sequences and their relationship. GSPs or RGSPs are not included. The CapFinder sequence was selected from the *Yersinia pestis* Genome sequence. BLASTN search results using it can be retrieved with request ID (RID) 1009985097-16592-24994. P_{total} is biotin labeled at the 5' end. This is a "lock-docking" degenerate primer that actually consists of three primers with different (A, G, or C) nucleotides at its 3' end. Its RID is 1009987520-16411-27741.

2. Heat 1 µg of polyA$^+$ RNA and 10 pmol P_{total} primer in 11.75 µL of water at 80°C for 3 min, cool rapidly on ice, and spin for 5 s in a microcentrifuge. Combine with the components from **step 1**.
3. Add 1 µL (200 U) of SuperScript II reverse transcriptase to the above mixture and incubate for 5 min at room temperature, 30 min at 42°C, 30 min at 45°C, and 10 min at 50°C.
4. Incubate at 70°C for 15 min to inactivate the reverse transcriptase. Add in magnetic streptavidin beads; use about five times the binding capacity required to complex the amount of biotinylated primer used. Wash with TE at 50°C three times to eliminate the free CapFinder adaptors.
5. Dilute the reaction mixture to 0.5 mL with TE and store at 4°C (cDNA pool).

3.2. Amplification of the cDNA (see Notes 8–13)

3.2.1. First Round

1. Add an aliquot (1 µL) of the cDNA pool (resuspend well) and primers (25 pmol each of GSP1 and P_o for 3'-RACE, or RGSP1 and U_o for 5'-RACE) to 50 µL of PCR cocktail in a 0.5-mL PCR tube.
2. Heat the mixture in the thermal cycler at 95°C for 5 min to denature the first-strand products and the streptavidin; add 2.5 U *Taq* polymerase and mix well (hot start). Incubate at appropriate annealing temperature for 2 min. Extend the cDNAs at 72°C for 40 min. It is not necessary to keep the magnetic streptavidin

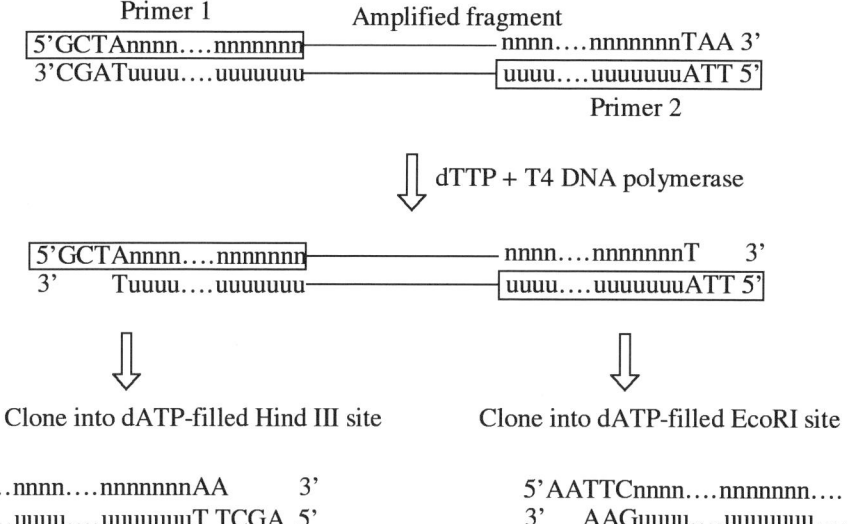

Fig. 3. A safe and easy cloning method. Details are described in **Subheading 3.3.**

beads resuspended during these incubations because the biotin should not interact with the denatured streptavidin. It has been reported that styrene beads are smaller, stay without agitation in solution, and are potentially better *(3)*. The respective performances have not been compared in the author's hands.

3. Carry out 30 cycles of amplification using a step program (94°C, 1 min; 52–68°C, 1 min; 72°C, 3 min), followed by a 15-min final extension at 72°C. Cool to room temperature. The extension time at 72°C needs to be adjusted according to the length of the product expected and the speed of the polymerase used.

3.2.2. Second Round (If Necessary)

1. Dilute 1 µL of the amplification products from the first round into 20 µL of TE.
2. Amplify 1 µL of the diluted material with primers GSP2 and P_i for 3′-RACE, or RGSP2 and U_i for 5′-RACE, using the first-round procedure, but eliminate the initial 2-min annealing step and the 72°C, 40-min extension step.

3.3. Safe and Easy Cloning Protocol (see Note 14)

1. Insert preparation: select a pair of restriction enzymes for which you can synthesize half-sites appended to PCR primers that can be chewed back to form appropriate overhangs, as shown for *Hind*III and *Eco*RI in **Fig. 3.** For example, add "TTA" to the 5′ end of P_i and add "GCTA" to the 5′ end of GSP2. Carry out PCR as usual.
2. After PCR, clean up using Qiagen PCR cleanup spin columns.

3. On ice, add the selected dNTP(s) (e.g., dTTP) to a final concentration of 0.2 mM, 1/10 vol of 10X T4 DNA polymerase buffer, and 1–2 U T4 DNA polymerase.
4. Incubate at 12°C for 15 min and then 75°C for 10 min to heat inactivate the T4 DNA polymerase. (Optional: gel-isolate DNA fragment of interest, depending on degree of success of PCR amplification.)
5. Vector preparation: digest vector (e.g., pGem-7ZF (Promega)) using the selected enzymes (e.g., *Hin*dIII and *Eco*RI) under optimal conditions, in a volume of 10 µL.
6. Add a 10-µL mixture containing the selected dNTP(s) (e.g., dATP) at a final concentration of 0.4 mM, 1 µL of the restriction buffer used for digestion, 0.5 µL Klenow, and 0.25 µL Sequenase.
7. Incubate at 37°C for 15 min and then 75°C for 10 min to heat inactivate the polymerases.
8. Gel-isolate the linearized vector fragment.
9. For ligation, use equal molar amounts of vector and insert.

Other manipulations of RACE PCR products are discussed in **Note 15**.

4. Notes

1. PolyA RNA is preferentially used for reverse transcription to decrease background, although total RNA can be used as well. An important factor in the generation of full-length cDNAs concerns the stringency of the reverse transcription reaction. Reverse transcription reactions were historically carried out at relatively low temperatures (37–42°C) using a vast excess of primer (approximately one-half the mass of the mRNA). Under these low-stringency conditions, a stretch of A residues as short as six to eight nucleotides will suffice as a binding site for an oligo(dT)-tailed primer. This may result in cDNA synthesis being initiated at sites upstream of the polyA tail, leading to truncation of the desired amplification product. One should be suspicious that this has occurred if a canonical polyadenylation signal sequence is not found near the 3' end of the cDNAs generated. This low-stringency problem can be minimized by controlling two parameters: primer concentration and reaction temperature. The primer concentration can be reduced dramatically without decreasing the amount of cDNA synthesized significantly and will begin to bind preferentially to the longest A-rich stretches present (i.e., the polyA tail). The recommended quantity in **Subheading 3.1.** represents a good starting point. It can be reduced fivefold further if significant truncation is observed.
2. The efficiency of cDNA extension is important, especially for 5'-RACE. In the described protocol, the incubation temperature is raised slowly to encourage reverse transcription to proceed through regions of difficult secondary structure. Synthesis of cDNAs at elevated temperatures should diminish the amount of secondary structure encountered in GC-rich regions of the mRNA. Because the half-life of reverse transcriptase rapidly decreases as the incubation temperature increases, the reaction cannot be carried out at elevated temperatures in its

entirety. Alternatively, the problem of difficult secondary structure (and non-specific reverse transcription) can be approached using heat-stable reverse transcriptases, which are now available from several suppliers (Perkin-Elmer-Cetus, Amersham, Epicentre Technologies, and others). Like PCR reactions, the stringency of reverse transcription can be controlled by adjusting the temperature at which the primer is annealed to the mRNA. Optimal temperature depends on the specific reaction buffer and reverse transcriptase used and should be determined empirically, but will usually be in the range 48–56°C for a primer terminated by a 17-nt oligo(dT).

3. In addition to synthesis of cDNAs at elevated temperature, there are several other approaches that encourage cDNA extension. First, use clean, intact RNA. Second, select a gene-specific primer for reverse transcription (GSP-RT) that is close to the 5′ end within the known sequences, thus minimizing difficult regions. A random hexamer (50 ng) can also be used to create a universal 5′-end cDNA pool. If using random hexamers, then a room-temperature 10-min incubation period is needed after mixing everything together.

4. The successfulness of 5′-RACE relies on the incorporation of the CapFinder Adaptor sequence at the beginning of the cDNA. As mentioned, this step depends on the addition of extra oligo(dC) at the end of the first strand of cDNA. It has been shown that the Mn^{2+} ion in the reverse transcription buffer greatly increases the percentage of oligo(dC) added to the end of the first strands of cDNAs *(5)*.

5. The presence of excess P_{total} and CapFinder adaptors during amplification will be detrimental to the reaction. Virtually all of the cDNA produced will contain the primer P_{total} at the 3′ end and (hopefully) the CapFinder sequence at the 5′ end. The physical presence of both primers will cause heavy background and failure of RACE. This phenomenon has been described even in the original RACE article *(1)*. Here, a semisolid-phase cDNA synthesizing protocol is adapted to deal with this problem. The primer P_{total} is biotin labeled, and after the reverse transcription reaction, streptavidin beads are used to separate to the cDNAs from unincorporated CapFinder adaptors.

6. The following discusses some issues regarding potential problems with the reverse transcription steps.

   a. Damaged RNA: Electrophorese RNA in a 1% formaldehyde minigel and examine the integrity of the 18*S* and 28*S* ribosomal bands. Discard the RNA preparation if ribosomal bands are not sharp.

   b. Contaminants: Ensure that the RNA preparation is free of agents that inhibit reverse transcription (e.g., lithium chloride and sodium dodecyl sulfate) *(6)*.

   c. Bad reagents: To monitor reverse transcription of the RNA, add 20 µCi of 32p-dCTP to the reaction, separate the newly created cDNAs using gel electrophoresis, wrap the gel in Saran Wrap™, and expose it to X-ray film. Accurate estimates of cDNA size can best be determined using alkaline agarose gels, but a simple 1% agarose minigel will suffice to confirm that reverse transcription took place and that cDNAs of reasonable length were generated. Note that adding 32p-dCTP to the reverse transcription reaction results in the detection of cDNAs synthesized both through the specific priming of mRNA and through RNA self-priming. When a gene-specific primer is used to prime transcription (5′-end RACE) or when total RNA is used as a template, the majority of the labeled cDNA will actually have been generated from RNA self-priming. To monitor extension of the primer used for reverse transcription, label the primer using T4 DNA kinase and 32p-γATP prior to reverse transcription. A much longer exposure time will be required to detect the labeled primer-extension products than when 32p-dCTP is added to the reaction.

7. To monitor reverse transcription of the gene of interest, one may attempt to amplify an internal fragment of the gene containing a region derived from two or more exons, if sufficient sequence information is available.

8. For 3′-end amplification, it is important to add the *Taq* polymerase after heating the mixture to a temperature above the T_m of the primers ("hot-start" PCR). Addition of the enzyme prior to this point allows one "cycle" to take place at room temperature, promoting the synthesis of nonspecific background products dependent on low-stringency interactions.

9. An annealing temperature close to the effective T_m of the primers should be used. Computer programs to assist in the selection of primers are widely available and should be used. An extension time of 1-min/kb expected product should be allowed during the amplification cycles. If the expected length of product is unknown, try 3–4 min initially.

10. Very little substrate is required for the PCR reaction. One microgram of polyA⁺ RNA typically contains approx 5×10^7 copies of each low-abundance transcript. The PCR reaction described here works optimally when 10^3–10^5 templates (of the desired cDNA) are present in the starting mixture; therefore, as little as 0.002% of the reverse transcription mixture suffices for the PCR reaction! The addition of too much starting material to the amplification reaction will lead to production of large amounts of nonspecific product and should be avoided. The RACE technique is particularly sensitive to this problem, as every cDNA in the mixture, desired and undesired, contains a binding site for the P_i and P_o primers.

11. It was found empirically that allowing extra extension time (40 min) during the first amplification round (when the second strand of cDNA is created) sometimes resulted in increased yields of the specific product relative to background amplification and, in particular, increased the yields of long cDNAs versus short cDNAs when specific cDNA ends of multiple lengths were present *(1)*. Prior treatment of cDNA templates with RNA hydrolysis or a combination of RNase H and RNase A infrequently improves the efficiency of amplification of specific cDNAs.

12. Some potential amplification problems are as follows:
    a. No product: If no products are observed for the first set of amplifications after 30 cycles, add fresh *Taq* polymerase and carry out an additional 15 rounds of

amplification (extra enzyme is not necessary if the entire set of 45 cycles is carried out without interruption at cycle 30). Product is always observed after a total of 45 cycles if efficient amplification is taking place. If no product is observed, carry out a PCR reaction using control templates and primers to ensure the integrity of the reagents.

b. Smeared product from the bottom of the gel to the loading well: There are too many cycles or too much starting material.

c. Nonspecific amplification, but no specific amplification: Check sequence of cDNA and primers. If all are correct, examine primers (using computer program) for secondary structure and self-annealing problems. Consider ordering new primers. Determine whether too much template is being added or if the choice of annealing temperatures could be improved. Alternatively, secondary structure in the template may block amplification. Consider adding formamide *(7)* or ^{7}aza-GTP (in a 1:3 ratio with dGTP) to the reaction to assist polymerization. ^{7}aza-GTP can also be added to the reverse transcription reaction.

d. Inappropriate templates: To determine whether the amplification products observed are being generated from cDNA or whether they derive from residual genomic DNA or contaminating plasmids, pretreat an aliquot of the RNA with RNase A.

13. The following describes the analysis of the quality of the RACE PCR products:

a. The production of specific partial cDNAs by the RACE protocol is assessed using Southern blot hybridization analysis. After the second set of amplification cycles, the first- and second-set reaction products are electrophoresed in a 1% agarose gel, stained with ethidium bromide, denatured, and transferred to a nylon membrane. After hybridization with a labeled oligomer or gene fragment derived from a region contained within the amplified fragment (e.g., GSP-Hyb in **Fig. 1B** or **1C**), gene-specific partial cDNA ends should be detected easily. Yields of the desired product relative to nonspecific amplified cDNA in the first-round products should vary from <1% of the amplified material to nearly 100%, depending largely on the stringency of the amplification reaction, the amplification efficiency of the specific cDNA end, and the relative abundance of the specific transcript within the mRNA source. In the second set of amplification cycles, approx 100% of the cDNA detected by ethidium bromide staining should represent specific product. If specific hybridization is not observed, then troubleshooting steps should be initiated.

b. Information gained from this analysis should be used to optimize the RT procedure. If low yields of specific product are observed because nonspecific products are being amplified efficiently, then annealing temperatures can be raised gradually (approx 2°C at a time) and sequentially in each stage of the procedure until nonspecific products are no longer observed. Alternatively, some investigators have reported success using the "touchdown PCR" procedure to optimize the annealing temperature without trial and error *(8)*. Optimizing the annealing temperature is also recommended if multiple species of specific products are observed, which could indicate that truncation of specific products is occurring. If multiple species of specific products are observed after the reverse transcription and amplification reactions have been fully optimized, then the possibility should be entertained that alternate splicing or promoter use is occurring.

   c. Look for TATA, CCAAT, and initiator element (Inr) sites at or around the candidate transcription site in the genomic DNA sequence if it is available. One should usually be able to find either TATA or an Inr.

14. Cloning of RACE products like any other PCR products.

   a. Option 1: To clone the cDNA ends directly from the amplification reaction (or after gel purification, which is recommended), ligate an aliquot of the products to plasmid vector encoding a one-nucleotide 3′ overhang consisting of a "T" on both strands. Such vector DNA is available commercially (Invitrogen's "TA Kit") or can be easily and inexpensively prepared (e.g., ref. **9**).

   b. Option 2: A safer and very effective approach is to modify the ends of the primers to allow the creation of overhanging ends using T4 DNA polymerase to chew back a few nucleotides from the amplified product in a controlled manner and Klenow enzyme (or Sequenase) to fill in partially restriction-enzyme-digested overhanging ends on the vector, as shown in **Fig. 3** and discussed in **Subheading 3.4.** (adapted from **refs. *10*** and ***11***).

   This approach has many advantages. It eliminates the possibility that the restriction enzymes chosen for the cloning step will cleave the cDNA end in the unknown region. In addition, vector dephosphorylation is not required because vector self-ligation is no longer possible, insert kinasing (and polishing) is not necessary, and insert multimerization and fusion clones are not observed. Overall, the procedure is more reliable than "TA" cloning.

15. Other manipulation of RACE PCR products:

   a. Sequencing: RACE products can be sequenced directly on a population level using a variety of protocols, including cycle sequencing, from the end at which the gene-specific primers are located. Note that 3′-RACE products cannot be sequenced on a population level using the P_i primer at the unknown end, because individual cDNAs contain different numbers of A residues in their polyA tails and, as a consequence, the sequencing ladder falls out of register after reading through the tail. Using the set of primers TTTTTTTTTTTTTTTTTA/G/C, 3′-end products can be sequenced from their unknown end. The non-T nucleotide at the 3′ end of the primer forces the appropriate primer to bind to the inner end of the polyA tail (*12*). The other two primers do not participate in the sequencing reaction. Individual cDNA ends, once cloned into a plasmid vector, can be sequenced from either end using gene-specific or vector primers.

   b. Hybridization probes: RACE products are generally pure enough that they can be used as probes for RNA and DNA blot analyses. It should be kept in mind that small amounts of contaminating nonspecific cDNAs will always be present. It is also possible to include a T7 RNA polymerase promoter in one or both primer sequences and to use the RACE products with in vitro transcription reactions to produce RNA probes. Primers encoding the T7 RNA polymerase promoter sequence do not appear to function as amplification primers as efficiently as the others listed in **Fig. 2** (personal observation). Therefore, the T7 RNA polymerase promoter sequence should not be incorporated into RACE primers as a general rule.

c. Construction of full-length cDNAs: It is possible to use the RACE protocol to create overlapping 5′ and 3′ cDNA ends that can later, through judicious choice of restriction enzyme sites, be joined together through subcloning to form a full-length cDNA. It is also possible to use the sequence information gained from acquisition of the 5′ and 3′ cDNA ends to make new primers representing the extreme 5′ and 3′ ends of the cDNA and to employ them to amplify a *de novo* copy of a full-length cDNA directly from the cDNA pool. Despite the added expense of making two more primers, there are several reasons why the second approach is preferred. First, a relatively high error rate can be associated with the PCR conditions for which efficient RACE amplification takes place (depending on the conditions used) and numerous clones may have to be sequenced to identify one without mutations. In contrast, two specific primers from the extreme ends of the cDNA can be used under less efficient but low-error-rate conditions *(13)* for a minimum of cycles to amplify a new cDNA that is likely to be free of mutations. Second, convenient restriction sites are often not available, making the subcloning project difficult. Third, by using the second approach, the synthetic polyA tail (if present) can be removed from the 5′ end of the cDNA. Homopolymer tails appended to the 5′ ends of cDNAs have, in some cases, been reported to inhibit translation. Finally, if alternate promoters, splicing, and polyadenylation signal sequences are being used and result in multiple 5′ and 3′ ends, it is possible that one might join two cDNA halves that are never actually found together in vivo. Employing primers from the extreme ends of the cDNA as described confirms that the resulting amplified cDNA represents a mRNA actually present in the starting population.

Acknowledgments

The author thanks Dr. Michael A. Frohman for the title of the chapter and critical comments, and Dr. James B. Bliska, Dr. Gloria Viboud, and Michelle B. Ryndak for editorial suggestions. Portions of this chapter have been adapted and reprinted by permission of the publisher from "Using Rapid Amplification of cDNA Ends (RACE) to Obtain Full-Length cDNAs" by Yue Zhang and Michael A. Frohman in *cDNA Library Protocols* (Cowell, I. and Austin, C., eds.), pp. 61–87, copyright 1997 by Humana Press, Totowa, NJ.

References

1. Frohman, M. A., Dush, M. K., and Martin, G. R. (1988) Rapid production of full-length cDNAs from rare transcripts: amplification using a single gene-specific oligonucleotide primer. *Proc. Natl. Acad. Sci. USA* **85(23)**, 8998–9002.
2. Zhang, Y. and Frohman, M. A. (1997) Using rapid amplification of cDNA ends (RACE) to obtain full-length cDNAs, in *Methods in Molecular Biology* (Cowell, I. G. and Austin, C. A., eds.), Vol. 69, pp. 61–87. Humana, Totowa, NJ.
3. Schramm, G., Bruchhaus, I., and Roeder, T. (2000) A simple and reliable 5′-RACE approach. *Nucleic Acids Res.* **28(22)**, E96.
4. Clontech Laboratories. (1996) CapFinder™ PCR cDNA Library Construction Kit. *Clontechniques* **11,** 1.
5. Schmidt, W. M. and Mueller, M. W. (1999) CapSelect: a highly sensitive method for 5′ CAP-dependent enrichment of full-length cDNA in PCR-mediated analysis of mRNAs. *Nucleic Acids Res.* **27(21),** e31.
6. Sambrook, J., Fritsch, E. F., and Maniatis, T. (1989) *Molecular Cloning: A Laboratory Manual*, Cold Spring Harbor Laboratory, Cold Spring Harbor, NY.
7. Sarkar, G., Kapelner, S., and Sommer, S. S. (1990) Formamide can dramatically improve the specificity of PCR. *Nucleic Acids Res.* **18(24),** 7465.
8. Don, R. H., Cox, P. T., Wainwright, B. J., Baker, K., and Mattick, J. S. (1991). "Touchdown" PCR to circumvent spurious priming during gene amplification. *Nucleic Acids Res.* **19(14),** 4008.
9. Mead, D. A., Pey, N. K., Herrnstadt, C., Marcil, R. A., and Smith, L. M. (1991) A universal method for the direct cloning of PCR amplified nucleic acid. *Bio/Technology* **9(7),** 657–663.
10. Stoker, A. W. (1990) Cloning of PCR products after defined cohesive termini are created with T4 DNA polymerase. *Nucleic Acids Res.* **18(14),** 4290.
11. Iwahana, H., Mizusawa, N., Ii, S., Yoshimoto, K., and Itakura, M. (1994) An end-trimming method to amplify adjacent cDNA fragments by PCR. *Biotechniques* **16(1),** 94–98.
12. Thweatt, R., Goldstein, S., and Shmookler Reis, R. J. (1990) A universal primer mixture for sequence determination at the 3′ ends of cDNAs. *Anal. Biochem.* **190(2),** 314–316.
13. Eckert, K. A. and Kunkel, T. A. (1990) High fidelity DNA synthesis by the Thermus aquaticus DNA polymerase. *Nucleic Acids Res.* **18(13),** 3739–3744.

3

cDNA Generation on Paramagnetic Beads

Zhaohui Wang and Michael G. K. Jones

1. Introduction

Synthesis of complementary DNA (cDNA) by reverse transcription (RT) is a key step in investigating specific gene expression of a single transcript by RT-PCR (polymerase chain reaction) or to study the more complex profiles of gene expression in a biological sample using cDNA library or other techniques. Solid-phase generation of cDNA *(1)*, based on immobilization of the synthesized cDNA on paramagnetic beads and the magnetic-bead separation technology, is a very useful method for studying gene expression particularly when handling limited amounts of starting materials (*see* **Note 1**). Using this method, analysis of gene expression can be carried out at a single-cell level *(2–4)*. This is not possible using conventional techniques of mRNA isolation and cDNA cloning.

There are two standard methods for adapting magnetic beads to the synthesis of cDNA. One method is to covalently link an oligo(dT) tail directly to the paramagnetic beads, and this oligo(dT) tail can be used to capture poly(A)$^+$ RNA (messenger RNA [mRNA]) and prime the RT reaction. Alternatively, an oligonucleotide arm can be inserted between the oligo(dT) site and the beads, and the oligonucleotide arm can serve as a 3′-end priming site in the subsequent PCR amplification *(5)*. Both methods share the benefits of isolation of mRNA and RT directly on beads, which simplifies the experimental process and minimizes DNA contamination (*see* **Note 2**). The cDNA synthesized on beads can be used directly in downstream molecular processes, such as construction of cDNA libraries *(6)*, subtractive libraries *(7,8)*, 5′-RACE *(9)* and cDNA-amplified fragment length polymorphism (cDNA–AFLP) analysis *(10)*.

From: *Methods in Molecular Biology, vol. 221: Generation of cDNA Libraries: Methods and Protocols*
Edited by: S.-Y. Ying © Humana Press Inc., Totowa, NJ

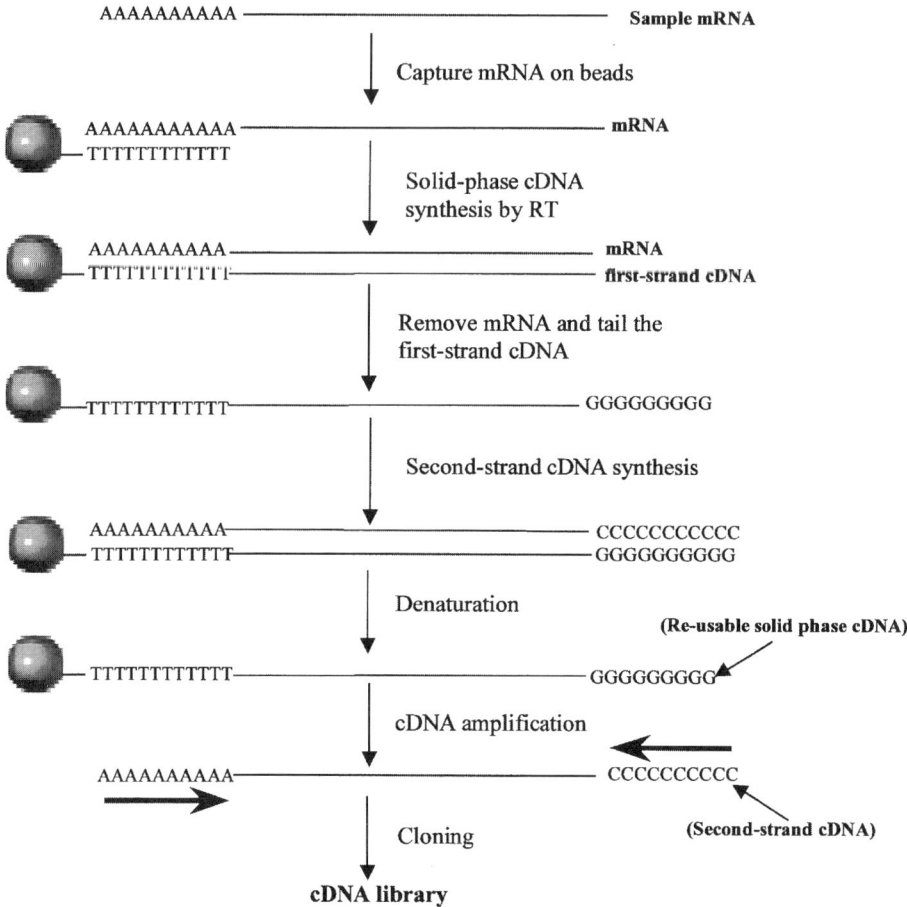

Fig. 1. Solid-phase generation of cDNA from sample mRNA and cDNA amplification.

The method described in this chapter to generate solid-phase cDNA libraries is illustrated in **Fig. 1**. Briefly, mRNA is purified from crude lysate of small tissue samples by annealing to the oligo(dT)$_{25}$ site on the paramagnetic beads. A magnetic particle concentrator (MPC) is used to attract the beads to the side of the reaction tube and enables rapid changes of buffers and solutions. First-strand cDNA is synthesized on beads primed by the oligo(dT)$_{25}$ tail. The uncoupled oligo(dT) sites are removed by T4 DNA polymerase treatment. The first-strand cDNA is then G-tailed by terminal transferase and second-strand cDNA is synthesized using an oligo(dC) primer in a PCR reaction. The first-strand cDNA on beads can be isolated and reused as template for RT-PCR.

The double-stranded cDNA is then amplified and cloned to construct a cDNA library (*see* **Note 3**).

2. Materials

1. Dynabeads oligo(dT)$_{25}$ (product no. 610.02, Dynal Inc.) and MPC (Dynal). Store at 4°C.
2. Diethyl pyrocarbonate (DEPC)-treated water. Use to prepare all solutions for mRNA isolation and first-strand cDNA synthesis.
3. Lysis/binding buffer: 100 mM Tris-HCl (pH 8.0), 500 mM LiCl, 10 mM EDTA (pH 8.0), 1% sodium dodecyl sulfate (SDS), 5 mM dithiothreitol (DTT). Store at 4°C.
4. Washing buffer A: 10 mM Tris-HCl (pH 8.0), 0.15 M LiCl, 1 mM EDTA, 0.1% lithium dodecyl sulfate (LiDS). Store at 4°C.
5. Washing buffer B: 10 mM Tris-HCl (pH 8.0), 0.15 M LiCl, 1 mM EDTA. Store at 4°C.
6. Superscript II reverse transcriptase (200 U/µL) (Life Technologies). Store at –20°C.
7. 5X First-strand buffer: 250 mM Tris-HCl (pH 8.3), 375 mM KCl, 15 mM MgCl$_2$. Store at –20°C.
8. Reverse transcriptase mix: 4 µL of 5X first-strand buffer, 2 µL of 0.1 M DTT, 0.5 µL RNasin (40 U/µL), 1 µL of 10 mM dNTPs (Promega), and 11.5 µL DEPC-treated water; prepare just before use.
9. Storage buffer: 10 mM Tris-HCl (pH 8.0).
10. T4 DNA polymerase (5 U/µL) (Gibco-BRL). Store at –20°C.
11. 5X T4 DNA polymerase buffer: 165 mM Tris-acetate (pH 7.9), 330 mM Na-acetate, 50 mM Mg-acetate, 2.5 mM DTT, 0.5 mg/mL bovine serum albumin (BSA). Store at –20°C.
12. T4 DNA polymerase reaction mix: 4 µL of 5X T4 DNA polymerase buffer, 1 µL T4 DNA polymerase, and 15 µL DEPC-treated water, prepare just before use.
13. 0.5 M EDTA (pH 8.0).
14. Terminal deoxynucleotidyl transferase (TdT, 20 U/µL) (Gibco-BRL). Store at –20°C.
15. 5X Tailing buffer: 500 mM cacodylate buffer (pH 7.2), 10 mM CoCl2, 1 mM DTT. Store at –20°C.
16. Tailing reaction mix: 4 µL of 5X tailing buffer, 5 µL of 20 µM dGTP (Promega), 1 µL of TdT, and 10 µL of DEPC-treated water; prepare just before use.
17. *Taq* DNA polymerase (5.5 U/µL) (Biotech International). Store at –20°C.
18. 10X PCR reaction buffer: 67 mM Tris-HCl (pH 8.8), 16.6 mM (NH$_4$)$_2$SO$_4$, 0.45% Triton X-100. Store at –20°C.
19. PCR primers: 10 pM of *Not*I-dC (5′-dCTCTCTATAGTCGAC$_{14}$-3′) and *Sal*I-dT (5′-dCTCTGCGGCCGCT$_{17}$-3′).
20. PCR reaction mix: 5 µL of 10X PCR buffer, 1 µL of *Taq* DNA polymerase, 2.5 µL each of *Not*I-dC and *Sal*I-dT primer, 2 µL of 25 mM MgCl$_2$, 2 µL of 10 mM dNTPs, 35 µL of water; prepare just before use.

3. Methods

3.1. mRNA Isolation Using Paramagnetic Beads

Plant root tissue was used as starting material for this protocol. However, other plant tissues, animal tissues, cultured cells, or extract of cytoplasmic contents from single cells can also be used as starting material. Upon collection, they must be stored at −80°C as quickly as possible. All manipulations must be carried out in an RNase-free environment.

1. Transfer 250 µL of Dynabeads oligo-(dT)$_{25}$ from stock suspension to an RNA-free tube placed in a MPC and remove the supernatant by pipetting.
2. Wash the beads once by resuspending in 200 µL lysis/binding buffer.
3. Grind 100 mg of frozen plant root tissue in liquid nitrogen.
4. Transfer the frozen powder to a 1.5-mL Eppendorf tube containing 1 mL lysis/binding buffer. Vortex the tube for 1–2 min to obtain complete lysis.
5. Centrifuge the lysate at 20,800g for 30 s.
6. Remove the lysis/binding buffer from the beads placed in the MPC. Transfer the supernatant of the lysate to the washed beads.
7. Resuspend the beads by pipetting and anneal the mRNA to the beads by rotating for 3–5 min at room temperature.
8. Place the tube in the MPC to settle the beads and discard the supernatant.
9. Wash the beads twice by resuspending in 500 µL washing buffer A at room temperature.
10. Repeat the washing step twice with 500 µL washing buffer B (*see* **Note 4**).

3.2. First-Strand cDNA Synthesis on Paramagnetic Beads

The isolated mRNA bound to Dynabeads can be directly used as a template in reverse transcription. Alternatively, the mRNA can be eluted from the beads and store at −80°C until use.

1. Wash the mRNA on beads once with 1X first-strand buffer. Discard the supernatant using the MPC.
2. Resuspend the beads in 19 µL of reverse transcriptase mix solution and heat to 42°C for 2 min.
3. Add 1 µL of Superscript II reverse transcriptase to the reaction tube and mix thoroughly.
4. Incubate the reaction tube at 42°C for 60 min.
5. Stop the reaction by heating at 70°C for 15 min.
6. Denature the mRNA–cDNA hybridization by heating up to 92°C for 2 min.
7. Immediately transfer the reaction tube into the MPC; discard the supernatant containing mRNA.
8. Add 20 µL of storage buffer into the tube. The first-strand cDNA is now attached to the beads.

3.3. Tailing of the First-Strand cDNA

To synthesis second-strand cDNA for further cDNA library construction, the first-strand cDNA is tailed with oligo-(dG) to create the binding site for *Not*I-dC primer. The free oligo-(dT) sites on the beads are removed by T4 DNA polymerase prior to tailing (*see* **Note 5**).

1. Remove the buffer from the first-strand cDNA on beads using the MPC.
2. Add the 20 μL T4 DNA polymerase reaction mix to the tube and incubate at 37°C for 10 min.
3. Stop the reaction by adding 1 μL of 0.5 *M* EDTA.
4. Wash the cDNA on beads twice with 100 μL of 1X tailing buffer.
5. Remove the supernatant using the MPC and add 20 μL of tailing reaction mix.
6. Incubate the reaction at 37°C for 30 min.
7. Stop the reaction by adding 1 μL of 0.5 *M* EDTA.

3.4. Second-Strand cDNA Synthesis and cDNA Amplification

The second-strand cDNA is synthesised on the beads. The first-strand cDNA linked to the beads is isolated using the MPC by denaturing the double-stranded cDNA and can be reused as template in second-strand cDNA generation or PCR amplification.

1. Wash the dG-tailed first-strand cDNA on beads twice with 1X PCR reaction buffer.
2. Remove the buffer and add 50 μL of the PCR reaction mix.
3. Create second-strand cDNA by one cycle of PCR: 94°C for 2 min, 52°C for 2 min, and 72°C for 3 min.
4. Denature the double-stranded cDNA at 94°C for 2 min.
5. Immediately place the reaction tube into the MPC. Transfer the supernatant to another PCR tube.
6. Continue the PCR amplification: 30–35 cycles of 94°C for 30 s, 52°C for 30 s, 72°C for 3 min, followed by 1 cycle of 72°C for 10 min.
7. The first-strand cDNA on beads can be kept in the storage buffer and reused.
8. The amplified cDNA can be purified and directly cloned into a TA vector to generate a small library of thousands of primary transformants. Alternatively, the two restriction sites (*Not*I and *Sal*I) at the 5′/3′ end of the cDNA can be used to clone the cDNA into a vector (*see* **Note 6**).

4. Notes

1. The protocol of mRNA isolation using Dynabeads is recommended for direct, high-purity, and intact poly(A⁺) RNA isolation from small amount of starting materials. A specific amount of tissue, such as 100 mg of plant tissue or 20–50 mg of animal tissue, should be used for each isolation using 250 μL of beads, because

an excess of tissue will reduce the mRNA yield and purity. For mRNA isolation and cDNA generation at the single-cell level, the amount of beads can be scaled down to 20–50 μL. We have used 20 μL of beads to capture the mRNA from extracts of cytoplasmic contents from single giant cells induced by root-knot nematodes in tomato root and generated first-strand cDNA on beads to detect different transcripts in giant cells *(4)*. However, cDNA generated on beads from a single cell may not be suitable for constructing a full cDNA library. Large variations in PCR amplification can occur when the PCR template amount falls below a certain threshold copy number *(2)*. This effect, termed "Monte Carlo," will directly decrease the reproducibility of cDNA amplification and the representation of the cDNA library constructed from single cells.

2. To carry out large-scale mRNA isolation, the beads can be reused up to four times to reduce preparation costs. After the first round of mRNA isolation, the mRNA bound on beads can be eluted from the beads by adding 10–20 μL of elution buffer (2 m*M* EDTA [pH 8.0]), keeping it at 65°C for 2 min. Immediately place the tube into the MPC and transfer the supernatant containing the mRNA to another tube. Resuspend the beads in 200 μL of reconditioning buffer (0.1 *M* NaOH), keep at 65°C for 2 min, and repeat the reconditioning step once. Wash the beads three times with storage solution (250 m*M* Tris-HCl [pH 8.0], 20 m*M* EDTA, 0.1% Tween-20, 0.02% sodium azide). The beads are then ready for another mRNA isolation. To avoid any cross contamination, reuse of the reconditioned beads for different tissue or cell samples is not recommended.

3. This protocol normally results in very pure mRNA with traces of ribosomal RNA. DNA contamination might be found for some cell types and tissues. For critical applications such as cDNA library construction, trace rRNA and DNA contamination should be avoided by carrying out an extra round of mRNA purification with the same protocol.

4. For all steps of changing the solution, the beads must be washed thoroughly and the supernatant removed properly by pipetting to eliminate any residue of reagents, such as salts, detergent, and enzymes, from the previous step.

5. It is necessary to conduct T4 DNA polymerase treatment on the first-strand cDNA on beads to remove the uncoupled oligo-(dT) sites on the beads. The terminal transferase can also tail the residual oligo-(dT) sites with oligo-(dG). These short oligo-(dT)-(dG) fragments may interfere in the following PCR amplification by competing for the primer binding sites.

6. The majority of the cloned cDNA in the cDNA library is relatively small-size fragments (<300 bp). This reflects selective amplification and cloning of smaller cDNAs, and this is a common limitation of all PCR-based cDNA libraries. However, the presence of full length cDNA can be confirmed by amplifying specific transcripts using first-strand cDNA on beads as PCR template with gene-specific primers, and full-length cDNA is always obtained. Size fractionation of the amplified cDNA before the cloning step should lead to increased insert sizes and thus provide more sequence information of the translated regions.

References

1. Raineri, I., Moroni, C., and Senn, H. P. (1991) Improved efficiency for single-sided PCR by creating a reusable pool of first-strand cDNA coupled to a solid phase. *Nucleic Acids Res.* **19,** 4010.
2. Karrer, E. E., Lincoln, J. E., Hogenhout, S., Benett, A. B., Bostock, R. M., Martineau, B., et al. (1995) *In situ* isolation of mRNA from individual plant cells—creation of cell-specific cDNA libraries. *Proc. Natl. Acad. Sci. USA* **92,** 3814–3818.
3. Schütze, K. and Lahr, G. (1998) Identification of expressed genes by laser-mediated manipulation of single cells. *Nat. Biotech.* **16,** 737–742.
4. Wang, Z., Potter, R. H., and Jones, M. G. K. (2001) A novel approach to extract and analyse cytoplasmic contents from individual giant cells in tomato roots induced by *Meloidogyne javanica. Int. J. Nematol.* **11,** 219–225.
5. Lambert, K. N. and Williamson, V. M. (2000) cDNA library construction using streptavidin-paramagnetic beads and PCR, in *The Nucleic Acid Protocols Handbook* (Rapley, R., ed), Humana, Totowa, NJ, pp. 289–294.
6. Lambert, K. N. and Williamson, V. M. (1993) cDNA library construction from small amounts of RNA using paramagnetic beads and PCR. *Nucleic Acids Res.* **21,** 775–776.
7. Heinrich, T., Washer, S., Marshall, J., Jones, M. G. K., and Potter, R. H. (1997) Subtractive hybridisation of cDNA from small amounts of plant tissue. *Mol. Biotech.* **8,** 8–12.
8. Sharma, P., Lönneburg, A., and Stougaard, P. (1993) PCR-based construction of subtractive cDNA library using magnetic beads. *BioTechniques* **15,** 610–611.
9. Rodriguez, I. R., Mazuruk, K., Schoen, T. J., and Chader, G. J. (1994) Structural analysis of the human hydroxyindole-*o*-methyltransferase gene. *J. Biol. Chem.* **269,** 31,969–31,977.
10. Bachem, C. W. B., Ommen, R. J. F. J., and Visser, R. G. F. (1998) Transcript imaging with cDNA-AFLP: a step-by-step protocol. *Plant Mol. Biol. Rep.* **16,** 157–173.

4

Construction of a Normalized cDNA Library by mRNA–cDNA Hybridization and Subtraction

Ye-Guang Chen

1. Introduction

The human genome project has predicted that the human genome encodes about 35,000 genes (1,2). These genes are not uniformly expressed in all of the cells. Some of them are expressed in most of the cells, but others are cell- or tissue-specific. It has been estimated that about 10,000 genes are expressed in a cell, but the abundance of their expression varies from 1 copy to 200,000 copies per cell (3). On average, the 10 most prevalent genes encode more than 5000 copies per each, whereas most others may be represented only by 1–15 copies (4,5). It is very difficult to identify the rarely represented mRNA from any tissue or cell type. Therefore, when constructing a cDNA library, a big challenge is how to reduce the number of the highly abundant species and, at the same time, to maintain the complexity of cDNA in the population (i.e., how to generate a normalized library). The generation of expressed sequence tags (ESTs) by single-pass sequencing of cDNA clones has been greatly accelerating gene discovery (6). One of the important applications of normalized cDNA libraries is to provide great sources for efficient large-scale generation of ESTs (7–9).

To bring the representation of each cDNA species in a population within a narrow range, various normalization procedures have been applied (3,7,10–14). All of those normalization procedures take advantage of the second-order kinetics of nucleic acid association: The highly abundant species associate faster than the low abundant ones (10,15,16). By complementary DNA (cDNA)–cDNA or messenger RNA (mRNA)–cDNA self-hybridization and double-strand exclusion, the high-abundant species are eliminated. In this way, the abundance

From: *Methods in Molecular Biology, vol. 221: Generation of cDNA Libraries: Methods and Protocols*
Edited by: S.-Y. Ying © Humana Press Inc., Totowa, NJ

of cDNAs in a library is brought to a narrow range and the chance to identify a rare species is increased.

The procedures described here are based on a method developed by Sasaki et al. *(11)*. They have reported a normalization procedure in which a cDNA library is constricted following removal of abundant mRNA species by sequential cycles of self-hybridization between a whole mRNA population and its corresponding cDNA immobilized on beads. This method involves relatively simple manipulations. It has been shown not only to achieve a reasonable normalization but, at the same time, to also conserve the original length of clones. Therefore, a cDNA library generated by this method has a high possibility to yield cDNAs with a full-length open reading frame and can be used for expression cloning. Another potential use of this method is for cloning of the genes differentially expressed in specific tissues or cell types or the genes differentially expressed in response to a treatment. This method has been verified by constructing a normalized cDNA library from the human brain. Abundant mRNA species such as those for cytochrome-*c* oxidase subunit III and NADH dehydrogenase subunit 2 were reduced approx 25-fold, whereas rare mRNA such as one for prohibitin was enriched eightfold in the normalized cDNA library when compared to a normal cDNA library *(17)*.

2. Materials

Because RNA is vulnerable for degradation, it has to be handled with extreme care. A RNase-free environment should be set up *(18)*. Gloves should be worn at all times. Plasticware should be autoclaved. All glassware used for isolation of RNA is treated with 0.02% diethyl pyrocarbonate (DEPC)-treated water and autoclaved before use. Aerosol-resistant pipet tips are recommended. All the buffers are stored at 4°C unless indicated.

2.1. Isolation of Total Cellular RNA

1. DEPC-treated water: Add DEPC (diethylpyrocarbonate) to 0.02% (v/v) and mix. Stand overnight at 37°C and then autoclave. Store at room temperature.
2. Trizol reagent (Invitrogen, cat. no. 15596026).
3. RNas-free DNase I (10 U/μL) (Invitrogen, cat. no. 18068015). Store at –20°C.
4. 10X DNas I buffer: 200 m*M* Tris-HCl (pH 8.3), 500 m*M* KCl, 20 m*M* MgCl$_2$. Store at –20°C.
5. Phenol : choloroform : isoamyl alcohol (25 : 24 : 1): purchased from Fisher (cat. no. BP1752I-400).
6. RNase inhibitor (20 U/μL): purchased from Promega (cat. no. N2511). Store at –20°C.

2.2. Preparation of mRNA

1. Oligotex suspension (Qiagen, cat. no. 79000). Keep at room temperature.
2. Buffer A: 20 mM Tris-HCl (pH 7.5), 1 M NaCl, 2 mM EDTA, 0.2% sodium dodecyl sulfate (SDS).
3. Washing buffer: 10 mM Tris-HCl (pH 7.5), 150 mM NaCl, 1 mM EDTA.
4. Elution buffer: 5 mM Tris-HCl (pH 7.5).

2.3. Conversion of mRNA to cDNA on Latex Beads

1. Buffer B: 50 mM Tris-HCl (pH 8.3), 10 mM MgCl$_2$, 100 mM KCl.
2. Buffer C: 50 mM Tris-HCl (pH 8.3), 3 mM MgCl$_2$, 75 mM KCl.
3. Reverse transcriptase Superscript II (Invitrogen, cat. no. 18064-022). Store at –20°C.
4. TE buffer: 10 mM Tris-HCl (pH 7.5), 1 mM EDTA. Store at room temperature.
5. dNTP mix (Promega, cat. no. U1330 or U1242). Store at –20°C.
6. [α-^{32}P]-dCTP (specific activity ~3000 Ci/mmol) (Amersham, cat. no. PB10205). Store at –20°C.

2.4. mRNA–cDNA Self-Hybridization

1. Oligo (dT) primer with XhoI and KpnI sites: 5′-GAA GAA GAA <u>CTC GAG GGT ACC</u> TTT TTT TTT TTT TTT-3′.
2. Oligo(dA) (25–30 mer): synthesized as custom oligonucleotide.
3. Hybridization buffer: 10 mM Tris-HCl (pH 7.5), 1 mM EDTA, 100 mM NaCl.

2.5. Construction of cDNA Libraries

1. Reverse transcriptase Superscript II (Invitrogen, cat. no. 18064-022). Store at –20°C.
2. *Escherichia coli* RNase H (2 U/µL) (Invitrogen, cat no. 18021-071). Store at –20°C.
3. *E. coli* DNA polymerase I (10 U/µL) (Invitrogen, cat. no. 18010025). Store at –20°C.
4. *E. coli* DNA ligase (10 U/µL) (Invitrogen, cat. no. 18052019). Store at –20°C.
5. β-NAD (Fisher, cat. no. BP2532). Store at –20°C.
6. T4 DNA polymerase (Invitrogen, cat. no. 18005017). Store at –20°C.
7. Sephacryl s-400 column (Promega, cat. no. V3181).
8. *Eco*RI-*Not*I-*Bam*HI adapter: synthesized as custom oligonucleotide or purchased from Takara (cat. no. TAK 4510).

3. Methods

3.1. Isolation of Total Cellular RNA

1. Lyse monolayer cells by adding 3.0 mL of Trizol Reagent to 1 of the 100-mm cell culture dishes. Pipet cell lysate up and down several times (*see* **Note 1**).

2. Incubate for 5 min at room temperature.
3. Add 0.2 mL of chloroform per 1 mL of Trizol. Shake tube vigorously by hand for 15 s and incubate 3 min at room temperature.
4. Centrifuge at 12,000g for 15 min at 4°C.
5. Transfer the colorless upper aqueous phase to a new tube and add 0.5 mL of isopropanol per 1 mL of Trizol used to precipitate RNA.
6. After incubating for 10 min at room temperature, centrifuge the sample at 12,000g for 10 min at 4°C.
7. Wash the RNA pellet with 75% ethanol. Use more than 1 mL of 75% ethanol per 1 mL of Trizol used (*see* **Note 2**).
8. Briefly air-dry the RNA pellet for 5–10 min at room temperature and dissolve RNA to 1 µg/µL in DEPC-treated water (*see* **Note 3**). Store RNA solution at –80°C.
9. Treat the total RNA with RNase-free DNase I:
   250 µL of total RNA (250 µg, 1 µg/µL);
   50 µL of 10X DNas I buffer (200 mM Tris-HCl 8.3, 500 mM KCl, 20 mM MgCl$_2$);
   2 µL of RNase inhibitor (20 U/µL);
   2.5 µL of DNase I (10 U/µL);
   195.5 µL of H$_2$O.
   Mix well and incubate at room temperature for 15 min (*see* **Note 4**).
10. Terminate the reaction by adding 2.5 µL of 500 mM EDTA and extract with 500 µL of phenol:choloroform:isoamyl alchohol (25:24:1). Transfer the supernatant and precipitate RNA with 50 µL of 3 M sodium acetate (pH 5.5) and 1 mL of 100% ethanol. Dissolve RNA to 1 µg/µL in DEPC-treated water and store at –80°C.

3.2. Isolation of Poly(A$^+$) mRNA

Isolation of poly(A$^+$) mRNA is accomplished with Oligotex beads that are covalently linked to oligo(dT)$_{30}$. Refer to the manual from the manufacturer for the details.

1. Warm up Oligotex suspension at 37°C before use. Heat elution buffer to 70°C.
2. To 250 µg total RNA, add 250 µL buffer A and 25 µL Oligotex suspension. Mix the contents thoroughly by pipetting.
3. Incubate the sample for 3 min at 70°C in a water bath to disrupt the secondary structure of the RNA.
4. Place the sample at room temperature for 10 min to allow hybridization between the oligo(dT)$_{30}$ of the Oligotex bead and the poly(A) tail of the mRNA.
5. Pellet the Oligotex:mRNA complex by centrifugation for 2 min at 14,000–18,000g and carefully remove the supernatant by pipetting. (Save the supernatant until certain that satisfactory binding and elution of poly(A$^+$) mRNA has occurred.)

6. Resuspend the Oligotex:mRNA pellet in 1 mL washing buffer by vortex or pipetting.
7. Pellet the Oligotex:mRNA complex by centrifugation for 2 min at 14,000–18,000*g* and carefully remove the supernatant with a pipet.
8. Repeat **steps 6** and **7** once.
9. Add 20–100 µL of hot (70°C) elution buffer. Pipet up and down three or four times to resuspend the resin and centrifuge for 2 min at 14,000–18,000*g*. Carefully transfer the supernatant, which contains the eluted poly(A⁺) mRNA, to another RNase-free tube (*see* **Note 5**).
10. For maximal yield, add another 20–100 µL of elution buffer to the Oligotex pellet and combine the eluates.

Approximately 5–10 µg of poly(A⁺) mRNA are yielded from 250 µg of total RNA.

3.3. Conversion of mRNA to cDNA on Latex Beads

1. Mix 20 µg poly(A)⁺ RNA with 2.5 mg Oligotex beads in 250 µL buffer B. Incubate at 37°C for 20 min.
2. Centrifuge at 15,000*g* for 10 min at room temperature.
3. Resuspend beads in 250 µL of buffer C.
4. Set up reverse transcription in a 500-µL reaction:
   50 m*M* Tris-HCl (pH 8.3);
   3 m*M* MgCl$_2$;
   75 m*M* KCl;
   1 m*M* of dNTP (dATP, dCTP, dGTP, dTTP);
   10 m*M* dithiothreitol (DTT);
   300 U RNase inhibitor;
   2000 U Superscript II.
   Incubate 90 min at 45°C.
5. Wash beads two times with TE.
6. Heat for 3 min at 95°C to remove RNA.
7. Spin and resuspend beads in TE.
8. Store at 4°C.
9. Monitor cDNA synthesis with 0.5 µL of [α-^{32}P]-dCTP (10 µCi/µL) in the 10-µL reaction and count bead-associated radioactivities.

3.4. mRNA–cDNA Self-Hybridization (see Note 6)

1. Suspend 2.5 mg cDNA–Oligotex beads in 90 µL of TE containing 100 µg of oligo(dA) (25–30 mer). Heat 5 min at 70°C.
2. Add 10 µL of 5 *M* NaCl and incubate for 10 min at 37°C to mask free oligo(dT) residues on beads.
3. After centrifugation, the beads are incubated with 2 µg of poly(A)⁺ RNA in 200 µL hybridization buffer. Incubate 15 min at 55°C with occasional agitation.

4. Remove the beads by centrifugation (15,000g) for 10 min.
5. The supernatant fraction is subjected to a second cycle of hybridization. Repeat hybridization (**steps 1–4**) three times using regenerated cDNA–Oligotex beads (*see* **step 7**). After four cycles of hybridization and subtraction, approx 2–4% of the input poly(A) RNA are left in the supernatant fraction.
6. Treat the supernatant with equal volume of phenol:chloroform. RNA is precipitated with ethanol, air-dried, dissolved with H_2O, and stored in –20°C; it can be used for construction of cDNA libraries.
7. Regeneration of cDNA–Oligotex beads:
   a. Resuspend the beads in 200 μL of TE.
   b. Heat 5 min at 70°C and then chill on ice.
   c. Wash two times with TE and resuspend with TE.

3.5. Construction and Characterization of Normalized cDNA Libraries

1. Anneal poly(A)$^+$ RNA to the oligo(dT) primer with *Xho*I and *Kpn*I sites by adding 0.5 μg of the primer, the normalized mRNA, and DEPC-treated water to 10 μL into an autoclaved RNase-free 1.5-mL microcentrifuge tube. Heat the mixture to 70°C for 10 min and quickly chill on ice. Collect the contents of the tube by brief centrifugation.
2. Synthesize the first strand of cDNA with Superscript II in a 20-μL reaction volume.
   50 mM Tris-HCl (pH 8.3);
   75 mM KCl;
   3 mM MgCl$_2$;
   10 mM DTT;
   500 μM each dATP, dCTP, dGTP, dTTP;
   200 U Superscript II.
   Mix gently and incubate at 45°C for 1 h. Place the tube on ice.
3. Synthesize second strand by adding the following reagents directly to the first-strand reaction mixture:
   5 μL of 1 M Tris-HCl (pH 6.9) (final 50 mM);
   4.5 μL of 1 M MgCl$_2$ (final 5 mM);
   12 μL of 1 M KCl (final 100 mM);
   3.3 μL of 1 M DTT (final 5 mM);
   1.2 μL of 1 M (NH$_4$)$_2$SO$_4$ (final 10 mM);
   1.2 μL of 10 mM β-NAD$^+$ (final 0.1 mM);
   6 μL of 5 mM dNTP (final 0.33 mM each);
   93.5 μL H$_2$O;
   1 μL of 2 U/μL *E. coli* RNase H (final 2 U);
   1 μL of 10 U/μL *E. coli* DNA ligase (final 10 U);
   4 μL of 10 U/μL *E. coli* DNA polymerase I (final 40 U).

Mix gently and incubate at 16°C for 2–4 h. Then, add 10 U of T4 DNA polymerase and incubate at 16°C for 5 min. Place reaction on ice and add 10 μL of 0.5 *M* EDTA. The product is blunt-ended, double-stranded cDNA.

4. Purify cDNA by phenol : chloroform extraction and ethanol precipitation.
5. Ligate to *Eco*RI-*Not*I-*Bam*HI adapters.
6. Purify DNA with Sephacryl s-400 column.
7. Clone cDNA to λgt10 vector via *Eco*RI site or to λZAPII vector via *Eco*RI/*Xho*I sites.
8. Normalization of cDNA species can be confirmed by DNA sequencing of randomly picked clones from the library or by plague hybridization.

4. Notes

1. For isolation of RNA from suspension-cultured cells or from tissues, refer to the Trizol Reagent instruction from the manufacturer. Samples can be stored in Trizol solution at –80°C for at least 1 mo.
2. RNA pellet can be stored at –20°C for 1 yr at this step.
3. Do not dry completely because it would result in low solubility. When dissolving RNA, incubation at 55°C for 10 min helps dissolution.
4. It is important not to exceed the 15-min incubation time or the room-temperature incubation. Higher temperatures and longer times could lead to Mg^{2+}-dependent hydrolysis of RNA.
5. The volume of elution buffer used depends on the expected or desired concentration of poly(A^+) mRNA. Ensure that elution buffer does not cool significantly during handling.
6. Normalization efficiency can be controlled by changing the number of hybridization cycles and the molar ratio of cDNA on the beads to mRNA in solution. When mRNA source is limited and, at the same time, a great degree of normalization needs to be achieved, DNA inserts in a phage λ cDNA library can be transcribed in vitro by T3 RNA polymerase, and the resulting transcripts can be used for hybridization with the cDNA–Oligotex beads (*see* **ref. *17***).

References

1. Lander, E. S., Linton, L. M., Birren, B., Nusbaum, C., Zody, M. C., et al. (2001) Initial sequencing and analysis of the human genome. *Nature* **409,** 860–921.
2. Venter, J. C., Adams, M. D., Myers, E. W., Li, P. W., Mural, R. J., et al. (2001) The sequence of the human genome. *Science* **291,** 1304–1351.
3. Patanjali, S. R., Parimoo, S., and Weissman, S. M. (1991) Construction of a uniform-abundance (normalized) cDNA library. *Proc. Natl. Acad. Sci. USA* **88,** 1943–1947.
4. Davidson, E. H. and Britten, R. J. (1979) Regulation of gene expression: possible role of repetitive sequences. *Science* **204,** 1052–1059.

5. Bishop, J. O., Morton, J. G., Rosbash, M., and Richardson, M. (1974) Three abundance classes in HeLa cell messenger RNA. *Nature* **250,** 199–204.
6. Adams, M. D., Dubnick, M., Kerlavage, A. R., Moreno, R., Kelley, J. M., Utterback, T. R., et al. (1992) Sequence identification of 2,375 human brain genes. *Nature* **355,** 632–634.
7. Bonaldo, M. F., Lennon, G., and Soares, M. B. (1996) Normalization and subtraction: two approaches to facilitate gene discovery. *Genome Res.* **6,** 791–806.
8. Hillier, L. D., Lennon, G., Becker, M., Bonaldo, M. F., Chiapelli, B., et al. (1996) Generation and analysis of 280,000 human expressed sequence tags. *Genome Res.* **6,** 807–828.
9. Soares, M. B., Bonaldo, M. F., Jelene, P., Su, L., Lawton, L., and Efstratiadis, A. (1994) Construction and characterization of a normalized cDNA library. *Proc. Natl. Acad. Sci. USA* **91,** 9228–9232.
10. Ko, M. S. (1990) An "equalized cDNA library" by the reassociation of short double-stranded cDNAs. *Nucleic Acids Res.* **18,** 5705–5711.
11. Sasaki, Y. F., Ayusawa, D., and Oishi, M. (1994) Construction of a normalized cDNA library by introduction of a semi-solid mRNA–cDNA hybridization system. *Nucleic Acids Res.* **22,** 987–992.
12. Diatchenko, L., Lau, Y. F., Campbell, A. P., Chenchik, A., Moqadam, F., et al. (1996) Suppression subtractive hybridization: a method for generating differentially regulated or tissue-specific cDNA probes and libraries. *Proc. Natl. Acad. Sci. USA* **93,** 6025–6030.
13. Neto, E. D., Harrop, R., Correa-Oliveira, R., Wilson, R. A., Pena, S. D., and Simpson, A. J. (1997) Minilibraries constructed from cDNA generated by arbitrarily primed RT-PCR: an alternative to normalized libraries for the generation of ESTs from nanogram quantities of mRNA. *Gene* **186,** 135–142.
14. Poustka, A. J., Herwig, R., Krause, A., Hennig, S., Meier-Ewert, S., and Lehrach, H. (1999) Toward the gene catalogue of sea urchin development: the construction and analysis of an unfertilized egg cDNA library highly normalized by oligonucleotide fingerprinting. *Genomics* **59,** 122–133.
15. Weissman, S. M. (1987) Molecular genetic techniques for mapping the human genome. *Mol. Biol. Med.* **4,** 133–143.
16. Galau, G. A., Klein, W. H., Britten, R. J., and Davidson, E. H. (1977) Significance of rare mRNA sequences in liver. *Arch. Biochem. Biophys.* **179,** 584–599.
17. Sasaki, Y. F., Iwasaki, T., Kobayashi, H., Tsuji, S., Ayusawa, D., and Oishi, M. (1994) Construction of an equalized cDNA library from human brain by semi-solid self-hybridization system. *DNA Res.* **1,** 91–96.
18. Sambrook, J., Fritsch, E. F., and Maniatis, T. (1989) *Molecular Cloning: A Laboratory Manual*, 2nd ed., Cold Spring Harbor Laboratory, Cold Spring Harbor, NY.

5

Amplification of cDNA Ends Using PCR Suppression Effect and Step-Out PCR

Mikhail V. Matz, Naila O. Alieva, Alex Chenchik, and Sergey Lukyanov

1. Introduction

In Chapter 10, we describe two methods of complementary DNA (cDNA) synthesis and amplification. In this chapter, we will refer to these two methods of cDNA synthesis and amplification as "method A" (the technique based on classical double-stranded cDNA synthesis and subsequent adapter ligation, implemented in the Marathon kit from Clontech) and "method B" (based on the template-switching effect, implemented in the SMART cDNA synthesis kit from Clontech). These methods not only allow for amplification of representative cDNA populations from microscopic tissue samples but also provide an excellent starting point for amplification of unknown flanks of a known cDNA fragment. There are many techniques designed for this purpose (*1–10*); among gene hunters, they go under the generic name RACE (rapid amplification of cDNA ends). It must be remembered, though, that this name formally belongs to a particular technique introduced by Frohman et al. (*8*).

The method of RACE described here is based on three key principles. First, to achieve required specificity (and therefore sensitivity), we employ two-step polymerase chain reaction (PCR) using nested gene-specific primers. The necessity of this is usually underestimated in RACE, because, historically, the heaviest nonspecific amplification in RACE stemmed from ubiquitous adapter-specific primers rather than gene-specific ones. Our own techniques were developed for single-step RACE (*4,10*), but they are only as good as the specificity of the single gene-specific primer upon which they rely. At the current stage of methodology development, it is nonspecific amplification

From: *Methods in Molecular Biology, vol. 221: Generation of cDNA Libraries: Methods and Protocols*
Edited by: S.-Y. Ying © Humana Press Inc., Totowa, NJ

originating from a gene-specific primer that often limits the power of the RACE, so we presently switched back to using the good old two-step nested PCR to eliminate this problem.

Second, the design of adapters and adapter-specific primers evokes so-called PCR-suppression effect, or PS effect, to prohibit amplification of molecules that do not contain annealing site for a gene-specific primer *(11)*. The PCR-suppression technology is based on the observation that molecules flanked by inverted terminal repeats at least 40 bases long are amplified very inefficiently when the single primer used for amplification corresponds to the distal half of the repeat *(10,12)* or when a long primer corresponding to the whole repeat sequence is used in low concentration *(13)*. The reason for this is the equilibrium between productive PCR primer annealing and nonproductive self-annealing of the fragment's complementary ends, which arises within each PCR cycle at the primer annealing stage and, in the above two cases, is markedly shifted toward self-annealing *(11)*.

The third principle is step-out PCR *(4)*. In our protocol, this trick is applied during the second stage of amplification. It consists in substituting the adapter-specific primer used during the first stage for another long primer of which only the 3′-half corresponds to the adapter, specifically, to its distal half. In the resulting product, the nonmatching 5′ portion of the new primer sequence is added to the distal end of the original adapter. This situation can be viewed as shifting of the primer annealing site to the outside of the original amplicon, hence the name "step-out." Although this procedure seems to be the contrary of nested PCR, it is, in fact, analogous to it when primer systems evoking the PS effect are concerned: It increases the length of inverted terminal repeats carried by the nonspecifically amplified molecules, which suppresses their amplification even further.

The protocol described here assumes simultaneous RACEs for 3′ and 5′ flanks, starting from a known fragment of cDNA sequence 500–1000 bases long. This situation is the most common in PCR-based gene hunting. The logistics and schematic protocol of this experiment are outlined in **Fig. 1**, typical results in **Fig. 2**. As depicted there, in this case the four gene-specific primers can be designed to allow amplification of the positive control (a part of the known fragment) at the first RACE stage, along with nested PCR required at the second RACE stage. If the known fragment is longer than 1000 bases, the RACE primers should be designed to keep 5′ and 3′ nested steps (*see* **Fig. 1** for the definition of the term "nested step") within the recommended limits. This may dictate the need to design two additional gene-specific primers (one for each RACE direction) for amplification of positive controls. These addi-

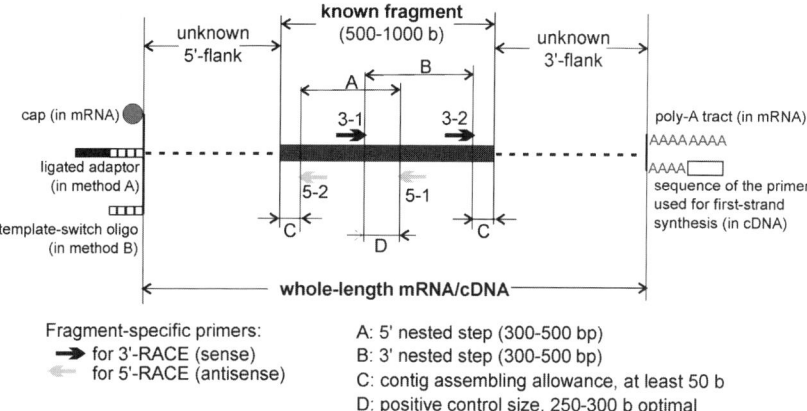

Fragment-specific primers:

➤ for 3'-RACE (sense)

← for 5'-RACE (antisense)

A: 5' nested step (300-500 bp)
B: 3' nested step (300-500 bp)
C: contig assembling allowance, at least 50 b
D: positive control size, 250-300 b optimal

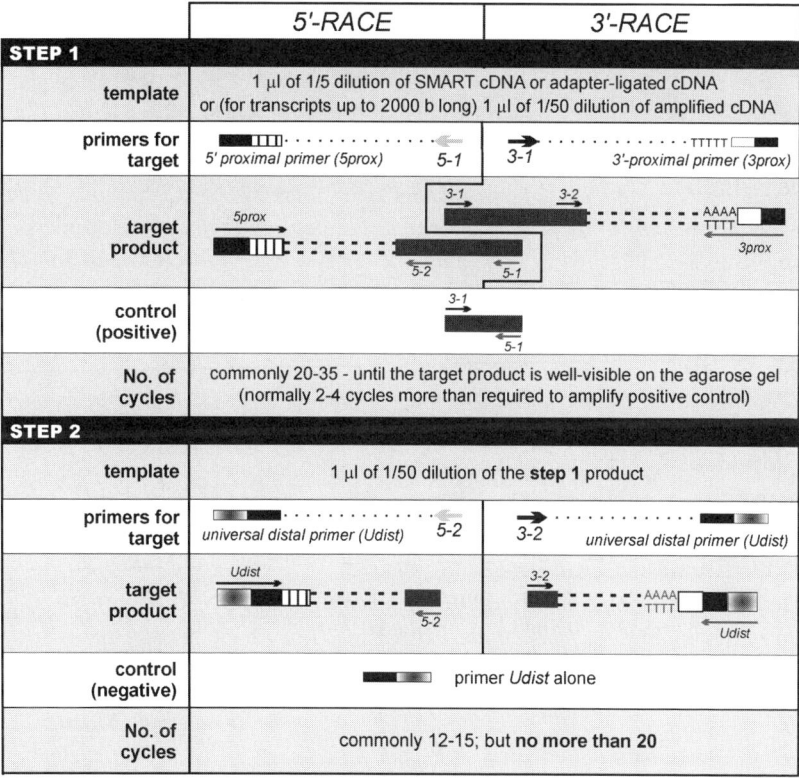

Fig. 1. RACE logistics and scheme of the protocol. The sequences of the adapter-specific primers are pattern coded in accordance with **Fig. 3**.

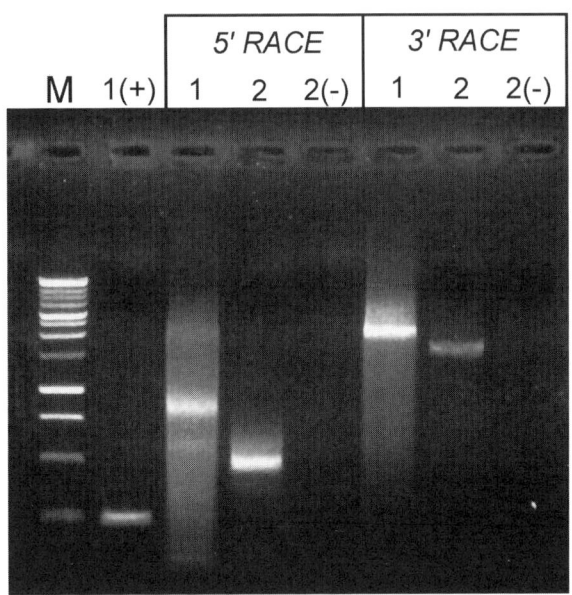

Fig. 2. Typical example of RACE results. In this experiment, 3′ and 5′ flanks of cDNA coding for cadherin-related protein expressed in central nervous system of sea hare *Aplysia californica* were obtained. The template was the amplified cDNA prepared from six identified neurons (metacerebral cells), according to method A (*see* Chapter 10). Lane M: 1-kb DNA ladder (Promega); lane 1(+): positive control from the first step. Three lanes correspond to each RACE direction: 1, first step; 2, second step; 2(–), negative control from the second step.

tional primers should produce amplicons 300–500 base pairs long when used with their correspondent first-stage RACE primers (5-1 and 3-1 in **Fig. 1**).

2. Materials

1. cDNA (amplified or nonamplified; *see* **Note 1**) synthesized according to one of the two methods described in Chapter 10.
2. cDNA dilution buffer: 10 m*M* Tris-HCl (pH 8.0), 10 ng/µL yeast tRNA (*see* **Note 2**).
3. Long-and-Accurate PCR enzyme mix with provided buffer (Advantage2 polymerase mix from Clontech, LA-PCR from Takara, Expand Taq by Boehringer or equivalent).
4. dNTP mix, 10 m*M* each.
5. Oligonucleotides: *see* **Fig. 3** and **Notes 3** and **4**.
6. Agarose gel (1%) containing ethidium bromide.

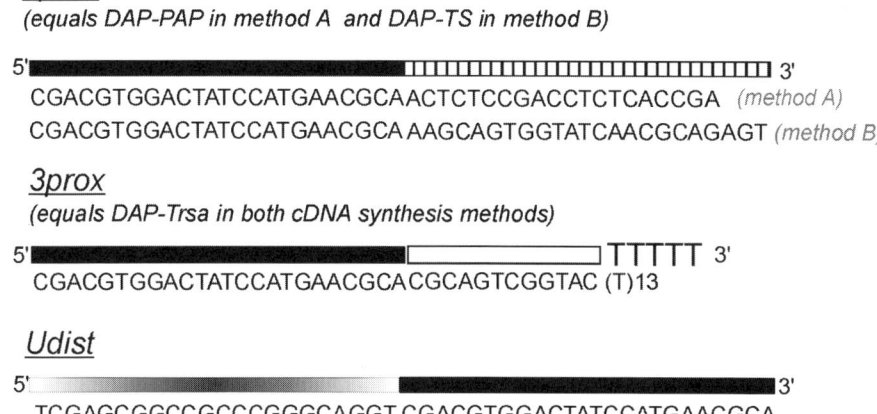

5prox
(equals DAP-PAP in method A and DAP-TS in method B)

5' ▰▰▰▰▰▰▰▰▰▰▰▰▱▱▱▱▱▱▱▱▱▱▱▱▱▱▱▱▱▱▱ 3'

CGACGTGGACTATCCATGAACGCAACTCTCCGACCTCTCACCGA *(method A)*

CGACGTGGACTATCCATGAACGCA AAGCAGTGGTATCAACGCAGAGT *(method B)*

3prox
(equals DAP-Trsa in both cDNA synthesis methods)

5' ▰▰▰▰▰▰▰▰▰▰▰▰▱▱▱▱▱▱▱ TTTTT 3'

CGACGTGGACTATCCATGAACGCACGCAGTCGGTAC (T)13

Udist

5' ░░░░░░░░░░░░░░░░ ▰▰▰▰▰▰▰▰▰▰▰▰ 3'

TCGAGCGGCCGCCCGGGCAGGT CGACGTGGACTATCCATGAACGCA

Fig. 3. Legend to primer sequences.

3. Methods

3.1. First Stage of RACE

1. Prepare the template by diluting the aliquot of nonamplified cDNA fivefold in the cDNA dilution buffer. If the use of amplified cDNA is feasible, dilute the aliquot of the product of cDNA amplification 50-fold in the cDNA dilution buffer.

2. Prepare three PCR mixtures, corresponding to 3′-RACE, 5′-RACE, and positive control. These mixtures will differ only in primers, so a master mixture without primers can be prepared: 6 µL of 10X PCR buffer (provided with the polymerase mixture), 1.5 µL of dNTP mix (10 mM of each), 3 µL of diluted cDNA template, polymerase mixture sufficient for 60 µL of PCR (*see* manufacturer's recommendations), and H$_2$O to 57 µL. Dispense the master mix into three tubes (18 µL to each tube); label the tubes. Add the following primers to the tubes (*see* **Fig. 1** for locations of primer annealing sites):

    a. For 5′-RACE, 1 µL of 2 µM 5prox primer and 1 µL of 2 µM fragment-specific primer 5-1.

    b. For 3′-RACE, 1 µL of 2 µM 3prox primer and 1 µL of 2 µM fragment-specific primer 3-1.

    c. For positive control, 1 µL of 2 µM primer 5-1 and 1 µL of 2 µM primer 3-1. Note that the final concentration of primers is 0.1 µM.

3. Perform cycling: For block-controlled thermocyclers, 94°C for 40 s (annealing temperature of the weakest gene-specific primer; *see* **Note 4**), 1 min – 72°C 2 min 30 s; for thermocyclers with tube-controlled temperature or simulated tube control, 95°C 10 s – [•••] 30 s – 72°C for 2 min 30 s. Perform as many cycles as required to amplify the known fragment from the particular cDNA. If this

value is unknown, do 22 cycles. Check 3 μL of the products on a 1% agarose gel, keeping the PCR tube at room temperature while the electrophoresis runs. If none of the tubes contains detectable product, put the tubes back into the thermal cycler and do five more cycles. If the products are barely visible, do only three more cycles. Repeat the checks on agarose gel and adding cycles according to these guidelines until the products in all three tubes are readily detectable on an agarose gel with standard EtBr staining. However, the total number of cycles should not be more than 40 (*see* **Note 5**).

3.2. Second Stage of RACE

1. Prepare the templates for second stage RACEs by diluting the product of the first stages 50-fold in cDNA dilution buffer.
2. Prepare four PCR mixtures, corresponding to nested/step-out 3′-RACE, nested/step-out 5′-RACE, and two negative controls, one for each RACE direction. A master mix for the four samples includes 8 μL of 10X PCR buffer (provided with the polymerase mixture), 2 μL of dNTP mix (10 mM of each), 4 μL of 2 μM primer *Udist*, polymerase mixture sufficient for 80 μL of PCR (*see* manufacturer's recommendations), and H$_2$O to 76 μL. Dispense the master mix into two tubes, 38 μL to each. To one tube, add 2 μL of diluted first-stage 3′-RACE; to the other add 2 μL of diluted first-stage 5′-RACE. Dispense each of the two mixtures into two tubes, 19 μL to each. One of these resulting mixtures will contain the *Udist* primer alone and will serve as a negative control; to the other, add 1 μL of 2 μM nested gene-specific primer corresponding to the RACE direction (3-2 for 3′ RACE and 5-2 for 5′ RACE). Perform cycling using the same program as described for the first stage, but with the fewer number of cycles. The number of cycles before the first check on agarose gel should be 12. After that, add more cycles if necessary using the above-described guidelines; the only difference is that the maximum allowed number of cycles in this case is 20 instead of 40 (*see* **Note 6**).
3. Examine the RACE results on agarose gel. Load the products of the first stage side-by-side with corresponding second-stage products and negative controls from the second stage, achieve a good resolution, and analyze the image. Normally, the products of RACE, especially if the flanks to be obtained are longer than 1000 basepairs, exhibit multiple bands of different sizes, of which some may be nonspecific (*see* **Note 7**). A specific band is the one that does not appear in the negative control (*see* **Note 8**); it also may correspond to a band in the product of the first stage that is longer by the size of the nested step (*see* **Fig. 1** for the term definition). The latter cannot be expected if the second-stage amplification took more than 15 cycles. When the product of RACE turns out to be complex, we recommend picking the specific bands out of the agarose gel, reamplifying them, and cloning one by one (*see* **Note 9**). Otherwise, the whole second-stage product can be cloned using any vector system suitable for cloning PCR products (such as the pGEM-T vector system from Promega), following the manufacturer's protocol.

4. Notes

1. Rapid amplification of cDNA ends can be performed either starting from amplified cDNA (obtained according to method A or B (*see* Chapter 10), or from raw nonamplified cDNA. In the case of method A, this raw cDNA is the ligation mixture of double-stranded cDNA with a pseudo-double-stranded adaptor, whereas in the case of method B, it is the first-strand synthesis reaction mixture. Using amplified cDNA is highly recommended for short transcripts (total length less than 1500 bases), as there is much less background amplification. For longer transcripts, the farthest 5′ flank may be underrepresented in the amplified cDNA, so the use of raw nonamplified cDNA is recommended.

2. Complementary DNA dilution buffer contains yeast transfer RNA to stabilize the DNA in low concentration. The diluted samples can be stored at –20°C for several months.

3. Adaptor-specific primers (**Fig. 3**) are designed to work on cDNAs synthesized with methods A and B (*see* Chapter 10). Note that the primer 5prox is different for these two methods due to the different 5′ flanking adapter sequence. The adapter-specific primers should be purified to the highest degree by high pressure liquid chromatography (HPLC) or polyacrylamide gel electrophoresis.

4. Gene-specific primers should be designed following the five simple guidelines:
   a. The 3′-terminal base should be A or T.
   b. The five 3′-terminal bases should include no more than two G's or C's.
   c. The four 3′-terminal bases should not find a perfect match within the primer being designed and within primers that are going to be used in PCR with this primer (adapter-specific primers and another gene-specific primer to amplify positive control, if applicable).
   d. The length of the primer should be at least 20 bases.
   e. The annealing temperature of the primer should be equal or higher than 60°C; calculated by the formula $4(G+C) + 2(A+T) + 3$.

   In RACE amplification reactions, the annealing should be set to the lowest of the temperatures calculated for the participating gene-specific primers using this formula.

5. If during the first stage of RACE the product in the positive control is not seen after 38–40 cycles, the gene-specific primers were poorly designed or synthesized or the particular cDNA sample does not contain the target transcript. If the positive control shows a well-detectable band and the RACEs are empty even at five to eight cycles more than the cycle number required to amplify the positive control, the problem may be the poor quality of adapter-specific primers (5prox and 3prox) or the unsuccessful cDNA synthesis.

6. If no bands discriminating the experimental samples from negative controls are detected after 20 PCR cycles during second stage of RACE, this means either that the first-stage amplification did not work properly or something is wrong with the nested gene-specific primers. Note that although adding more cycles may eventually produce some bands, these, most probably, would not correspond to the transcript of interest.

7. It is very common for specific RACE products to appear as several bands ("RACE ladders"). In 5'-RACE, this is explained by partial degradation of the original mRNA or falloffs of the reverse transcriptase during first-strand synthesis. In 3'-RACE, multiple bands correspond to the sites of nonspecific annealing of oligo(dT)-containing primer within the mRNA rather than at the poly(A) tract. In addition to these artifactual sources, there may be natural causes of "RACE ladders" such as the presence of multiple splice forms or the presence of high-similarity repeats within the target transcript to which the gene-specific primers anneal.

8. The negative control may show some product of amplification, especially when the number of cycles performed during the second stage of RACE approaches 20. This product looks like a condensed smear with fragment lengths from 4 to 8 kb and represents the longest cDNA fragments that were present in the sample. Fragments that long are affected the least of all by the PCR-suppression effect and have a chance to be amplified to the detectable concentration, especially in the absence of amplification of the target product.

9. To pick the band out of agarose gel, stab the gel in the middle of the band with a pipetman tip (cut the tip to make an opening approximately 1 mm wide). Avoid exposing the gel to ultraviolet for prolonged periods of time. Squeeze the piece of agarose out of the tip into 20 µL of Milli-Q water and either leave it overnight at 4°C or incubate it at 55°C for 1 h; then, take 1 µL of the liquid to reamplify the band using the same primers, controls, program, and cycle number as at the second stage of RACE.

References

1. Yu, Y. P., Lin, F., Dhir, R., Krill, D., Becich, M. J., and Luo, J. H. (2001) Linear amplification of gene-specific cDNA ends to isolate full-length of a cDNA. *Anal. Biochem.* **292,** 297–301.
2. Tellier, R., Bukh, J., Emerson, S. U., and Purcell, R. H. (1996) Amplification of the full-length hepatitis A virus genome by long reverse transcription–PCR and transcription of infectious RNA directly from the amplicon. *Proc. Natl. Acad. Sci. USA* **93,** 4370–4373.
3. Shi, X. Z. and Kaminskyj, S. G. W. (2000) 5' RACE by tailing a general template-switching oligonucleotide. *Biotechniques* **29,** 1192.
4. Matz, M., Shagin, D., Bogdanova, E., Britanova, O., Lukyanov, S., Diatchenko, L., and Chenchik, A. (1999) Amplification of cDNA ends based on template-switching effect and step-out PCR. *Nucleic Acids Res.* **27,** 1558–1560.
5. Kruger, M., Beger, C., Welch, P. J., Barber, J. R., and Wong-Staal, F. (2001) C-SPACE (cleavage-specific amplification of cDNA ends): a novel method of ribozyme-mediated gene identification. *Nucleic Acids Res.* **29,** U44–U50.
6. Koehler, D. R., Tellier, R., Hu, J., and Post, M. (1999) System for PCR identification of cDNA ends (SPICE). *Biotechniques* **27,** 46.
7. Huang, S. H., Hu, Y. Y., Wu, C. H., and Holcenberg, J. (1990) A simple method for direct cloning cDNA sequence that flanks a region of known sequence from

total RNA by applying the inverse polymerase chain-reaction. *Nucleic Acids Res.* **18,** 1922–1922.

8. Frohman, M. A., Dush, M. K., and Martin, G. R. (1988) Rapid production of full-length cDNAs from rare transcripts—amplification using a single gene-specific oligonucleotide primer. *Proc. Natl. Acad. Sci. USA* **85,** 8998–9002.

9. Eyal, Y., Neumann, H., Or, E., and Frydman, A. (1999) Inverse single-strand RACE: an adapter-independent method of 5′ RACE. *Biotechniques* **27,** 656.

10. Chenchik, A., Diachenko, L., Moqadam, F., Tarabykin, V., Lukyanov, S., and Siebert, P. D. (1996) Full-length cDNA cloning and determination of mRNA 5′ and 3′ ends by amplification of adaptor-ligated cDNA. *Biotechniques* **21,** 526–534.

11. Lukyanov, K. A., Gurskaya, N. G., Bogdanova, E. A., and Lukyanov, S. A. (1999) Selective suppression of polymerase chain reaction. *Bioorg. Khim.* **25,** 163–170.

12. Siebert, P. D., Chenchik, A., Kellogg, D. E., Lukyanov, K. A., and Lukyanov, S. A. (1995) An improved PCR method for walking in uncloned genomic DNA. *Nucleic Acids Res.* **23,** 1087–1088.

13. Shagin, D. A., Lukyanov, K. A., Vagner, L. L., and Matz, M. V. (1999) Regulation of average length of complex PCR product. *Nucleic Acids Res.* **27,** e23.

6

Use of Inverse PCR to Clone cDNA Ends

Sheng-He Huang, Steven H. M. Chen, and Ambrose Y. Jong

1. Introduction

Since the first report on complementary DNA (cDNA) cloning in 1972 *(1)*, the technology has been developed into a powerful and universal tool in the isolation, characterization, and analysis of both eukaryotic and prokaryotic genes. However, the conventional methods of cDNA cloning require much effort to generate a cDNA library and then screen for a large number of recombinant phages or plasmid clones. There are three major limitations in these methods. First, a substantial amount of purified mRNA (at least 1 µg) is needed as starting material to generate libraries of sufficient diversity *(2)*. Second, the intrinsic difficulty of multiple sequential enzymatic reactions required for cDNA cloning often leads to low yields and/or truncated clones *(3)*. Finally, screening of a library with hybridization technique is time-consuming.

Polymerase chain reaction (PCR) technology can simplify and improve cDNA cloning. Using PCR with two gene-specific primers, a piece of known sequence cDNA can be specifically and efficiently amplified and isolated from very small numbers ($<10^4$) of cells *(4)*. Previously, it was difficult to isolate full-length cDNA copies of mRNA on the basis of very limited sequence information. The unknown sequence flanking a small stretch of the known sequence of DNA cannot be amplified by the conventional PCR. Then, anchored PCR *(5–7)* and inverse PCR *(8–10)* have been developed to resolve this problem. Anchored-PCR techniques have the common point: DNA cloning goes from a small stretch of known DNA sequence to the flanking unknown sequence region with the aid of a gene-specific primer at one end and a universal primer at other end. Because of only one gene-specific primer in

From: *Methods in Molecular Biology, vol. 221: Generation of cDNA Libraries: Methods and Protocols*
Edited by: S.-Y. Ying © Humana Press Inc., Totowa, NJ

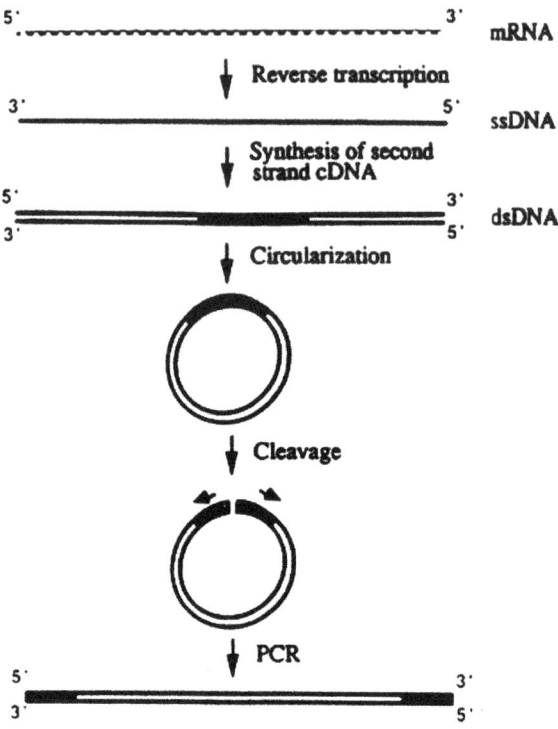

Fig. 1. Diagram of IPCR for cDNA cloning. The procedure consists of five steps: reverse transcription, synthesis of second-strand cDNA, circularization of double-strand cDNA, reopening of the circle DNA, and amplification of reverse DNA fragment. The solid and open bars represent the known and unknown sequence regions of double-stranded cDNA, respectively.

the anchored PCR, it is easier to get a high level of nonspecific amplification by PCR than with two gene-specific primers *(10,11)*. The major advantage of inverse PCR (IPCR) is to amplify the flanking unknown sequence by using two gene-specific primers. Recently, many genomic DNA sequences have been completed. PCR-related techniques become even more useful for various cloning works.

At first, IPCR was successfully used in the amplification of genomic DNA segments that lie outside the boundaries of a known sequence *(8,9)*. We have developed a procedure that extends this technique to the cloning of unknown cDNA sequence from total RNA *(10)*. Double-stranded cDNA is synthesized from RNA and ligated end to end (*see* **Fig. 1**). Circularized cDNA is nicked by selected restriction enzyme or denatured by NaOH treatment *(12,13)*. The

A **B**

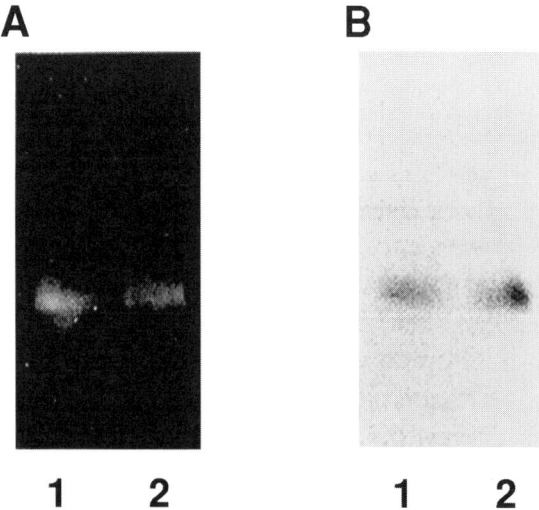

1   2   **1   2**

Fig. 2. Application of IPCR to amplifying the joining region (280 bp) from 5′ (160 bp) and 3′ (120 bp) sequences of human ERp72 cDNA. Amplified DNA from CCRF/CEM cells sensitive (lane 1) and resistant (lane 2) to cytosine arabinoside stained by ethidium bromide (**A**) or hybridized with ^{32}P-labeled ERp72 cDNA (**B**). *See* text for the sequences of the primers and the parameters of IPCR.

reopened or denatured circular cDNA is then amplified by two gene-specific primers. This technique has been efficiently used in cloning several full-length cDNAs *(14–16)*. The following protocol was used to amplify cDNA ends for the human stress-related protein ERp72 *(10)* (*see* **Fig. 2**).

2. Materials

2.1. First-Strand cDNA Synthesis

1. Total RNA prepared from human CCRF/CEM leukemic lymphoblast cells *(17,18)*.
2. dNTP mix (10 mM of each dNTP).
3. Random primers (Boehringer-Mannheim, Indianapolis, IN). Prepare in sterile water at 1 µg/µL. Store at –20°C.
4. RNasin (Promega, Madison, WI).
5. Actinomycin D (1 mg/mL). Actinomycin D is light sensitive and toxic. It should be stored in a foil-wrapped tube at –20°C.
6. MMLV reverse transcriptase.
7. 5X First-strand buffer: 0.25 M Tris-HCl (pH 8.3), 0.375 M KCl, 50 mM MgCl$_2$, 50 mM dithiothreitol (DTT), and 2.5 mM spermidine. The solution is stable at –20°C for more than 6 mo.

2.2. Second-Strand Synthesis

1. 10X Second-strand buffer: 400 mM Tris-HCl (pH 7.6), 750 mM KCl, 30 mM MgCl$_2$, 100 mM (NH$_4$)$_2$SO$_4$, 30 mM DTT, and 0.5 mg/mL of bovine serum albumin (BSA). The solution is stable at –20°C for at least 6 mo.
2. NAD (1 mM).
3. RNase H (2 U/µL).
4. *Escherichia coli* DNA polymerase I (5 U/µL).
5. *E. coli* DNA ligase (1 U/µL).
6. Nuclease-free H$_2$O.
7. T4 DNA polymerase.
8. 200 mM EDTA (pH 8.0).
9. GeneClean (Bio 101 Inc., La Jolla, CA).
10. TE buffer: 10 mM Tris-HCl (pH 7.6), 1 mM EDTA. Sterile filter.
11. DNA standards. Prepare 1-mL aliquots of a purified DNA sample at 1, 2.5, 5, 10, and 20 µg/mL in TE buffer. Store at –20°C for up to 6 mo.
12. TE/ethidium bromide: 2 µg/mL of ethidium bromide in TE buffer. Store at 4°C for up to 6 mo in a dark container.

2.3. Circularization and Cleavage or Denaturation

1. 5X Ligation buffer (supplied with T4 DNA ligase).
2. T4 DNA ligase (1 U/µL).
3. T4 RNA ligase (4 µg/µL).
4. Hexaminecobalt chloride (15 µM).
5. Phenol:CHCl$_3$:isoamyl alcohol (25:24:1).
6. 3 M sodium acetate (pH 7.0).
7. Absolute ethanol.
8. 70% Ethanol.

2.4. Inverse PCR

1. 10X PCR buffer: 100 mM Tris-HCl (pH 8.3), 500 mM KCl, 15 mM MgCl$_2$, 0.01% (w/v) gelatin.
2. 15 mM MgCl$_2$.
3. Deoxyoligonucleotides were synthesized on an Applied Biosystems (Foster City, CA) 380B DNA synthesizer and purified by an OPEC column from the same company. The primer pairs were selected from the 5′ and 3′ sequence of the cDNA coding for human ERp72 stress-related protein (5′-primer: 5′-TTCCTCC TCCTCCTCCTCTT-3′; 3′-primer: 5′-ATCTAAATGTCTAGT-3′) *(10)*.
4. Light mineral oil.
5. *Taq* DNA polymerase.

3. Methods

3.1. First-Strand cDNA Synthesis (19)

Perform reverse transcription in a 25-µL reaction mixture, adding the following components:

5X First-strand buffer	5.0 μL
dNTP mix	2.5 μL
Random primers	2.5 μL
RNasin	1.0 U
Actinomycin D	1.25 μL
MMLV reverse transcriptase	250 U
RNA	15–25 μg of total RNA (heat denature RNA at 65°C for 3 min prior to adding to reaction)
Nuclease-free H20	to 25 μL final vol.

3.2. Second-Strand Synthesis (20)

1. Add components to the first-strand tube on ice in the following order:

10X Second strand buffer	12.5 μL
1 mM NAD	12.5 μL
RNase H (2 U/μL)	.5 μL
E. coli DNA polymerase I (5 U/μL)	5.75 μL
E. coli ligase (1U/μL)	1.25 μL
Nuclease-free water	67.5 μL

2. Incubation at 14°C for 2 h.
3. Heat the reaction mix to 70°C for 10 min, spin for few seconds, and then put on ice.
4. Add 4 U of T4 DNA polymerase and incubate at 37°C for 10 min to blunt the ends of double-stranded cDNA.
5. Stop the reaction with 12.5 μL of 0.2 M EDTA and 200 μL sterile H_2O.
6. Concentrate and purify the sample with Geneclean. Resuspend the DNA in 100–200 μL of sterile H_2O.
7. Estimate the DNA concentration by comparing the ethidium bromide fluorescent intensity of the sample with that of a series of DNA standards on a sheet of plastic wrap *(21)*. Dot 1–5 μL of sample onto plastic wrap on an ultraviolet (UV) transilluminator. Also, dot with 5 μL of DNA standards. Add an equal volume of TE buffer containing 2 μg/mL of ethidium bromide; mix by pipetting up and down. Use proper UV shielding for exposed skin and eyes.

3.3. Circularization and Cleavage (see Notes 1–4)

1. Set up the circularization reaction mix containing the following components: 100 μL (100 ng DNA) of the purified sample, 25 μL of 5X ligation buffer, and 6 μL of T4 DNA ligase. Finally, add 2 μL of T4 RNA ligase or 15 μL of 15 μM hexaminecobalt chloride (*see* **Note 5**).
2. Incubate at 18°C for 16 h.
3. Boil the ligated circular DNA for 2–3 min in distilled water or digest with an appropriate restriction enzyme to reopen circularized DNA.
4. Purify the DNA sample with Geneclean as described in **step 6** in **Subheading 3.2.** or extract with water-saturated phenol/CHCl$_3$ and then precipitate with ethanol *(20)*.

3.4. Inverse PCR (see Note 6)

1. Add 1/10 of the purified cDNA to 100 µL of amplification mix *(22)*:

10X PCR buffer	10 µL
15 m*M* MgCl$_2$	10 µL
dNTP mix (2.5 m*M* of each)	10 µL
5′ Primer (10 pmol/µL)	10 µL
3′ Primer (10 pmol/µL)	10 µL
cDNA	10 µL
Nuclease-free H$_2$O	39.5 µL
Taq DNA polymerase (2.5 U/µL)	0.5 µL

2. Cap and vortex the tubes to mix. Spin briefly in a microfuge. Cover each reaction with a few drops of light mineral oil to prevent evaporation.
3. Put a drop of mineral oil into each well of the thermal cycler block that will hold a tube. Load the reaction tubes.
4. Amplify by PCR using the following cycle profile (25 cycles):
   94°C, 1 min (denaturation);
   65°C, 2 min (annealing);
   72°C, 4 min (elongation).

4. Notes

1. For maximum efficiency of intramolecular ligation, low concentration of cDNA should be used in the ligation mix. A high density of cDNA may enhance the level of intermolecular ligation, which creates nonspecific amplification.
2. Cleavage or denaturation of circularized double-strand cDNA is important because circular double-strand DNA tends to form supercoil and is a poor template for PCR *(23)*. Circularized double-strand DNA is only good for amplification of a short DNA fragment.
3. The following three ways can be considered to introduce nicks in circularized DNA. Boiling is a simple and common way. Because of the unusual secondary structure of some circular double-strand DNA, sometimes this method is not sufficient in nicking and denaturing circular double-strand DNA. A second method is selected restriction enzyme digestion. The ideal restriction site is located in the known sequence region of cDNA. In most cases, it is difficult to make the right choice of a restriction enzyme because the restriction pattern in the unidentified region of cDNA is unknown. If an appropriate enzyme is not available, EDTA–oligonucleotide-directed specific cleavage may be tried *(24,25)*. An oligonucleotide linked to EDTA–Fe at T can bind specifically to double-stranded DNA by triple-helix formation and produce double-stranded cleavage at the binding site.
4. Alkali denaturation has been successfully used to prepare plasmid DNA templates for PCR and DNA sequencing *(12,13,26)*. This method should be feasible in denaturing circularized double-strand cDNA.

5. Inclusion of T4 RNA ligase or hexamine cobalt chloride can enhance the efficiency of blunt-end ligation of double-strand DNA catalyzed by T4 DNA ligase *(27)*.

6. Inverse PCR can be used to efficiently and rapidly amplify regions of unknown sequence flanking any identified segment of cDNA or genomic DNA. This technique does not need construction and screening of DNA libraries to obtain additional unidentified DNA sequence information. Some recombinant phage or plasmid may be unstable in bacteria and amplified libraries tend to lose them *(23)*. IPCR eliminates this problem.

References

1. Verma, I. M., Temple, G. F., Fan, H., and Baltimore, D. (1972) In vitro synthesis of double-stranded DNA complimentary to rabbit reticulocyte 10S RNA. *Nature* **235,** 163–169.

2. Akowitz, A. and Mamuelidis, L. (1989) A novel cDNA/PCR strategy for efficient cloning of small amounts of undefined RNA. *Gene* **81,** 295–306.

3. Okayama, H., Kawaichi, M., Brownstein, M., Lee, F., Yokota, T., and Arai, K. (1987) High-efficiency cloning of full-length cDNA; construction and screening of cDNA expression libraries for mammalian cells. *Methods Enzymol.* **154,** 3–28.

4. Brenner, C. A., Tam, A. W., Nelson, P. A., Engleman, E. G., Suzuki, N., Fry, K. E., et al. (1989) Message amplification phenotyping (MAPPing): a technique to simultaneously measure multiple mRNAs from small numbers of cells. *BioTechniques* **7,** 1096–1103.

5. Frohman, M. A. (1990) RACE: rapid amplification of cDNA ends, in *PCR Protocols: A Guide to Methods and Applications* (Innis, M. A., Gelfand, D. H., Sninsky, J. J., and White, T. J., eds.), Academic, San Diego, CA, pp. 28–38.

6. Shyamala, V. and Ames, G. F.-L. (1989) Genome walking by single-specific-primer polymerase chain reaction: SSP-PCR. *Gene* **84,** 1–8.

7. Huang, S.-H., Jong, A. Y., Yang, W., and Holcenberg, J. (1993) Amplification of gene ends from gene libraries by PCR with single-sided specificity. *Methods Mol. Biol.* **15,** 357–363.

8. Ochman, H., Gerber, A. S., and Hartl, D. L. (1988) Genetic applications of an inverse polymerase chain reaction. *Genetics* **120,** 621–625.

9. Triglia, T., Peterson, M. G., and Kemp, D. J. (1988) A procedure for in vitro amplification of DNA segments that lie outside the boundaries of known sequences. *Nucleic Acids Res.* **16,** 8186.

10. Huang, S.-H., Hu, Y. Y., Wu, C.-H., and Holcenberg, J. (1990) A simple method for direct cloning cDNA sequence that flanks a region of known sequence from total RNA by applying the inverse polymerase chain reaction. *Nucleic Acids Res.* **18,** 1922.

11. Delort, J., Dumas, J. B., Darmon, M. C., and Mallet, J. (1989) An efficient strategy for cloning 5′ extremities of rare transcrips permits isolation of multiple

5′-untranslated regions of rat tryptophan hydroxylase mRNA. *Nucleic Acids Res.* **17,** 6439–6448.

12. Cusi, M. G., Cioé, L., and Rovera, G. (1992) PCR amplification of GC-rich templates containing palindromic sequences using initial alkali denaturation. *BioTechniques* **12,** 502–504.

13. Lau, E. C., Li, Z.-Q., and Slavkin, S. C. (1993) Preparation of denatured plasmid templates for PCR amplification. *BioTechniques* **14,** 378.

14. Green, I. R. and Sargan, D. R. (1991) Sequence of the cDNA encoding ovine tumor necrosis factor-α: problems with cloning by inverse PCR. *Gene* **109,** 203–210.

15. Zilberberg, N. and Gurevitz, M. (1993) Rapid isolation of full length cDNA clones by "inverse PCR": purification of a scorpion cDNA family encoding α-neurotoxins. *Anal. Biochem.* **209,** 203–205.

16. Austin, C. A., Sng, J.-H., Patel, S., and Fisher, L. M. (1993) Novel HeLa topoisomerase II is the IIβ isoform: complete coding sequence and homology with other type II topoisomerases. *Biochim. Biophys. Acta* **1172,** 283–291.

17. Delidow, B. C., Lynch, J. P., Peluso, J. J., and White, B. A. (1993) Polymerase chain reaction: basic protocols. *Methods Mol. Biol.* **15,** 1–29.

18. Davis, L. G., Dibner, M. D., and Battey, J. F. (1986) *Basic Methods in Molecular Biology*, Elsevier Science, New York.

19. Kru, M. S. and Berger, S. L. (1987) First strand cDNA synthesis primed by oligo(dT). *Methods Enzymol.* **152,** 316–325.

20. Promega (1991) *Protocols and Applications*, 2nd ed., Promega, pp. 199–238.

21. Sambrook, J., Fritch, E. F., and Maniatis, T. (1989) *Molecular Cloning*, 2nd ed., Cold Spring Harbor Laboratory, Cold Spring Harbor, N.Y.

22. Saiki, R. K., Gelfand, D. H., Stoffel, S., Scharf, S. J., Higuchi, R., Horn, G. T., et al. (1988) Primer-directed enzymatic amplification of DNA with a thermostable DNA polymerase. *Science* **239,** 487–491.

23. Moon, I. S. and Krause, M. O. (1991) Common RNA polymerase I, II, and III upstream elements in mouse 7SK gene locus revealed by the inverse polymerase chain reaction. *DNA Cell Biol.* **10,** 23–32.

24. Strobel, S. A. and Dervan, P. B. (1990) Site-specific cleavage of a yeast chromosome by oligonucleotide-directed triple-helix formation. *Science* **249,** 73–75.

25. Dreyer, G. B. and Dervan, P. B. (1985) Sequence-specific cleavage of single-stranded DNA: oligodeoxynucleotide–EDTA•Fe(II). *Proc. Natl. Acad. Sci. USA* **82,** 968–972.

26. Zhang, H., Scholl, R., Browse, J., and Somerville, C. (1988) Double strand DNA sequencing as a choice for DNA sequencing. *Nucleic Acids Res.* **16,** 1220.

27. Sugino, A., Goodman, H. M., Heynecker, H. L., Shine, J., Boyer, H. W., and Cozzarelli, N. R. (1977) Interaction of bacteriophage T4 RNA and DNA ligases in joining of duplex DNA at base-paired ends. *J. Biol. Chem.* **252,** 3987.

7

Construction of Size-Fractionated cDNA Library Assisted by an In Vitro Recombination Reaction

Osamu Ohara

1. Introduction

In the past, the construction of complementary DNA (cDNA) libraries has served as a key technology for the discovery of biologically interesting genes *(1)*. For mammalian genes, this is true even in the postgenome sequencing era. The prediction of protein-coding sequences solely from the genomic sequence is still difficult because protein-coding sequences are divided into multiple small pieces by introns. Therefore, currently, cDNA libraries are frequently used to complement genomic information for the comprehensive analysis of cDNAs, rather than the discovery of particular genes. For this purpose, a large number of randomly sampled cDNA clones are usually analyzed using sequencing, whereas cDNA clones are subjected to characterization after screening processes for the purpose of gene discovery. In practice, comprehensive analysis differs from that of conventional gene discovery in that comprehensive analysis is most frequently carried out on a single-clone-for-single-gene basis, whereas gene discovery is done through the isolation of multiple cDNA clones for a single gene. Thus, current requirements for high-quality cDNA libraries suitable for comprehensive cDNA analysis are (1) high rate of full-length and authentic copies of mRNAs, (2) low population bias of cDNA clones, (3) high complexity, (4) low occurrence rate of artificial clones such as chimera, and (5) low occurrence rate of immature mRNA copies. As for requirement (1), many groups have made a tremendous effort to date and have used their methods in practice *(2–7)*. Tricks in the enrichment of full-length cDNA are mainly in the process of converting mRNA to cDNA in a clonable form. On the other hand, requirement (5) will be satisfied only

From: *Methods in Molecular Biology, vol. 221: Generation of cDNA Libraries: Methods and Protocols*
Edited by: S.-Y. Ying © Humana Press Inc., Totowa, NJ

by modifying the RNA preparation method. For example, construction of cDNA libraries from cytoplasmic RNA is at an advantage for meeting this requirement. Requirements (2–4) are related to the cloning process (i.e., retrieval of in vitro synthesized double-stranded cDNA as a clone, rather than in cDNA synthesis). Although there are many reports regarding cDNA cloning methods in the literature *(1)*, a simple and robust cloning step of directional cDNA library construction most often depends on ligation-assisted cloning (LC) of cDNAs bearing two different restriction sites at their ends to plasmid vectors *(8)*. Although ligation has so far been the most reliable reaction for joining cDNA to vectors with high efficiency in vitro, another technology has recently emerged and become commercially available: an in vitro site-specific recombination technology based on the integrase–excisionase system of bacteriophage λ known as "gateway cloning system" from Invitrogen *(9)*. This in vitro recombination cloning (RC) system requires a set of recombination sites in each of the donor and acceptor DNAs. The recombination sites for this system are classified as attB (25 bp), attP (200 bp), attL (100 bp), and attR (125 bp). attB and attP sites specifically recombine to yield attL and attR sites, and vice versa. An enzyme mix that catalyzes the recombination reactions between attB and attP or between attL and attR is called the BP or LR clonase enzyme mix, respectively. The recombination reactions between attB and attP and between attL and attR are called the BP and LR reactions, respectively. DNA clones yielded by the BP and LR recombination reactions are termed "entry clones" and "expression clones," respectively. Detailed information on RC can be found at Invitrogen's website (http://www.invitrogen.com/content.cfm? pageid=2497&cfid=1336112&cftoken=69698350). In principle, RC is specific and efficient enough to apply to cDNA library construction. In fact, it has recently been demonstrated that RC-assisted directional cDNA library construction has many advantages over the conventional LC-assisted method *(10)*. Thus, I herein describe the details of RC-assisted directional cDNA library construction for the purpose of comprehensively analyzing cDNA clones. In addition, the usefulness of a set of size-fractionated cDNA libraries for comprehensive cDNA analysis has been demonstrated through the Kazusa human cDNA sequencing project, which focuses sequencing efforts on large cDNAs *(11,12)*. In this context, the generation of a set of strictly size-fractionated cDNA libraries is also described in this chapter. The resultant set of size-fractionated cDNA libraries yielded by RC enables us to analyze cDNA clones with sizes of interest on a random sampling basis. More importantly, because the resulting cDNA clones are compatible with further RC, they serve as versatile and powerful reagents for the functional analysis of genes.

2. Materials

1. Water, autoclaved Milli-Q water (Millipore Corp., MA).
2. TE: 10 mM Tris-HCl and 1 mM ethylenediamine tetracetic acid (pH 8.0).
3. attB2-dT adapter primer [5′-GCGAAGCCCACCACTTTGTACAAGAAAG CTGGGCGGCCGC(T)$_{20}$-3′], purified by polyacrylamide gel electrophoresis before use.
4. attB1 adapter (upper strand, 5′-TCGACGCGTACAAGTTTGTACAAAAAAG CAGGCTCTTC-3′; lower strand, 5′-pGAAGAGCCTGCTTTTTTGTACAAACT TGTACGCG-3′), purified by high-performance liquid chromatography and annealed by 30-min gradual cooling to room temperature after heating at 95°C for 5 min in TE.
5. 5X First-strand buffer: 250 mM Tris-HCl (pH 8.3), 375 mM KCl, and 15 mM MgCl$_2$.
6. 5X Second-strand buffer: 100 mM Tris-HCl (pH 6.9), 450 mM KCl, 23 mM MgCl$_2$, 0.75 mM β-NAD$^+$, and 50 mM (NH$_4$)$_2$SO$_4$.
7. 5X T4 DNA ligase buffer: 250 mM Tris-HCl (pH 7.6), 50 mM MgCl$_2$, 5 mM ATP, 5 mM dithiothreitol (DTT), and 25% (w/v) polyethylene glycol 8000.
8. 10 mM dNTP mix: 10 mM each of dATP, dGTP, dCTP, and dTTP.
9. 1X Gel loading solution: 0.1% sodium dodecyl sulfate (SDS), 5% glycerol, and 0.005% bromophenol blue.
10. attP donor plasmid harboring ampicillin-resistance marker gene (e.g., attP pSP73 donor plasmid or attP pSPORT-1 donor plasmid, or a down-sized attP pSPORT-1, described in **refs. *10*** and ***12***).
11. SYBR Green I (Molecular Probes, Inc., OR).
12. BP clonase enzyme mix (Invitrogen, CA).
13. Phenol/chloroform/isoamyl alcohol (25:24:1).
14. 2XYT medium: 1.6% tryptone, 1.0% yeast extract, and 0.5% NaCl.
15. SOC medium: 0.5% yeast extract, 2% tryptone, 10 mM NaCl, 2.5 mM KCl, 10 mM MgCl$_2$, 10 mM MgSO$_4$, and 20 mM glucose.
16. LB agar plate containing ampicillin: 1.5% agar containing 1% tryptone, 0.5% yeast extract, 0.5% NaCl, and 50 µg/mL ampicillin.
17. Enzymes: Superscript II reverse transcriptase (200 U/µL; Invitrogen), *Escherichia coli* DNA polymerase I (10 U/µL), *E. coli* DNA ligase (10 U/µL), T4 DNA polymerase (5 U/µL), *E. coli* RNase H (2 U/µL), T4 DNA ligase (5 U/µL), β-agarase.
18. Polyadenylic acid (Sigma Chemical Co., MO) used after phenol treatment followed by ethanol precipitation.
19. PEG/NaCl solution: 20% (w/v) polyethylene glycol 6000 and 2.5 M NaCl.
20. ElectroMax DH10B (Invitrogen).
21. *E. coli* Pulser equipped with *E. coli* Pulser Cuvette with 0.1-cm electrode gap (Bio-Rad Laboratories, CA).
22. Dark reader™ (Non-ultraviolet transilluminator, Clare Chemical Research, CO).

23. DNA size markers: bacteriophage λ DNA digested with *Hin*dIII (available from many companies) and supercoiled DNA ladders (Invitrogen).
24. FluoreImager 595 (Amersham Biosciences Corp., NJ).

3. Methods

The cDNA library construction method described below is essentially according to that reported by Ohara and Temple *(10)*. However, the core part of this method originates from the method by Gubler and Hoffman *(8)*, which is widely used in a kit for cDNA library construction because of its simplicity and robustness. RNA and DNA manipulations were performed by standard methods *(13,14)* and are not described here in detail because of space limitations, except for those critical in this method. The method described in this section outlines in vitro synthesis of double-stranded cDNA, purification of the double-stranded cDNA, in vitro recombination reaction and transformation, and size fractionation of the cDNA library. The outline is schematically shown in **Fig. 1**.

3.1. Synthesis of Double-Stranded cDNA Ligated with attB1 Adapter

Poly(A)$^+$ RNA, not total cellular RNA, should be used as a template for cDNA synthesis as far as possible. Because many excellent kits are commercially available, it is not difficult to obtain a high-quality RNA on your own (*see* **Note 1**). However, it is highly recommended that you check the integrity of your poly(A)$^+$ RNA before cDNA synthesis by RNA blotting analysis. A cDNA fragment for cytoplasmic dynein (about 16 kb) is our favorite probe for quality control of poly(A)$^+$ RNA by RNA blotting analysis because RNA blotting analysis becomes more sensitive to the integrity of the RNA preparation when longer mRNA is examined. Even when commercially available poly(A)$^+$ RNA is used for cDNA library construction, this quality check should be done unless the amount of the RNA is seriously limited. As a rule of thumb, it is recommended that cDNA library construction should start with more than 1 µg of poly(A)$^+$ RNA if the expected number of primary cDNA clones is larger than 10^6.

The method described here is a modification of a commercially available kit from Invitrogen (http://www.invitrogen.com:80/Content/Tech-Online/molecular_biology/manuals_pps/18267.pdf), which is essentially based on the method of Gubler and Hoffman *(8)*.

1. Mix poly(A)$^+$ RNA (1–5 µg) with 100 pmol of attB2-dT primer in a sterile 0.5-mL microcentrifuge tube. Heat the mixture to 70°C for 10 min and then quick chill on ice.

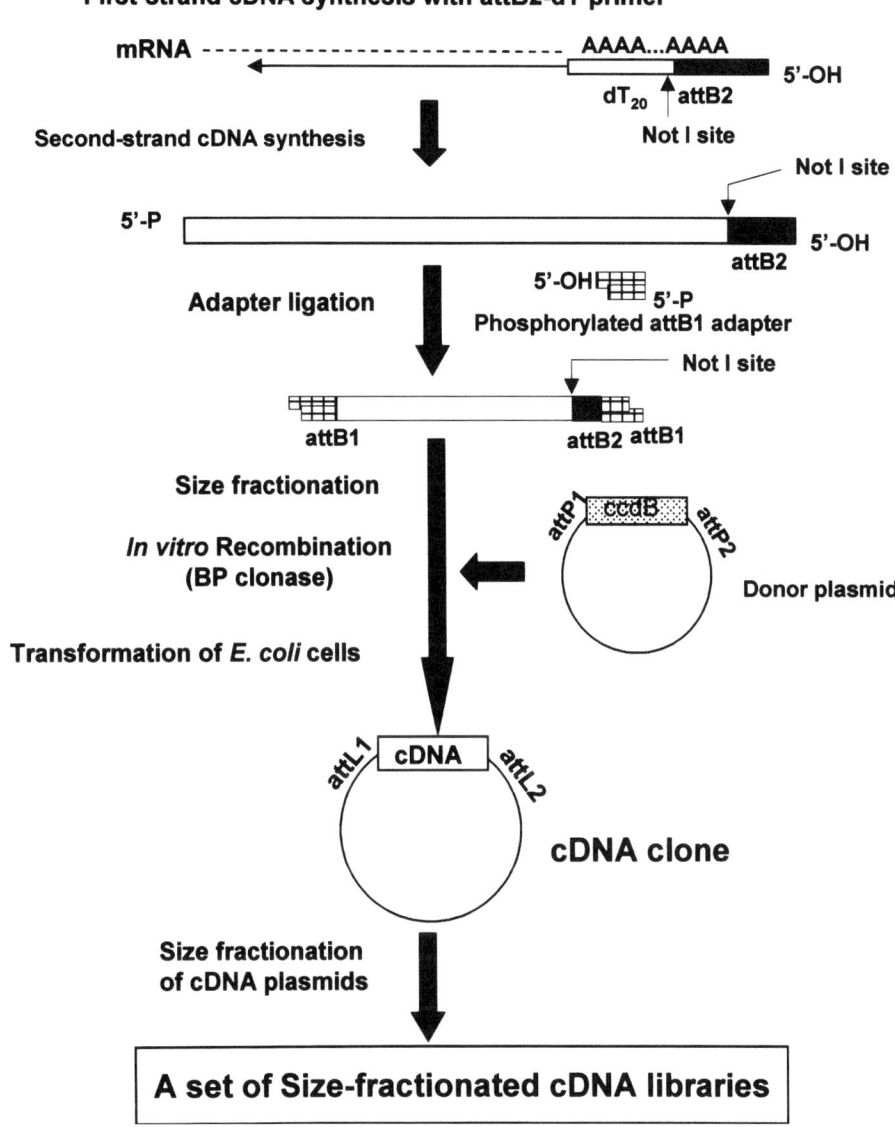

Fig. 1. Overall procedures for cDNA library construction and size selection.

2. After collecting the contents of the tube by brief centrifugation, add 4 μL of 5X first-strand buffer, 2 μL of 0.1 *M* DTT, and 1 μL of 10 m*M* dNTP mix to give a reaction mixture of 15 μL (for 5 μg of poly(A)+ RNA; this final volume is determined according to the amount of SuperScript II reverse transcriptase ([RT] finally added). Place the tube containing the reaction mixture at 37°C for 2 min.

3. Add SuperScript II RT to the reaction mixture (1 μL [200 U] of SuperScript II RT for each of 1 μg of poly(A)$^+$ RNA). Regardless of the amount of starting poly(A)$^+$ RNA, the total volume should be kept at 20 μL. Mix gently and incubate at 37°C for 30 min, and then at 42°C for an additional 30 min. If necessary, the reaction temperature may be elevated to 50°C (*see* **Note 2**).

4. Place the tube on ice to temporarily terminate the first-strand synthesis reaction. Add 91 μL of water, 30 μL of 5X second-strand buffer, 3 μL of 10 m*M* dNTP mix, 1 μL of *E. coli* DNA ligase, 4 μL of *E. coli* DNA polymerase, and 1 μL of *E. coli* RNase H to the reaction mixture for the first-strand synthesis (final volume, 150 μL). The second-strand synthesis reaction allowed to proceed for 2 h at 16°C.

5. Add 10 U of T4 DNA polymerase and continue the incubation at 16°C for 5 min. Because ligation of attB1 adapter will fail without this blunt-ending step, this step must not be skipped. Add 5 μg of polyadenylic acid to the reaction mixture. Then, purify the reaction mixture by extraction with phenol/chloroform/isoamyl alcohol, followed by ethanol precipitation. Dissolve the resultant cDNA with 50 μL of TE; then add 30 μL of PEG/NaCl. Incubate the tube containing the mixture on ice for at least 15 min, then centrifuge it at 18,500*g* for 20 min with a conventional microcentrifuge at 4°C. Wash the pellet with 70% ethanol. Dissolve the pellet in 20 μL of TE, then add 10 μL of 5X T4 DNA ligase buffer, 150 pmol of annealed upper and lower strands of attB1 adapter, and 5 μL of T4 DNA ligase to give a final volume of 50 μL. Incubate the reaction mixture at 16°C at least for 16 h.

3.2. Purification and Size Fractionation of the Synthesized cDNA

To prepare a high-quality cDNA library with high complexity, it is critical to purify the cDNA before inserting it into the plasmid vector. Because small cDNA fragments have a higher cloning efficiency than large ones in RC as well as LC, unnecessary small cDNAs (<100 bp) must be removed prior to RC. In addition, if a cDNA library containing large cDNAs (>3 kb) is required, it is also wise to size fractionate the synthesized cDNAs prior to RC. Although gel filtration is widely used for this purpose, size fractionation of the synthesized cDNAs on low-melting-temperature agarose gel is described here. In our hands, this method is most robust and efficient in recovery yield and strictness of size selection, particularly when a small amount of cDNA is handled.

1. Purify the reaction mixture by extraction with phenol/chloroform/isoamyl alcohol followed by ethanol precipitation. After washing with 70% ethanol, the pellet is briefly air-dried.

2. Dissolve the pellet with 50 μL of TE and add 30 μL of PEG/NaCl. Mix well and incubate the tube containing the reaction mixture on ice at least for 15 min. Centrifuge the 0.5-mL tube containing the TE/PEG/NaCl mixture at 18,500*g* at 4°C for 20 min with a conventional microcentrifuge. Wash the pellet with 70%

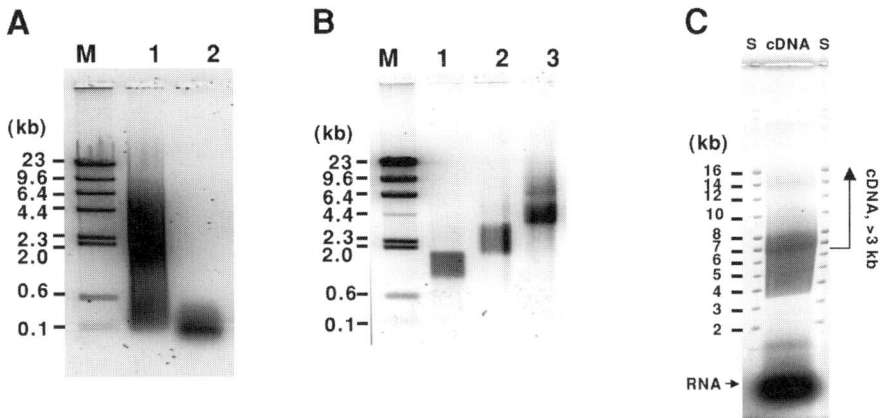

Fig. 2. Typical results of cDNA preparation and size fractionation. The results were obtained during the construction of the human brain cDNA library. cDNAs are shown as typical ones. cDNA synthesis was performed using 5 µg of human brain poly(A)$^+$ RNA (Clontech, CA). DNA was visualized by staining with SYBR-Green I (**panels A** and **B**) or with ethidium bromide (**panel C**). **(A)** One-fortieth of the PEG/NaCl pellet of synthesized double-stranded cDNA (before attB1 ligation) was run on a 0.7% agarose gel (lane 1). DNA remaining in the PEG/NaCl solution was recovered by ethanol precipitation, and one-fortieth was also run in the neighboring lane (lane 2). λ Phage DNAs digested with *Hin*dIII were run in lane M as size markers. **(B)** The recovered size-fractionated cDNAs (lane 1, 1–2 kb; lane 2, 2–3 kb; lane 3, >3 kb) were run on a 0.7% agarose gel to check their amounts and purity. Lane M contained DNA size markers of *Hin*dIII-digested λ phage DNA. **(C)** cDNA plasmids recovered from more than 10^6 transformants originating from cDNAs larger than 3 kb were run on a 0.7% agarose gel as described in **Subheading 3.4.** Because they were mostly in supercoiled form, their size was estimated using supercoiled DNA size markers from Invitrogen (lane S). Because the size of the vector portion used in this experiment plus two attL sites was about 3.5 kb, the size of cDNA inserts was roughly estimated from the size of cDNA plasmid minus 3.5 kb. A bias in the population of cDNA clones toward a small size was evidently introduced during RC, although it seemed to be less than that caused by LC.

ethanol and dissolve the pellet with 20.5 µL of 1X gel loading solution. Run 0.5 µL of the products onto 0.7% agarose gel and check the size and the amount of the synthesized cDNA. A typical result is shown in **Fig. 2A**. After confirming the integrity of the products, run the remaining products (20 µL) in a 1% low-melting-temperature agarose gel in parallel with size markers (λ phage DNA digested with *Hin*dIII) until bromophenol blue moves about 5 cm. Visualize DNA bands by staining with SYBR-green I using a dark reader. At this stage, it is best to avoid ultraviolet illumination for visualization of DNA bands.

3. Separate cDNAs smaller than 1 kb, between 1 kb and 3 kb, and larger than 3 kb as three agarose blocks along the lane. Recover the cDNAs from the agarose blocks using β-agarase as described by the instructions provided by the supplier. Final recovery of cDNAs is done by ethanol precipitation using 5 µg of polyadenylic acid as a carrier. The quantity of the recovered cDNAs was checked by agarose gel electrophoresis and SYBR-green I staining. A typical result is shown in **Fig. 2B**. The amounts of cDNAs are determined by comparing the signal intensity of cDNAs with those of known amounts of DNA size markers on a FluoroImager 595.

3.3. In Vitro Recombination Reaction Followed by Electroporation of E. coli *Cells*

After purification and size fractionation, the recovered cDNAs are subjected to RC. For RC, the protocols are well documented in detail in a manual of the GATEWAY system (Invitrogen), the commercial name of the RC system. You can access this manual at the Invitrogen's website (http://www.invitrogen.com/content.cfm?pageid=2497&cfid=1336112&cftoken=69698350). The method described here is essentially described in the manual of the GATEWAY system, but with slight modifications for application to cDNA library construction.

1. Mix 10–30 ng of cDNA with 400 ng of attP donor plasmid carrying the ampicillin-resistance marker gene (e.g., attP-pSP73 or attP-pSPORT-1 or a down-sized attP-pSPORT-1; *see* **Note 3**). You can easily generate your own attP donor plasmid using a kit from Invitrogen (Gateway™ vector conversion kit). The in vitro recombination reaction is carried out in 20 µL as instructed by Invitrogen. In brief, allow the recombination reaction to proceed for 16 h at room temperature. Add 2 µL of proteinase K solution appended with the BP clonase enzyme mix and incubate the mixture at 37°C for 10 min to quench the reaction.
2. After the addition of 5 µg of polyadenylic acid as a carrier, the reaction mixture is purified by extraction with phenol/chloroform/isoamyl alcohol, followed by ethanol precipitation. After washing the pellet with 70% ethanol, the pellet is dissolved in 10 µL of water.
3. Mix 1 µL of the reaction mixture with a 20-µL suspension of ElectroMax DH10B *E. coli* cells. Electroporate the *E. coli* cells with the reaction mixture at a setting of 16.6 kV/cm and a pulse length of 5 ms. Add 1 mL of SOC medium to the electroporated cells and incubate them at 37°C for 1 h with vigorous shaking. After the incubation, the electroporated *E. coli* cells allow to grow on agar LB plates containing 50 µg/mL of ampicillin at 30°C overnight. In parallel, 10 pg of pUC19 is also introduced to ElectroMAX DH10 B *E. coli* cells to check the competence of the lot of the DH10B *E. coli* cells used for cDNA library construction. Usually, the electroporation of pUC19 gives a transformation efficiency of more than 10^{10} colony formation units (cfu)/µg of pUC19. Using the same electroporation-competent cells, the transformation efficiency of cDNA is usually on the order of 10^8 cfu/µg of cDNA. If the transformation efficiency

is expressed as cfu/μg of donor plasmid, it is approximately on the order of 10^7 cfu/μg of donor plasmid and the transformation efficiency of donor plasmid alone is on the order of 10^4 cfu/μg.

3.4. Size Selection of cDNA Plasmids

Although the synthesized cDNAs are size fractionated prior to RC, RC followed by transformation of *E. coli* cells results in a bias toward small sizes of cDNA in the library *(10)*. Although the degree of this size bias is considerably lower than that in LC, the size fractionation of cDNA plasmids is required to make it possible to analyze cDNAs with the sizes of interest on a random sampling basis. In general, cDNA clones harboring cDNA inserts smaller than 3 kb are easily size fractionated by various conventional methods. However, the method described here is particularly designed to obtain size-fractionated cDNA libraries containing cDNAs larger than 3 kb (*see* **Note 4**).

1. Recover cDNA plasmids from transformants grown on agar plates containing ampicillin at 30°C according to a standard NaOH/SDS method *(13,14)*. Keep in mind that the number of the collected transformants eventually determines the complexity of cDNA library. For comprehensive cDNA analysis, a typical primary cDNA library should consist of more than 5×10^6 transformants.
2. The recovered cDNA plasmids in supercoiled form are run on 0.7% agarose gel 14 cm in length at 2 V/cm overnight. A typical result of this preparative electrophoresis is shown in **Fig. 2C**. The plasmids are visualized with ethidium bromide dissolved in water on a FluoroImager 595, and then size fractionated as gel blocks.
3. The size-fractionated cDNA plasmids are retrieved from the gel block with β-agarase after being electrophoretically transferred to a 1% low-melting-temperature agarose gel. When the amount of the size-fractionated cDNA plasmids is large enough (>100 ng), it is acceptable to use other conventional methods for recovery of DNA from gels *(15)*.
4. Each of the fractions of the size-fractionated cDNA plasmids is again introduced into DH10B *E. coli* cells by electroporation. Recover the plasmids from at least 10^6 transformants as described in **step 2**. The resultant cDNA plasmids are again separated and purified on agarose gel in the same way as described in **step 2**.
5. Repeat the procedures described in **steps 1** and **2** until the recovered cDNA plasmids appear as a single broad band on agarose gel. Although the fractions containing cDNA inserts smaller than 3 kb usually become pure enough after two rounds of size fractionation, three rounds of size fractionation are usually required for the fractions consisting of large cDNA plasmids.
6. Transform DH10B *E. coli* cells with the purified cDNA plasmids and recover cDNA plasmids from more than 10^6 cells.
7. Evaluate the purity of the final size-selected cDNA plasmids in supercoiled form by agarose gel electrophoresis. A typical result is shown in **Fig. 3**.

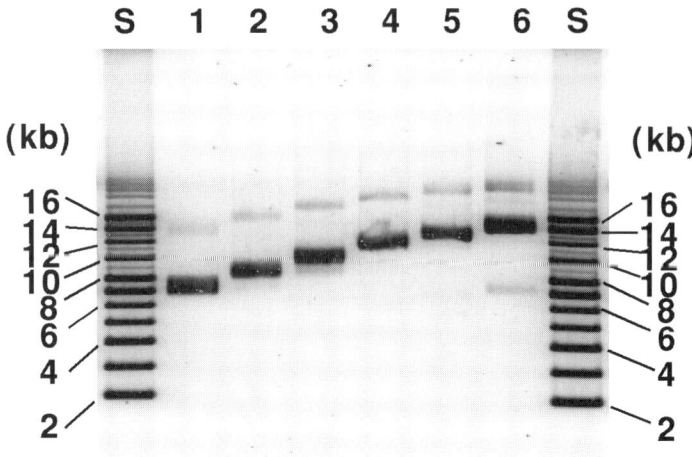

Fig. 3. A typical gel image of a set of size-fractionated cDNA plasmids in supercoiled form. A mixture of cDNA plasmids shown in **Fig. 2C** was size-separated as described in **Subheading 3.4.**, and the resultant size-separated cDNA plasmids recovered from *E. coli* cells were run in supercoiled form on a 0.7% agarose gel. In this case, size-fractionated human brain cDNA plasmids harboring cDNA inserts larger than 4 kb were retrieved and analyzed in lanes 1–6. Size markers for supercoiled plasmid DNAs were run in lane S to estimate the size of cDNA plasmids. Weak bands in each of lanes 1–6 next to strong bands corresponding to supercoiled cDNA plasmids are considered to be open-circular cDNA plasmids. As long as the size-fractionated cDNA plasmids were examined by agarose gel electrophoresis in supercoiled form, they were sufficiently size-selected for further characterization.

4. Notes

1. To obtain a high-quality cDNA library, it is essential to use poly(A)$^+$ RNA with high integrity as a template. For this purpose, many excellent kits are commercially available at present. Two additional commercial options have recently emerged. One is an agent that stabilizes RNA in tissues and cultured cells prior to RNA extraction, commercially named RNA *later*™ (available from many distributors; e.g., from Ambion, Inc., TX); the other is a cytoplasmic RNA extraction kit from Invitrogen. These two options are expected to be quite powerful in some cases, although their performance should be carefully examined prior to particular applications.

2. The use of SuperScript II, a modified MMLV RT lacking RNase H activity, is described in this chapter because this enzyme is widely accepted to have high performance. Recently, a different type of thermostable RT originating

from AMV RT becomes commercially available and is named ThermoScript™ (Invitrogen). Because a dT_{20} primer flanked by a relatively long attB2-*Not*I sequence (40 nucleotides long) is used in the method described in this chapter, the initial reaction temperature is set below 42°C for efficient priming of cDNA synthesis with this primer. However, recent articles have described that the elevated reaction temperature increased not only the product size but also the fidelity *(16)*. Furthermore, the inclusion of trehalose in the RT reaction mixture considerably improves the performance of RT at high temperature according to the report by Carninci et al. *(17)*. Thus, the use of a different RT and the elevation of temperature during reverse transcription might make sense if you have some concerns in the performance of RT.

3. A donor plasmid conferring ampicillin-resistance to *E. coli* cells is used as a vehicle of cDNA in this chapter, although a commercially available attP donor plasmid has kanamaycin-resistance. This is because the transformation efficiency of plasmids by electroporation is known to be dependent on the type of antibiotics used for selecting transformants *(18)*. Although some controversy still remains regarding the reported dependence of the electroporation transformation efficiency on the selection antibiotic marker, it is safe to use a donor plasmid harboring the ampicillin-resistance marker gene if you want to have the highest cloning efficiency for cDNA library construction. Although the attP donor plasmid with the ampicillin-resistant gene marker is not commercially available at present, one can prepare this type of new donor plasmid from any vector using the attR cassette DNA fragment available from Invitrogen (Gateway™ vector conversion kit). Another important factor in preparing a cDNA library with high complexity is to use competent *E. coli* cells with an extremely high transformation efficiency by electroporation (>10^{10} cfu/µg of pUC19). In addition, it is critical to use the same *E. coli* strain to propagate cDNA clones after yielding primary cDNA clones during the size-fractionation procedures because we have observed that a change in strain causes a serious bias in the population of cDNA clones in the library.

4. The size selection of cDNA plasmid in super coiled form is described in this chapter. In these protocols, a fresh plasmid sample should be subjected to size fractionation because the number of cDNA plasmids in open-circle form tends to increase during storage. Because no plasmid preparation can be completely free of the open-circle form, multiple rounds of size fractionation are usually required to obtain a plasmid mixture containing cDNA in a narrow range of sizes. The size separation of linearized DNA by gel electrophoresis is easier and more efficient than this, but it has some practical problems; that is, cDNA plasmids cannot be linearized with restriction enzymes without the risk of losing particular clones that have internal site(s) for the restriction enzyme and the recircularization step using ligase imposes a bias in the cDNA mixture toward small sizes. The size selection of cDNA plasmids in supercoiled form is the most robust of all those examined so far.

5. Future Developments

The resulting size-selected cDNA libraries serve as useful resources for comprehensive cDNA sequencing analysis based on random sampling. However, there is still some room for improvement in this method. First, because the cDNA synthesis steps in this method are essentially the same as those in conventional methods for LC-assisted cDNA library construction, it is easy to combine this method with other methods for full-length cDNA enrichment *(2–5)*. Second, although this method yields cDNA clones in a form of "entry" clone (i.e., carrying attL sites), clones in which cDNAs are flanked by shorter recombination sites (i.e., in a form of "expression" clone containing attB sites) are more convenient for sequencing and protein production. Third, because this method is found to generate a small but considerable number of artificial cDNA clones *(8)*, preventing the occurrence of these artificial clones is another important concern. These points are quite important to further improve the quality and performance of cDNA libraries and many efforts are currently being made to solve these problems. For example, the generation of cDNA libraries in a form of expression clone is recently reported *(12)*.

Acknowledgments

I thank Dr. Gary Temple and his colleagues in Life Technologies Division of Invitrogen Corporation for providing numerous helpful technical suggestions. I also thank Mr. Takashi Watanabe for his excellent technical assistance. This study was supported mainly by a grant from the Kazusa DNA Research Institute and in part by a grant from Special Coordination Funds of the Ministry of Education, Culture, Sports, Science and Technology, the Japanese Government.

References

1. Kimmel, A. R. and Berger, S. L. (1987) Preparation of cDNA and the generation of cDNA libraries: overview. *Methods Enzymol.* **152,** 307–316.
2. Maruyama, K. and Sugano, S. 1994, Oligo-capping: a simple method to replace the cap structure of eukaryotic mRNAs with oligoribonucleotides. *Gene* **138,** 171–174.
3. Edery, I., Chu, L. L., Sonenberg, N. and Pelletier, J. (1995) An efficient strategy to isolate full-length cDNAs based on an mRNA cap retention procedure (CAPture). *Mol. Cell Biol.* **15,** 3363–3371.
4. Zhu, Y. Y., Machleder, E. M., Chenchik, A., Li, R., and Siebert, P. D. (2001) Reverse transcriptase template switching: a SMART approach for full-length cDNA library construction. *Biotechniques* **30,** 892–897.
5. Carninci, P., Kvam, C., Kitamura, A., et al. (1996) High-efficiency full-length cDNA cloning by biotinylated CAP trapper. *Genomics* **37,** 327–336.

6. Wiemann, S., Weil, B., Wellenreuther, R., et al. (2001) Toward a catalog of human genes and proteins: sequencing and analysis of 500 novel complete protein coding human cDNAs. *Genome Res.* **11,** 422–435.

7. Kawai, J., Shinagawa, A., Shibata, K., et al. (2001) Functional annotation of a full-length mouse cDNA collection. *Nature* **409,** 685–690.

8. Gubler, U. and Hoffman, B. J. (1983) A simple and very efficient method for generating cDNA libraries. *Gene* **25,** 263–269.

9. Hartley, J. L., Temple, G. F., and Brasch, M. A. (2000) DNA cloning using in vitro site-specific recombination. *Genome Res.* **10,** 1788–1795.

10. Ohara, O. and Temple, G. (2001) Directional cDNA library construction assisted by the in vitro recombination reaction. *Nucleic Acids Res.* **29,** e22.

11. Nomura, N., Miyajima, N., Sazuka, T., Tanaka, A., Kawarabayashi, Y., Sato, S., et al. (1994) Prediction of the coding sequence of unidentified human genes. I. The coding sequences of 40 new genes (KIAA0001-KIAA0040) deduced by analysis of randomly sampled cDNA clones from human immature myeloid cell line KG-1. *DNA Res.* **1,** 27–35.

12. Ohara, O., Nagase, T., Mitsui, G., Kohga, H., Kikuno, R., Hiraoka, S., et al. (2002) Characterization of size-fractionated cDNA libraries generated by the in vitro recombination-assisted method. *DNA Res.* **9,** 47–57.

13. Sambrook, J. and Russell, D. W. (2001) *Molecular Cloning*, 3rd ed., Cold Spring Harbor Laboratory, Cold Spring Harbor, NY.

14. Ausubel, F. M., Brent, R., Kingston, R. E., Moore, D. D., Seidman, J. G., Smith, J. A., et al. (eds.) (1994) *Current Protocols in Molecular Biology*, Wiley, New York.

15. Kurien, B. T. and Scofield, R. H. (2002) Extraction of nucleic acid fragments from gels. *Anal. Biochem.* **302,** 1–9.

16. Malboeuf, C. M., Isaacs, S. J., Tran, N. H., and Kim, B. (2001) Thermal effects on reverse transcription: improvement of accuracy and processivity in cDNA synthesis. *BioTechniques* **30,** 1074–1078.

17. Carninci, P., Nishiyama, Y., Westover, A., et al. (1998) Thermostabilization and thermoactivation of thermolabile enzymes by trehalose and its application for the synthesis of full length cDNA. *Proc. Natl. Acad. Sci. USA* **95,** 520–524.

18. Steele, C., Zhang, S., and Shillitoe, E. J. (1994) Effect of different antibiotics on efficiency of transformation of bacteria by electroporation. *BioTechniques* **17,** 360–365.

8

Construction of a Full-Length Enriched and a 5′-End Enriched cDNA Library Using the Oligo-Capping Method

Yutaka Suzuki and Sumio Sugano

1. Introduction

With the completion of the draft sequence of the human genome *(1,2)*, it is now essential to extract biological information from the large volumes of human genomic sequence data. A number of attempts, which can be comprehensively termed "functional genomics," are being carried out to decipher which parts of the human genome are transcribed, how the transcripts are spliced and translated, and what functions the eventual protein products conduct. For these purposes, a full-length complementary DNA (cDNA) that contains the entire sequence of the mRNA from the cap structure to the poly(A) tail is a unique resource because a variety of information about the gene functions is contained in a full-length cDNA sequence. The intensive analysis of a full-length cDNA would enable us to identify the following:

1. The exact position of the mRNA transcriptional start site, which is indispensable for the identification of the adjacent promoter.
2. The sequence of the complete 5′ untranslated region (5′ UTR), which is related to the translation efficiency and the cellular localization of mRNA.
3. The continuous protein coding region (CDS), which is required for producing a recombinant protein.
4. The sequence of the complete 3′ untranslated region (3′ UTR), which is related to the translation efficiency, the cellular localization, and the stability of mRNA.

Therefore, a collection of full-length cDNAs should serve as a valuable resource for the functional analysis of the human genome.

From: *Methods in Molecular Biology, vol. 221: Generation of cDNA Libraries: Methods and Protocols*
Edited by: S.-Y. Ying © Humana Press Inc., Totowa, NJ

However, a serious drawback exists in the cDNA libraries that are widely used for current cDNA analyses. In many cases, reverse transcriptase cannot make a full cDNA copy of an mRNA; instead, drops off somewhere in the middle, yielding an incomplete copy. Thus, cDNA libraries made by the conventional methods contain many incomplete cDNA clones that usually lack the 5'-end sequences of the template mRNA. This, at least, in part accounts for the current situation that most cDNA data cover mainly the 3' ends of messenger RNA (mRNA) and the information about sequences near the 5' ends remains limited

In order to eliminate this drawback, we considered it essential to develop new technology to construct "a full-length cDNA library," which is a cDNA library consisting preferentially of full-length cDNAs. In order to make a full-length cDNA library, we need to devise a type of selection procedure to pick up full-length cDNAs from a cDNA pool predominantly occupied by truncated ones. "Selection of a full-length cDNA" is synonymous to "selection of a cDNA that contains both ends of the mRNA." Thus, in order to select a full-length cDNA, the features characteristic of the 3' end and the 5' end of an mRNA should be tagged for the selection. A full-length cDNA could be selected via selection steps for both the 3'-end and the 5'-end "tags."

The poly(A) stretch is a characteristic feature of the 3' end of an mRNA. Conventional methods have generally utilized the poly(A) as a "sequence tag" to select the 3'-end of an mRNA. In the conventional methods, the first strand cDNA is usually synthesized starting from an oligo(dT) primer. Because dT primers mostly hybridize at the poly(A) tail, most of the cDNA is selectively synthesized from the 3' end of the mRNA. Therfore, the conventional methods include the selection step for the 3'-end "tag" of the mRNA.

In contrast, conventional methods include no selection step for the 5' end of the mRNA. As a result, the largest part of the cDNA library constructed by the conventional methods consists of cDNAs that lack the 5' end of the mRNA. The main reason for this lies, in our view, in the fact that mRNA does not originally have a "sequence tag" at the 5' end. The 5' end of an mRNA does have a characteristic structure, called the cap structure, but, unfortunately, it is not a "sequence tag." Unlike the poly(A) at the 3' end, the cap structure cannot be used for hybridization as it is. If the 5'-end "tag" of an mRNA were also a "sequence tag," it would be easy to utilize the tag to select the 5' end of an mRNA.

In order to introduce a "sequence tag" to the 5' end of an mRNA, we have developed a new method, which we named the "oligo-capping" method *(3)*. This method allowed us to enzymatically replace the cap structure of an mRNA with a synthetic oligonucleotide (*see* **Fig. 1**). Each mRNA product of the "oligo-capping" should contain the sequence tags at both its ends, namely poly(A) at the 3' end and the cap-replacing oligo at the 5' end. With the "oligo-

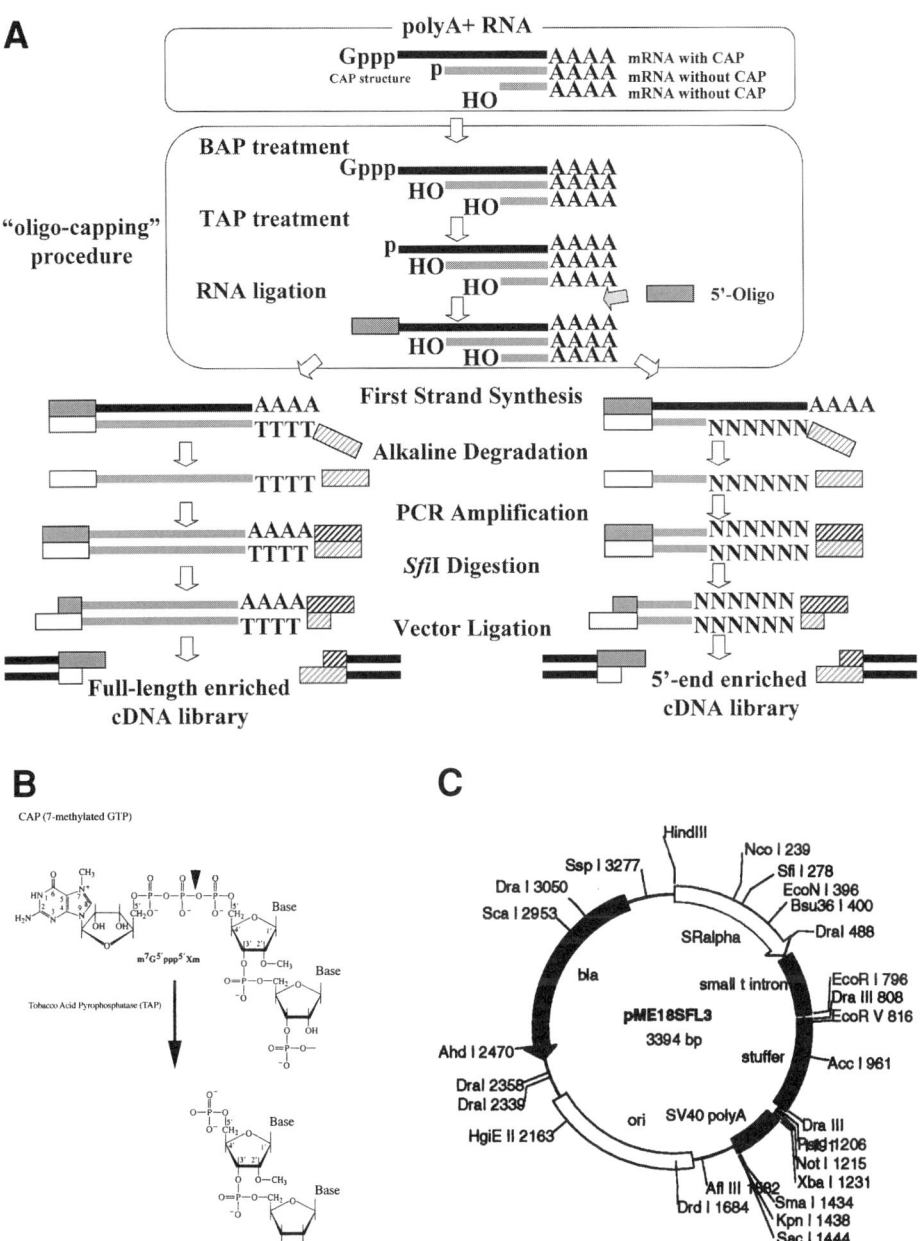

Fig. 1. **(A)** Schematic presentation of the construction of a full-length enriched and a 5′-end enriched cDNA library. According to the scheme shown, a full-length enriched and a 5′-end enriched cDNA library are constructed. Gppp: the cap structure; p: phosphate; OH: hydroxyl; AAAA: poly(A). **(B)** Enzymatic activity of TAP. TAP cleaves the triphosphate at the position shown by the arrowhead, leaving a phosphate at the 5′ end of the mRNA. **(C)** Plasmid map of the cloning vector pME18S-FL3. The cDNAs are inserted between the *Dra*III sites of the plasmid.

capped" mRNAs as starting material, we developed a new system to selectively clone the cDNAs that contain both of the sequence tags at the respective ends. Following the scheme shown in **Fig. 1**, we were able to construct a cDNA library in which the content of "full-length" cDNA is significantly enriched ("full-length enriched" cDNA library) *(4,5)*.

It is also possible that the full-length enriched cDNA library may not include cDNAs of long mRNAs, because the distance between the cap structure and the poly(A) may be beyond the limit that the reverse transcriptase can copy. In such cases, the cDNAs could not be cloned because of the lack of the 5'-end tags. This is especially likely when an mRNA longer than 5 kb is used as a template. In order to address this issue, we also developed a system to construct a "5'-end enriched cDNA library" to cover the 5' ends of long mRNAs (*see* **Fig. 5**). In case the full-length cDNA could not be directly obtained from the full-length enriched library, the 5'-end enriched library and the conventional cDNA library, which is enriched for the 3' end of the mRNA, could be used complimentarily with each other to isolate the full-length cDNA *(4)*.

1.1. Principles of the Construction of a Full-Length Enriched and a 5'-End Enriched cDNA Library

Figure 1A illustrates the scheme for the construction of a full-length enriched and a 5'-end enriched cDNA library. The "oligo-capping" procedure consists of three steps of enzymatic reactions. First, bacterial alkaline phosphatase (BAP) hydrolyzes the phosphate from the 5' ends of truncated mRNAs that are noncapped. The cap structure on full-length mRNAs remains intact during this reaction. Second, tobacco acid pyrophosphatase (TAP) cleaves the cap structure itself at the position indicated by an arrow in **Fig. 1B**, leaving a phosphate at the 5' ends of mRNAs. Finally, T4 RNA ligase selectively ligates the synthetic oligoribonucleotide to the phosphate at the 5' end. As a result, the oligoribonucleotide is introduced only to the 5' ends of mRNAs that originally had the cap structure.

With "oligo-capped" mRNA as the starting material, first-strand cDNA is synthesized using an oligo-(dT) adaptor primer for a full-length enriched cDNA library or using a random hexamer adaptor primer for a 5'-end enriched cDNA library. After first-strand cDNA synthesis, the template mRNA is subjected to alkaline degradation. Then, polymerase chain reaction (PCR) is performed with 3'- and 5'-end primers that contain part of the adaptor primer sequence and the cap-replaced oligonucleotide sequence, respectively. The amplified cDNA fragments are digested with restriction enzymes, size fractionated and cloned into a plasmid vector in an orientation-specific manner.

2. Materials (*see* Notes 1 and 2)

1. Thermal cycler: Zymoreactor II (Atto, Tokyo, Japan) for library construction and GeneAmp PCR System 9700 (ABI, Foster City, CA) for colony PCR and sequencing reaction.
2. Sequencer: 3700 DNA Analyzer (ABI).
3. Centrifuge: GS-6KR (Beckman) for centrifugation of 50-mL tubes and M150-IVD (Sakuma, Tokyo, Japan) for 1.5-mL microtubes. Any type of centrifuge with a refrigerator may be used.
4. Tubes: We have confirmed that the following tubes are RNase free when newly opened. They should be used without autoclaving or diethyl pyrocarbonate (DEPC) treatment: 50-mL tubes (cat. no. 227 261; Greiner, Frickenhausen, Gernamy); 1.5-mL tubes (cat. no. 0030 102.002; Eppendorf, Hamburg, Gernamy).
5. Mechanical homogenizer: Polytron (cat. no. PT10-35, Kinematica, Luzern, Switzerland).
6. Carrier for the ethanol precipitation (RNase-free): ethachinmate (cat. no. 312-01791; WAKO, Tokyo, Japan).
7. RNA extraction kit: RNeasy (cat. no. 75163; Qiagen, Chatsworth, CA); Trizol (cat. no. 15596-018; Invitrogen, Carlsbad, CA).
8. Poly(A) selection: Oligo-(dT) cellulose (cat. no. OT-125-B; Molecular Research Center, Cincinnati, OH).
9. Polypropylene column: Poly-Prep (cat. no. 731-1550; Bio-Rad, Hercules, CA).
10. RNasin (40 U/µL; cat. no. N2111; Promega, Madison, WI).
11. Bacterial alkaline phosphatase (0.25 U/µL; cat. no. 2110; TaKaRa, Kyoto, Japan).
12. Tobacco acid pyrophosphatase (20 U/µL; purified from tobacco cells, BY-2, following the procedure described in **ref. 6**). Alternatively, TAP is now commercially available (cat. no. 313-04021; Wako).
13. T4 RNA ligase (25 U/µL; cat. no. 2050; TaKaRa).
14. Polyethylene glycol (PEG) 8000: 50% (w/v) PEG 8000 (cat. no. P2139; Sigma, St. Louis, MO). Add dH_2O to PEG 8000 so that the concentration is 50% (w/v). Dissolve the PEG 8000 at 65°C. Sterilize the solution by filtration through a 0.22-µM membrane: MILLEX-GV (cat. no. SLGV025LS, Millipore, Molsheim, France).
15. DNase I (RNase-free) (5.0 U/µL; cat. no. 2215; TaKaRa).
16. Spin column: S-400HR (cat. no. 27-5140; Amersham Pharmacia Biotech, Piscataway, NJ).
17. Superscript II (200 U/µL; cat. no. 18064-014; Invitrogen).
18. PCR kit: GeneAmp (cat. no. N808-0192; Perkin-Elmer, Norwalk, CT, USA) for library construction; ExTaq (cat. no. RR001A; TaKaRa) for colony PCR.
19. Restriction enzymes: *Sfi*I (20 U/µL; New England Biolabs Beverly, MA); *Dra*III (20 U/µL; New England Biolabs).
20. Agarose (cat. no. 312-01193; WAKO).

21. DNA ligation kit (cat. no. 6021; TaKaRa).
22. MgCl$_2$ (cat. no. 135-00165; WAKO), ATP (cat. no. A2, 620-9; Sigma), dATP, dCTP, dGTP, dTTP (cat. no. 4026-4029; TaKaRa), NaOH (cat. no. 197-02125; WAKO), mineral oil (cat. no. M5904, Sigma).
23. Competent cells: TOP10 (cat. no. C4040-50; Invitrogen).
24. 2X Loading buffer for poly(A) selection: 40 mM Tris-HCl (cat. no. T-1503; Sigma) (pH 7.0), 1 M NaCl (cat. no. 191-01665; WAKO), 2 mM EDTA (cat. no. 345-01865; WAKO) (pH 8.0), 0.2% (w/v) sodium dodecyl sulfate (SDS) (cat. no. 191-07145; WAKO).
25. 5X BAP buffer: 500 mM Tris-HCl (pH 7.0), 50 mM of 2-mercaptoethanol (cat. no. 137-06862; WAKO). Do not use the supplied buffer because the pH of the buffer is so high that it is hazardous to RNA.
26. 5X TAP buffer: 250 mM sodium acetate (pH 5.5), 50 mM of 2-mercaptoethanol, 5 mM EDTA (pH 8.0).
27. 10X Ligation buffer: 500 mM Tris-HCl (pH 7.0), 100 mM of 2-mercaptoethanol.
28. 10X STE: 100 mM Tris-HCl (pH 7.0), 1 M NaCl, 10 mM EDTA (pH 8.0).
29. QIAquick Gel Extraction kit (cat. no. 28704; QIAGEN, Hilden, Germany).
30. DNA ligation kit (cat. no. 6021; TaKaRa).
31. Cloning vector: pME18S-FL3 (Genbank acc. no. AB009864; **Fig. 1C**).
32. Buffers and columns marked with asterisks in the text are included in the corresponding kit.
33. 5'-Oligoribonucleotide A: 5'-AGCAUCGAGUCGGCCUUGUUGGCCUACU GG-3' (100 ng/µL; custom order, TaKaRa).
34. Oligo-(dT) adapter primer B: 5'-GCGGCTGAAGACGGCCTATGTGGCC(T)$_{17}$-3' (5 pmol/µL; custom order, Invitrogen).
35. Random hexamer adapter primer C: 5'-GCGGCTGAAGACGGCCTATGTG GCCNNNNNNC-3' (10 pmol/µL; custom order, Invitrogen).
36. 5' Primer D: 5'-AGCATCGAGTCGGCCTTGTTG-3' (10 pmol/µL; custom order, Invitrogen).
37. 3' Primer E: 5'-GCGGCTGAAGACGGCCTATGT-3' (10 pmol/µL; custom order, Invitrogen).
38. 3' Primer F for the EF1-α amplification: 5'-ACGTTCACGCTCAGCTTTCAG-3' (10 pmol/µL; custom order, Invitrogen).
39. 3' Primer G for the EF1-α amplification: 5'-AACACCAGCAGCAACAAT CAG-3' (10 pmol/µL; custom order, Invitrogen).
40. Colony PCR primer H (forward): 5'-TCAGTGGATGTTGCCTTTAC-3' (3.2 pmol/µL; custom order, Invitrogen).
41. Colony PCR primer I (reverse): 5'-TGTGGGAGGTTTTTTCTCTA-3' (3.2 pmol/µL; custom order, Invitrogen).
42. Sequencing primer J (forward read): 5'-GGATGTTGCCTTTACTTCTA-3' (3.2 pmol/µL; custom order, Invitrogen).
43. Sequencing primer K (reverse read): 5'-CGACCTGCAGCTCGAGCACA-3' (3.2 pmol/µL; custom order, Invitrogen).

3. Method

3.1. RNA Isolation from Tissues or Cultured Cells (see Note 3)

3.1.1. Total RNA Isolation Using Trizol

1. Put 20 mL of Trizol* in a 50-mL tube.
2. Add up to 1 g of tissue or sample material to the tube and crash the sample using a Polytron (Kinematica). If a mechanical homogenizer is not available, grind the sample to a fine powder in a mortar filled with liquid nitrogen.
3. Centrifuge the tube at 3500*g* for 10 min at room temperature (RT) to remove the tissue that has remained undissolved.
4. Decant the supernatant to a new tube and discard the debris.
5. Add 4 mL of chloroform and rock the tube for 3 min at RT.
6. Centrifuge at 3500*g* for 20 min at RT.
7. Transfer the upper layer gently to a fresh tube (*see* **Note 4**).
8. Add 10 mL of 100% (v/v) isopropanol and let it stand for 10 min at RT.
9. Centrifuge at 3500*g* for 20 min.
10. Discard the supernatant and wash the pellet with 5 mL of 80% (v/v) ethanol.
11. Briefly spin down the pellet and remove the supernatant.
12. Dissolve the pellet in 1 mL of dH$_2$O.
13. To check the RNA, take 1 µL of the solution, add 4 µL of dH$_2$O and apply to a 2 % agarose gel in TAE for electrophoresis (*see* **Note 5**).

3.1.2. Purification of the Total RNA Using RNeasy

1. Add 15 mL of RLT* to the sample prepared in **Subheading 3.1.1.**
2. Add 15 mL of 70 % (v/v) ethanol to the tube and mix well.
3. Apply the solution to the RNeasy column* and let it pass through by brief centrifugation at 2300*g* at RT. Discard the flow through and repeat this step until the entire sample has been applied.
4. To wash the column, apply 10 mL of RW1* to the column and centrifuge briefly at 3000 rpm at RT. Discard the flowthrough.
5. To wash the column further more, apply 10 mL of RPE* to the column and centrifuge briefly at 2300*g* at RT. Discard the flow through and repeat this step once more.
6. Use a fresh tube to collect the eluate. To elute the sample, apply 1.2 mL of dH$_2$O to the column and let it stand for 1 min at RT. Centrifuge at 2300*g* for 1 min at RT. Collect the eluate.
7. To check the RNA, take 3 µL of the sample, add 2 µL of dH$_2$O and apply to a 2% agarose gel in TAE for electrophoresis (*see* **Note 5**).

3.1.3. Poly(A) Selection of the RNA (see **Notes 6** and **7**)

1. Transfer the oligo-(dT) powder from two prepacked columns* to a polypropylene Poly-Prep column (Biorad). The bed volume of the powder should be approx 0.5 mL when the powder from two columns is used (*see* **Note 6**).

2. Denature the dT powder by washing with 3 mL of 0.1 N NaOH.
3. Wash out the alkaline solution with 5 mL of dH_2O.
4. Pre-equilibrate the column with 5 mL of 1X loading buffer.
5. Set a fresh tube to collect the flowthrough.
6. Add an equal volume (1.2 mL) of 2X loading buffer to the sample, mix well, and apply to the column.
7. Collect the flow through and apply to the column. Repeat this step two more times.
8. Wash the column with 5 mL of 1X loading buffer.
9. Set a fresh collection tube and elute the sample by applying 3 mL of dH_2O.
10. Add 8 mL of 100% ethanol and 330 µL of 3 M sodium acetate (pH 5.5) and centrifuge for ethanol precipitation.
11. Dissolve the sample in 100 µL of dH_2O and ethanol precipitate once more (as described in **Subheading 3.2.1., steps 4** and **5**).
12. Dissolve the sample with 72.3 µL of dH_2O.
13. Take 5 µL of the solution and apply to a 2% agarose gel in TAE for electrophoresis (*see* **Note 5**).

3.2. BAP Reaction

1. Set up a BAP reaction by combining: 67.3 µL of the sample [10–20 µg of poly(A)$^+$ RNA], 20.0 µL of 5X BAP buffer, 2.7 µL of RNasin, and 10.0 µL of BAP.
2. Incubate at 37°C for 60 min.
3. Add an equal volume of phenol:chloroform (1:1) to the sample and mix well. Centrifuge at 12,000g briefly at 4°C. Transfer the upper aqueous layer to a fresh tube.
4. Ethanol precipitate the RNA by adding 1 µL of ethachinmate, 11 µL of sodium acetate (pH 5.5), and 275 µL of 100% ethanol. Centrifuge at 12,000g at 4°C for 10 min.
5. Remove the supernatant and rinse the pellet with 150 µL of 80% (v/v) ethanol. Drying is not necessary (*see* **Note 8**).
6. Dissolve the sample in 75.3 µL of dH_2O.

3.3. TAP Reaction

1. Set up a TAP reaction by combining: 75.3 µL of the sample, 20.0 µL of 5X TAP buffer, 2.7 µL of RNasin, and 2.0 µL of TAP.
2. Incubate at 37°C for 60 min.
3. Extract with phenol:chloroform (1:1) and ethanol precipitate (as described in **Subheading 3.2., steps 3–5**).
4. Resuspend the sample in 11.0 µL of dH_2O.

3.4. RNA Ligation

1. Ligate the BAP/TAP-treated poly(A)$^+$ RNA to the 5′-oligoribonucleotide (sequence A) by combining 11.0 µL of the sample, 4.0 µL of the 5′-oligoribonucleotide, 10.0 µL of 10X ligation buffer, 10.0 µL of 50 mM MgCl$_2$, 2.1 µL of

24 mM ATP, 2.5 µL of RNasin, 10.0 µL of T4 RNA ligase, and 50.0 µL of 50% (w/v) PEG 8000.
2. Incubate at 20°C for 3 h.
3. Add 200 µL of dH$_2$O.
4. Extract with 300 µL of phenol : chloroform (1 : 1) (*see* **Note 9**). Ethanol precipitate by adding 1 µL of ethachinmate, 33 µL of sodium acetate (pH 5.5), and 825 µL of 100% ethanol (the rest of the procedure is as described in **Subheading 3.2., steps 3–5**).
6. Dissolve the sample in 70.3 µL of dH$_2$O.

3.5. DNase I Treatment

1. Remove the residual DNA with DNase I by combining 70.3 µL of the sample, 16.0 µL of 50 mM MgCl$_2$, 4.0 µL of 1 M Tris-HCl (pH 7.0), 5.0 µL of 0.1 M dithiothreitol (DTT) (use DTT supplied with SuperScript II), 2.7 µL of RNasin, and 2.0 µL of DNase I.
2. Incubate at 37°C for 10 min.
3. Extract with phenol : chloroform (1 : 1) and ethanol precipitate (as described in **Subheading 3.2., steps 3–5**).
5. Dissolve the sample in 45 µL of dH$_2$O.

3.6. Spin-Column Purification

Remove excess 5′-oligoribonucleotide and fragmented DNA by spin-column chromatography.

1. Resuspend the resin of the column S400HR* thoroughly by vortexing.
2. Centrifuge the column for 1 min at RT at 735g (*see* **Note 10**) to remove the pre-equilibrated water. Discard the flow through.
3. In order to completely remove the water, repeat the previous step once more.
4. Add 5 µL of 10X STE to the sample, mix briefly, and gently, and apply to the column.
5. Set a fresh tube for collecting the eluate and let the sample pass through the column by centrifugation at 735g for 2 min at RT.
6. Add 50 µL of dH$_2$O.
7. Ethanol precipitate (as described in **Subheading 3.2., steps 4** and **5**).
8. Dissolve the sample in 21 µL of dH$_2$O.

3.7. First-Strand cDNA Synthesis

1. Synthesize first-strand cDNA with reverse transcriptase, SuperScript II, by combining: 21 µL of the sample, 10.0 µL of 5X first-strand buffer,* 8.0 µL of the 4 dNTPs at 5 mM each, 6.0 µL of 0.1 M DTT,* 2.5 µL of the oligo-(dT) adapter primer (sequence B), 1.0 µL of RNasin, and 2.0 µL of SuperScript II for the full-length cDNA library. For the 5′-end enriched library, use 2.5 µL of the dR (random hexamer) adapter primer (sequence C) instead of the oligo-(dT) adapter primer.

2. Incubate at 42°C for more than 3 h for the full-length cDNA library. For the 5′-end library, incubate at 12°C for 1 h and 42°C for more than 3 h (*see* **Note 11**).
3. Add 50 μL of dH₂O and extract the solution with phenol:chloroform (1:1) (as described in **Subheading 3.2., step 3**).
4. Add 2 μL of 0.5 *M* EDTA (pH 8.0) to stop the reaction completely.

3.8. Alkaline Degradation of the Template mRNA

1. Degrade the template RNA by adding 15 μL of 0.1 *M* NaOH. Incubate at 65°C for 40 min.
2. Add 20 μL of 1 *M* Tris-HCl (pH 7.0) to neutralize.
3. In order to remove the fragmented RNA, ethanol precipitate the first-strand cDNA by adding 1 μL of ethachinmate, 70 μL of 7.5 *M* ammonium acetate (*see* **Note 12**), and 500 μL of ethanol (the rest of the procedure is as described in **Subheading 3.2., steps 3–5**).
4. Dissolve the sample in 50 μL of dH₂O.

3.9. Confirmation of the First-Strand cDNA

In order to confirm the integrity of the first-strand cDNA and the rate of success of the "oligo-capping," PCR amplify the 5′ end of the EF1-α mRNA.

1. Combine 1 μL (1/50) of the first-strand cDNA (for the full-length library and the 5′-end library) in 52.4 μL of dH₂O with 30.0 μL of 3.3X reaction buffer II,* 8.0 μL of the 4 dNTPs at 2.5 m*M* each,* 4.4 μL of 25 m*M* magnesium acetate,* 1.6 μL of the 5′ primer (sequence D), 1.6 μL of the 3′ primer (sequence F) or 1.6 μL of the 3′ primer (sequence G), and 2.0 μL of rTth DNA polymerase. All of the reaction components except for the primers are included in the GeneAmp kit. Overlay with 125 μL of mineral oil.
2. Thermocycle for 30 cycles at 94°C, 1 min; 52°C, 1 min; 72°C, 1 min.
3. Take 5 μL (1/10–1/20) of the PCR products and apply to a 2% (w/v) agarose gel in TAE for electrophoresis. Confirm the PCR products of 312 bp and 474 bp for primer pairs D + F and D + G, respectively (*see* **Fig. 2**).

3.10. PCR Amplification of the cDNA

1. Use 10–25 μL (1/5–1/2; depending on the quality and quantity of the starting RNA; usually 10 μL is sufficient) of the first strand cDNA in 52.4 μL of dH₂O with 30.0 μL of 3.3X reaction buffer II,* 8.0 μL of the 4 dNTPs at 2.5 m*M* each,* 4.4 μL of 25 m*M* magnesium acetate,* 1.6 μL of 5′ Primer (sequence D), 1.6 μL of 3′ Primer (sequence E), and 2.0 μL of rTth DNA polymerase. Overlay with 125 μL of mineral oil.
2. Thermocycle for 12–15 cycles (depending on the quality and quantity of the library; usually 12 cycles is adequate) at 94°C, 1 min; 58°C, 1 min; 72°C, 10 min.
3. Extract with phenol:chloroform (1:1) and ethanol precipitate (as described in **Subheading 3.2., steps 3–5**).
4. Dissolve the sample in 89 μL of dH₂O.

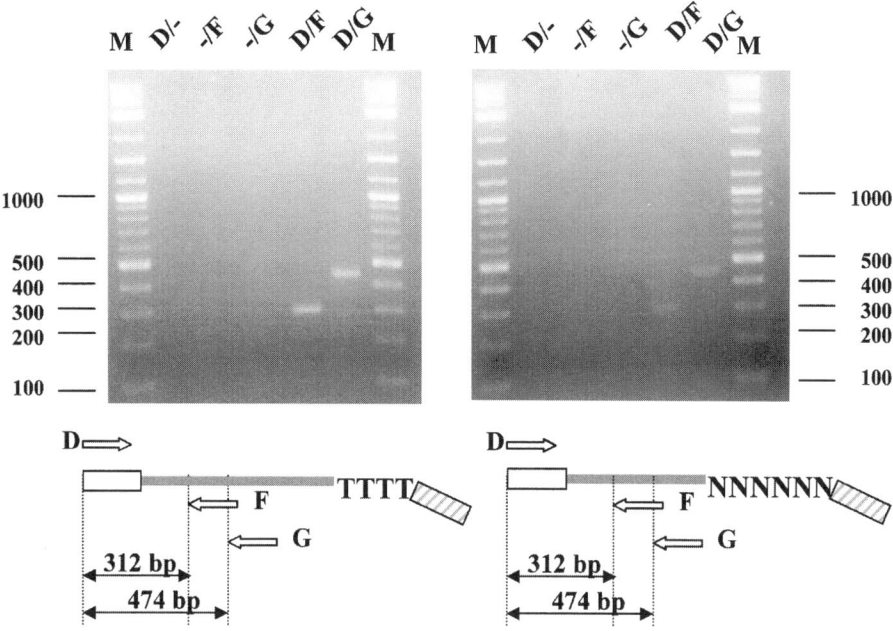

Fig. 2. Confirmation of the first-strand cDNA. The 5′ end of the EF1-α mRNA were amplified using a primer complementary to the cap-replacing oligo (D) and the EF1-α-specific primer (F or G). The first-strand cDNA was synthesized from the dT primer using "oligo-capped" RNA prepared from human intestinal mucosa tissue **(left)** or from the dR primer using "oligo-capped" RNA prepared from HEK293 cells **(right)**. The PCR products of the expected lengths (312 bp for primers B and H and 474 bp for primers B and I) were observed in both cases. The 5′ end of genes of interest could be amplified by a similar method. M: molecular weight marker, 2-Log DNA Ladder (cat. no. 3200S; New England Biolabs).

3.11. Sfi*I Digestion of the PCR Products*

1. Digest the PCR products by combining: 89 μL of the sample, 10 μL of NEBuffer 2,* 1 μL of 100X BSA,* and 2 μL of *Sfi*I.
2. Incubate at 50°C overnight.
3. Extract with phenol:chloroform (1:1) and ethanol precipitate (as described in **Subheading 3.2., steps 3–5**).

3.12. *Size Fractionation of the Double-Stranded cDNAs*

1. Electrophorese the *Sfi*I-digested PCR products in a 1% (w/v) agarose gel in TAE.
2. Excise the part of the gel containing the DNA fraction longer than 1.5 kb (*see* **Note 13** and **Fig. 3**).
3. Crush and completely dissolve the gel in 800 μL of QG* at RT. Incubation at 50°C and adding isopropanol (as instructed by the supplier) are not necessary.

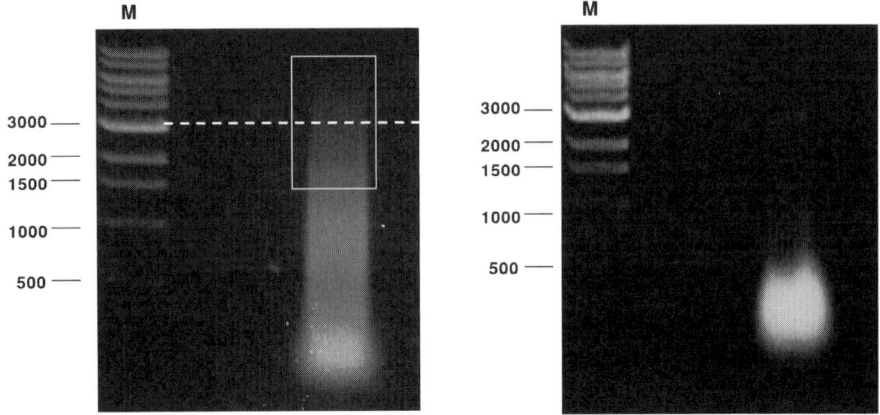

Fig. 3. Image of the smear of the PCR products after gel electrophoresis. The left panel shows the typical smeared profile of the amplified cDNAs. The open box indicates where the gel should be excised. The broken line represent the bottom line of the excision when enrichment of longer cDNAs is attempted (*see* **Note 13**). When the smear is observed only at the lower part of the gel (right), the PCR reaction has failed in amplifying the first strand cDNA. M: Molecular weight marker, 2-Log DNA Ladder (cat. no. 3200S, New England Biolabs).

4. Apply the solution to the column* and centrifuge briefly at RT. Repeat this step until the entire solution has been applied.
5. To wash the column, apply 500 µL of QG* and centrifuge briefly at RT. Discard the flowthrough.
6. To wash the column further more, apply 750 µL of PE* and centrifuge for 1 min at RT. Discard the flowthrough.
7. To elute the DNA, set a fresh tube to collect the eluate and apply 100 µL of dH$_2$O. Let it stand still for 1 min at RT.
8. Centrifuge for 1 min at RT and collect the elution.
9. Ethanol precipitate (as described in **Subheading 3.2., steps 4** and **5**).
10. Dissolve the sample in 9 µL of dH$_2$O.

3.13. Preparation of the Cloning Vector (see Note 14)

1. Digest the plasmid vector pME18S-FL3 with *Dra*III by combining 10 µg of pME18S-FL3 in 89 µL of dH$_2$O, 10 µL of NEBuffer 3,* 1 µL of 100X BSA,* and 2 µL of *Dra*III.
2. Incubate at 37°C overnight.
3. Extract with phenol:chloroform (1:1) and ethanol precipitate (as described in **Subheading 3.2., steps 3–5**).
4. Dissolve the sample in 89 µL of dH$_2$O and repeat the digestion (**Subheading 3.11., steps 1–3**) two more times, so that the residual uncut vector is completely digested.

5. Electrophorese the *Dra*III-digested vector in a 1% (w/v) agarose gel in TAE. Recover the 3.0-kb vector fragment (as described in **Subheading 3.12.**). Also, keep the 0.4-kb stuffer fragment to use as a mock insert.
6. Before using the prepared vector for the cloning of the PCR-amplified cDNA fragments, estimate the background level of undigested vector. Using the stuffer recovered in **step 5** instead of the cDNAs, follow the procedure described in **Subheading 3.12.** Compare the number of transformed bacterial colonies obtained using the "mock insert plus" and the "mock insert minus" preparations. Until the ratio of the numbers becomes more than 100:1, repeat the digestion and purification.

3.14. Cloning and Transformation of the Library (see Notes 15–18)

1. Using the vector prepared in **Subheading 3.13.**, ligate the PCR-amplified cDNA fragments to the vector by combining: 9 μL of the cDNA fragments, 1 μL of the prepared vector (10 ng/μL), 80 μL of Solution A,* and 10 μL of Solution B* (DNA ligation kit, TaKaRa).
2. Incubate at 16°C for 3 h.
3. Extract with phenol:chloroform (1:1) and ethanol precipitate (as described in **Subheading 3.2., steps 3–5**).
5. Dissolve the sample in 50 μL of dH$_2$O.
6. Use 1 μL of the library to transform competent *E. coli* cells TOP10 by electroporation according to the standard method.

3.15. Evaluation of the Quality of the Library

3.15.1. Checking the Distribution of the Insert Size

1. Randomly pick up the colonies and culture overnight in 185 μL of LB with ampicilin at 50 ng/mL in a 96-well microtiter plate (MTP) at 37°C.
2. Add 45 μL of 80% (v/v) glycerol and mix well. This solution can be stored at –80°C and used as a glycerol stock.
3. Set up the colony PCR by combining 1 drop of the glycerol stock in 4.93 μL of dH$_2$O with 1.0 μL of 10X ExTaq Buffer,* 0.9 μL of the four dNTPs at 2.5 m*M* each,* 1.56 μL of each colony PCR primer (sequences H and I), and 0.05 μL of ExTaq DNA polymerase.
4. Thermocycle for 30 cycles at 95°C, 15 s; 55°C, 15 s; 72°C, 4 min.
5. Electrophorese 2 μL of the PCR products in a 1% agarose gel in TAE.
6. Evaluate the length distribution of the inserts (*see* **Fig. 4A**).

3.15.2. Sequencing Analysis and Characterization of the Library

1. Set up the sequencing reaction by combining: 2.0 μL of the PCR products, 3.0 μL of dH$_2$O, 4.0 μL of BigDye buffer,* and 1.0 μL of the sequencing primer (sequence J or K).
2. Thermocycle for 25 cycles at 96°C, 10 s; 50°C, 5 s; 60°C, 4 min, using a 9700 ABI thermal cycler.

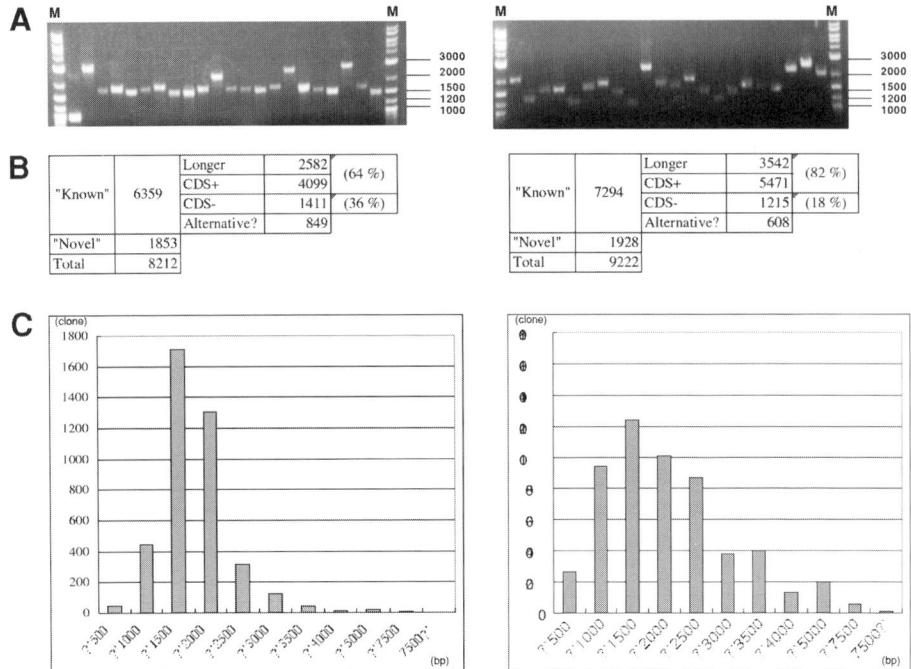

Fig. 4. Characterization and evaluation of a full-length enriched and a 5′-end enriched cDNA library. A full-length enriched and a 5′-end enriched cDNA library from human ileum (right panel) and HEK293 cells (left panel), respectively. **(A)** Results of the colony PCR; **(B)** results of the BLAST search against RefSeq; **(C)** length distribution of the isolated cDNAs of known genes. It is noteworthy that the population of long cDNAs (>5 kb), which is missing from the full-length library, is present in the 5′-end enriched cDNA library.

3. Determine the sequences using a 3700 ABI sequencer.
4. Base-call and trim the vector sequence. Using the processed sequence text data, perform the BLAST search against RefSeq, which is a database of reference human mRNA sequences maintained at NCBI (http://www.ncbi.nlm.nih.gov/LocusLink/refseq.html).
5. Evaluate the population of full-length cDNAs. When the 5′ end of a cDNA covers the translation initiator codon of the known gene according to the RefSeq data, the cDNA is tentatively categorized as "full length." Based on this criterion, the population of the "full-length" cDNAs in a library is usually 50–80% (*see* **Fig. 4B**).

3.15.3. Examples of a Full-Length Enriched and a 5′-End Enriched cDNA Library

A full-length enriched cDNA library and a 5′-end enriched cDNA library were constructed according to the above-described procedure. For the full-length library, HEK293 cells cultured in twenty 10-cm dishes, each of which contained 1×10^6 cells (2×10^7 cells in total) were used. For the 5′-end library, 1 g of frozen human intestinal mucosa tissue was used. After the first-strand cDNA synthesis, one-fifth of the sample was PCR-amplified and cloned into the vector in both cases. **Figure 4** exemplifies the results of the insert size check by colony PCR and the evaluation of the population of full-length cDNAs by BLAST search. Assessment using the cDNAs matched known genes (**Subheading 3.15., steps 4** and **5**) revealed that the populations of full-length cDNAs constituted 64% and 82% of the full-length and the 5′ end cDNA library, respectively. The distribution of the mRNA size was calculated using the cDNAs of known genes and the reported mRNA lengths. It is noteworthy that mRNAs longer than 5.0 kb is included in the 5′-end library. For further details, *see* **ref. 4**).

3.16. Genomewide Analysis of the "Oligo-Capped" cDNA Libraries

3.16.1. Large-Scale Sequencing of the Library

The full-length cDNA library construction technology enabled us to obtain full-length cDNAs of human genes efficiently. We have constructed more than 150 kinds of full-length enriched and 5′-end enriched cDNA libraries from a wide variety of human tissues and cultured cells. Using these cDNA libraries, we have been performing large-scale one-pass sequencing. So far, we have collected 20,000 different putative full-length cDNAs corresponding to candidate novel genes and more than 20,000 of these cDNAs have been completely sequenced. The analyzed cDNA data and the progress of the project are published at our website (http://cdna.ims.u-tokyo.ac.jp/). The cDNA sequence data have also been deposited in public databases through DNA Data Bank of Japan (DDBJ).

3.16.2. Large-Scale Identification of the Transcriptional Start Sites and Adjacent Promoters

Although motifs important for understanding the transcriptional regulation of human genes are embedded in the promoter, the number of genes whose promoter structure has been determined so far is quite limited. According to the Eukaryotic Promoter Database (http://www.epd.isb-sib.ch/), which accumulates reported promoter sequences, only several hundred human promoters have been characterized (273 human genes in Release 62). In part this may be the result of the fact that the exact mRNA start sites have not been identified for most human genes. The conventional methods for identifying the mRNA start site, such as S1 mapping, primer extension, and 5′-RACE, are technically difficult and often lead to the inaccurate identification of the mRNA start sites.

In order to elucidate the genomewide features of transcriptional regulation, we have initiated a large-scale identification of promoters of human genes. For that purpose, the "oligo-capped" cDNA libraries are good resources, because the 5′ end of a full-length cDNA corresponds to the transcriptional start site (TSS) and, in many cases, the promoter overlaps or is just proximal to the TSS. We computationally aligned the 5′ ends of the full-length cDNAs onto the human draft genomic sequences and obtained the positional information about the TSSs on the genome. Adjacent sequences were retrieved and considered to be the promoters. The latest information on the genomewide identification of TSSs and promoters is available from our website (http://elmo.ims. u-tokyo.ac.jp/dbtss/). For further details, refer to **ref. 7**.

4. Notes

1. Since the "oligo-capping" procedure consists of multi-step enzymatic reactions with long incubation times, the utmost care should be taken for every manipulation to avoid RNase contamination and ensure that all the reagents are kept in an RNase-free condition.
   1. Wear disposable gloves and change them at each step of the procedure.
   2. Use sterile, disposable tubes, tips, and pipets.
   3. Bake glassware in which dH_2O and other reagents will be stored at 150°C for 4 h.
   4. Take special care not to touch inside the cap of the tube whenever it is opened or closed.
   5. Every manipulation should be carried out on ice.
   6. The pH of every reagent should be strictly adjusted.
2. We also avoid using an autoclave because the autoclave is used for sterilizing culture medium in many cases, and airborne particles highly contaminated with RNases could easily be absorbed into the reagents. In addition, RNases cannot be inactivated by prolonged autoclaving. Some procedures employ the DEPC treatment of the plastic and glassware before use. This aims to inactivate the RNase activity by alkylating the histidine residues in the active sites of RNaseA-type enzymes. However, we do not recommend DEPC treatment because it has not been fully assessed how much influence the residual DEPC would have on the enzymatic activities of BAP, TAP, and RNA ligase.
3. The starting RNA material must be of the highest quality obtainable. One of the most popular methods for extracting total RNA is the AGPC method. This is a convenient method, which is applicable for a wide variety of tissues. However, the total RNA isolated with AGPC method contains a lot of fragmented RNA and genomic DNA. RNeasy (Qiagen) contains a column to remove such undesirable fractions. Therefore, we combine these two kits for the isolation of total RNA. If cultured cells are used as an RNA source, the most highly recommended method is the NP-40 method *(8)*. Using this method, only the cytoplasmic RNA can be isolated.

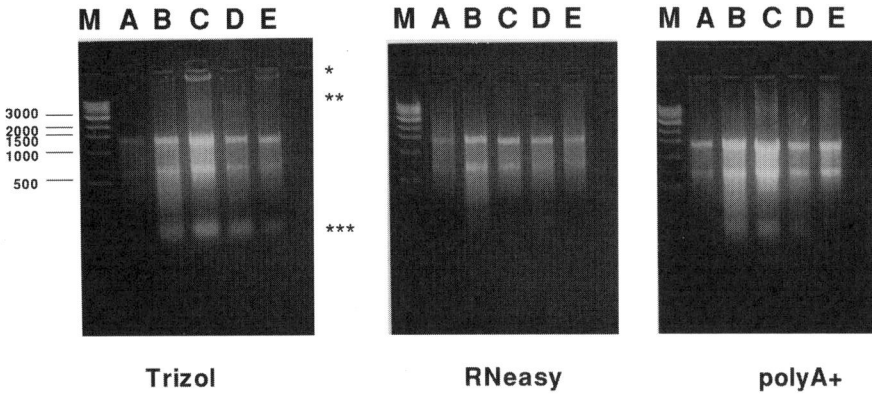

Fig. 5. Assessment of the RNA quality. The panels show gel electrophoretic profiles of the RNAs sampled after Trizol extraction *(left)*, RNeasy purification *(middle)*, and poly(A) selection *(right)*. The RNAs were isolated from human muscle, human ileum, monkey liver, monkey heart, and dog brain (lanes A–E, respectively). In some samples, contaminating genomic DNA (**) is observed. The slow-mobility fractions (*) might indicate that a certain amount of protein is included in the samples. It is noteworthy that these undesirable fractions and fragmented RNA (***) could be removed by RNeasy purification and poly(A) selection. M: Molecular weight marker, 1 kb DNA Ladder (cat. no. N3232S, New England Biolabs).

4. Be fully aware that a lot of genomic DNA is included near the borders of the layers. Residual genomic DNA could interfere with the oligo-capping reaction and/or could be cloned into the vector as an erroneous product. Take approximately two-thirds of the upper layer (or stop taking the solution at a position sufficiently distant from the border) and discard the rest.
5. Before applying the sample to the gel, heat-denature at 65°C for 2 min.
6. For poly(A) selection, many kits, which use latex or magnetic beads for the oligo-(dT) support (e.g., Oligo-Tex [cat. no. W9021B, Nippon-Roche, Tokyo, Japan] and FastTrak [cat. no. 1593-02; Invitrogen]) are commercially available. However, it is difficult to purify large quantities of poly(A)$^+$ RNA in high quality with these kits. We use oligo-(dT) cellulose powder so that we could adjust the bed volume and the washing conditions more flexibly according to the quality and quantity of the sample RNA.
7. In **Fig. 5**, RNA samples from each isolation and purification step are shown. For further details, refer to the figure legend. Now, more precise assessment of RNA quality is possible using Lab-Chip (Agilent 2100 bioanalyzer and RNA 6000 Nano Assay; cat. no. 5065-4476, Agilent Technologies, Waldbronn, Germany), which is a unit for performing the electrophoresis using a microcircuit. For further details, please refer to the instructions for this unit.

8. Drying the pellet is often hazardous to RNA, because RNase may get into the tube. Moreover, sometimes it becomes difficult to redissolve the pellet for the next reaction after the extensive drying.

9. At this step, the intermediate layer should be extremely sticky because of PEG 8000. Do not take this layer. Otherwise, it would be difficult to resuspend the sample in dH$_2$O for the next reaction.

10. Centrifugation at a higher g-force would destroy the column support resin and decrease the yield.

11. In order to avoid the misannealing of the oligo-(dT) primer, we omit the annealing step for the dT primer. However, when the dR primer is used, the incubation at 12°C is indispensable. In both cases, use a long extension time so that the reverse transcription can extend the cDNA maximally.

12. Use ammonium acetate at this step to remove fragmented RNA efficiently. Do not use the ammonium ion for ethanol precipitation until RNA ligation is completed, as ammonium ion interferes with T4 RNA ligase activity.

13. It is possible to excise the part of the gel containing the DNA fraction longer than 3 kb to construct a cDNA library in which a longer population of cDNAs of is enriched.

14. Regarding the vector plasmid pME18S-FL3, cDNA would be inserted downstream of the eukaryotic promoter, SRα, which is derived from the promoter of SV40 large T antigen. Thus, the full-length cDNA could be directly expressed if the vector was to be introduced into mammalian cells.

15. Usually the library size is 10^5–10^6 for 10–20 µg of starting poly(A)$^+$ RNA.

16. Use of PCR is a drawback for this procedure because it sometimes introduces a mutation into a cDNA. We estimate the frequency of such mutations as 1/2,000 bp for substitutions and 1/10,000 bp for insertion/deletion mutations. Also, PCR can cause a bias in the relative abundance of cDNAs because of differences in the efficiency of PCR for different cDNAs. Thus, information about the expression profile of mRNAs in cells may not be maintained in an "oligo-capped" cDNA library.

17. The restriction enzyme *Sfi*I, used for cDNA cloning, could cleave inside a cDNA, resulting in the loss of the cDNA from the library. However, *Sfi*I sites are expected to be rare in cDNAs because their recognition site consists of eight mers 5′GGCCNNNNNGGCC3′.

18. Other methods to construct a full-length enriched cDNA library using a cap-targeted selection step have also been reported by several groups (*9–11*).

Acknowledgments

The "oligo-capping" method was originally developed in collaboration with K. Maruyama. We thank T. Ota, J. M. Sugano, and T. Isogai for helpful discussions and suggestions, M. Shirota, H. Hata, K. Nakagawa, K. Abe, T. Mizuno, M. Morinaga, M. Ishizawa, and M. Kawamura for their excellent sequencing work, and M. Hida and M. Sasaki for their technical support.

This work was supported by a Grant-in-Aid for Scientific Research on Priority Areas from the Ministry of Education, Science, Sports and Culture of Japan.

References

1. Lander, E. S., Linton, L. M., Birren, B., Nusbaum, C., Zody, M. C., Baldwin, J., et al. (2001) Initial sequencing and analysis of the human genome. *Nature* **409,** 860–921.

2. Venter, J. C., Adams, M. D., Myers, E. M., Li, P. W., Mural, R. J., Sutton, G. G., et al. (2001) The sequence of the human genome. *Science* **291,** 1304–1351.

3. Maruyama, K. and Sugano, S. (1994) Oligo-capping: a simple method to replace the cap structure of eucaryotic mRNAs with oligoribonucleotides. *Gene* **138,** 171–174.

4. Suzuki, Y., Yoshitomo, K., Maruyama, K., Suyama, A., and Sugano, S. (1997) Construction and characterization of a full length-enriched and a 5′-end-enriched cDNA library. *Gene* **200,** 149–156.

5. Suzuki, Y. and Sugano, S. (2001) Construction of full-length-enriched cDNA libraries. The oligo-capping method. *Methods Mol. Biol.* **175,** 143–153.

6. Shinshi, H., Miwa, M., Kato, K., Noguchi, M., Matushima, T., and Sugimura, T. (1976) A novel phosphodiesterase from cultured tobacco cells. *Biochemistry* **15,** 2185–2190.

7. Suzuki, Y., Yamashita, R., Nakai, K., and Sugano, S. (2002) DBTSS: DataBase of human Transcriptional Start Sites and full-length cDNAs. *Nucleic Acids Res.* **30,** 328–331.

8. Sambrook, J., Fritsch, E. F., and Maniatis, T. (1989) *Molecular Cloning: A Laboratory Manual*, 2nd ed., Cold Spring Harbor Laboratory, Cold Spring Harbor, NY.

9. Carninci, P., Kvam, C., Kitamura, A., Ohsumi, T., Okazaki, Y., Itoh, M., et al. (1996) High-efficiency full-length cDNA cloning by biotinylated CAP trapper. *Genomics* **37,** 327–336.

10. Kato, S., Sekine, S., Oh, S. W., Kim, N. S., Umezawa, Y., Abe, N., et al. (1994) Construction of a human full-length cDNA bank. *Gene* **150,** 243–250.

11. Edery, I., Chu, L. L., Sonenberg, N., and Pelletier, J. (1995) An efficient strategy to isolate full-length cDNAs based on an mRNA cap retention procedure (CAPture). *Mol. Cell. Biol.* **15,** 3363–3371.

9

cDNA Library Construction Using In Vitro Transcriptional Amplification

Shi-Lung Lin and Henry Ji

1. Introduction

The generation of a complementary DNA (cDNA) library using transcriptional amplification has been developed to offer better linear amplification than polymerase chain reaction (PCR)-based methods. By incorporating a RNA promoter element during reverse transcription of messenger RNA (mRNA), a cDNA library can be preserved and amplified in the form of antisense RNA (aRNA) construct. In brief, the aRNA amplification procedure (*see* **Fig. 1**) is based on (1) reverse transcription of poly(A)$^+$ RNAs with promoter-linked oligo-(dT) primers, (2) double-stranding the resulting cDNA, (3) vitro transcription from the promoter element of the cDNA to synthesize the aRNA sequence up to 2000-fold increase, (4) reverse transcription of the aRNA, (5) denaturation and then double-stranding the resulting cDNA with promoter-linked oligo-(dT) primers, and (6) repeating steps 3–5 to achieve the desired cDNA or aRNA amount for the library preparation.

The advantages of this aRNA amplification are: First, single copy mRNAs can be increased to 2000-fold in one round of amplification without misreading mistakes (*see* **Note 1**). Second, the aRNA amplification is linear and does not result in preferential amplification of abundant RNA species. Finally, a limited RNA sample can be preserved using fixed cells as starting material. The aRNA-derived cDNA library prepared from a single live neuron was reported to cover average 50–75% of total intracellular mRNA population *(1,2)*. Although the aRNA amplification method has led to the identification of several diseased RNA markers from single cells, the rare mRNAs may not be assessable by the current aRNA method *(3)*.

From: *Methods in Molecular Biology, vol. 221: Generation of cDNA Libraries: Methods and Protocols*
Edited by: S.-Y. Ying © Humana Press Inc., Totowa, NJ

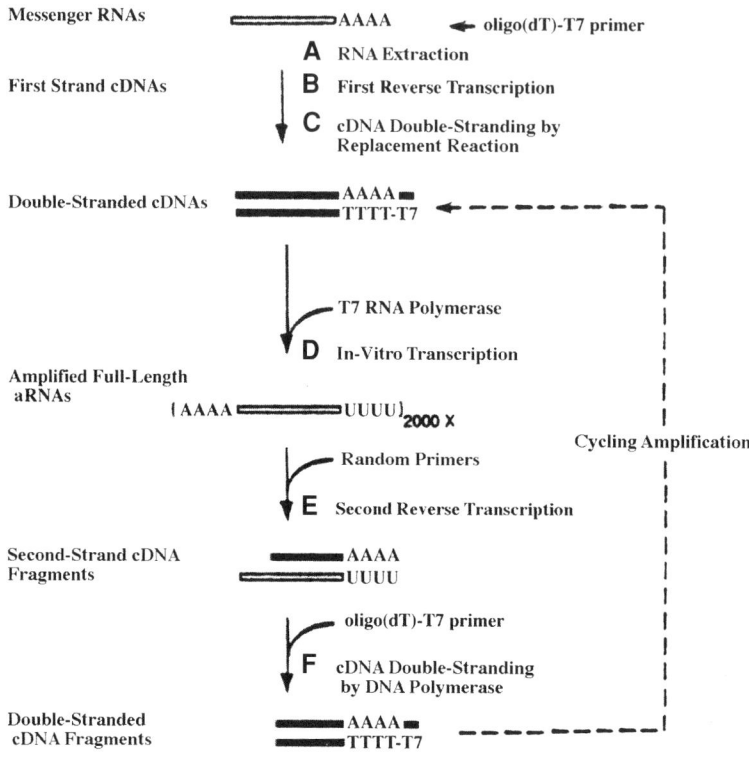

Fig. 1. aRNA amplification procedure.

The aRNA amplification has been widely used in probe preparation for the modern microarray (or called genechip) technology. Because most current microarrays use either sense RNA or denatured cDNA sequences on dot blots, it is necessary to prepare aRNA probes for providing higher-affinity hybridization. In general, RNA–RNA interaction is twice as strong as that of DNA–DNA. Under stringent washing condition, the aRNA probes usually generate less background noises, stronger positive signals, and more reliable reading data. Using Northern blot analysis, we have confirmed that the aRNA probed microarray results in much less deviation than those of cDNA probes. There is no doubt that the aRNA amplification has facilitated the practicability of microarray technology.

The original procedure has been improved using sequence replacement reaction (*4*) for cDNA double-stranding. The replacement reaction was introduced by Okayama and Berg in 1982 and modified by Gubler and Hoffman in 1983. Seeing that first-strand cDNA–mRNA hybrid as a template for a nick-translation reaction, the utilization of RNase H activity produces a nicked and gapped mRNA sequence, serving as a series of RNA primers that are used by Klenow fragment of *Escherichia coli* DNA polymerase I for the synthesis of the second strand of cDNA. Compared to the original random priming procedure, such improvement indeed provides better full-length conformation and a more correct populationary ratio for individual RNA species. However, it is impossible to reiterate the replacement reaction for cycling amplification of full-length cDNAs. The use of random primers for the reverse transcription of aRNAs usually disrupts the advantages of the replacement reaction.

In the most updated method, the above problem has been overcome by combining aRNA amplification with a template-switching effect *(5,6)*. It has been observed that Moloney murine leukemia virus reverse transcriptase (MMLV RT) usually adds a few nontemplate nucleotides (mostly dC) to the 3′ end of a newly reverse-transcribed cDNA upon reaching the 5′ end of the RNA template (*see* **Note 2**). A 3′-poly(dG)-containing specific primer can, therefore, be used to attach the protruding poly(dC) tail and guide the MMLV RT to add a specific sequence in the 3′ end of the cDNA. The specific sequence then serves as a primer binding site for cDNA double-stranding. Because the cDNA so generated is full length and flanked by a specific primer and a poly(dA) tail in termini, the cycling amplification of aRNA has been tested to yield approx 10^3-fold the estimated amount of starting mRNA in one round, and approx 10^5-fold after two rounds of amplification *(7)*. Conventional aRNA amplification usually introduces bias products because of a possible 5′-end underrepresentation and random priming. In conjunction with template-switching effect, the current method has prevented both problems.

From bulk RNA source (>5 µg total RNAs), the method can generate more than 100 µg aRNAs after 5–7 h of incubation. The resultant aRNA can be directly used in microarray analysis or for further reverse transcription to cDNAs. It is noted that RNA degradation occurs significantly after over 2–3 h of incubation in an in vitro transcription reaction. We therefore recommended a 3-h incubation for about 30–40 µg aRNA generation with higher quality. A much lower rate of false signals is always detected using the higher-quality aRNAs. For cycling amplification of aRNA, minimal 10 ng mRNA or aRNA can be used with the template-switching protocol *(7)* or poly(dA)-tailing method *(8)* described in Chapter 11.

2. Materials

2.1. Generation of First-Strand cDNA Using Reverse Transcription with Oligo-(dT)-Promoter Primers

1. Diethyl pyrocarbonate (DEPC) H2O: Stir double-distilled water with 0.1% DEPC for more than 12 h and then autoclave at 120°C under about 1.2 kgf/cm^2 for 20 min, twice.

2. Oligo-$(dT)_{24}$-T7 promoter primer: dephosphorylated 5′-dAAGCTTAGAT ATC TAATACG ACTCACTATA GGGAATTTTT TTTTTTTTTT TTTTTTTTT-3′ (100 pmol/μL).

3. AMV reverse transcriptase (50 U/μL) and 10X first-strand buffer (500 m*M* Tris-HCl (pH 8.5) at 25°C, 80 m*M* $MgCl_2$, 300 m*M* KCl, 10 m*M* dithiothreitol [DTT]).

4. First reverse transcriptase mix: 7 μL DEPC-treated ddH_2O, 2 μL 10X first-strand buffer, 2 μL of 10 m*M* dNTP mix (10 m*M* each of dATP, dGTP, dCTP, and dTTP), 1 μL RNasin (25 U/μL), and 2 μL AMV reverse transcriptase; prepare just before use.

5. Incubation chamber: 65°C, 42°C, and 50°C.

2.2. cDNA Double-Stranding Using a Replacement Reaction with DNA Polymerase–RNase H–Ligase Blend Cocktail

1. 10X Second-strand cDNA synthesis buffer: 250 m*M* Tris-HCl (pH 7.0) at 25°C, 1.35 *M* KCl, 65 m*M* $MgSO_4$, 85 m*M* $(NH_4)_2SO_4$, 2.25 *M* β-NAD, and 80 m*M* DTT; prepare fresh.

2. Second-strand reaction mix: 10 μL DEPC-treated ddH_2O, 4 μL 10X second-strand cDNA synthesis buffer, 3 μL *E. coli* DNA polymerase I, endonuclease-free (5 U/μL), 1 μL *E. coli* RNase H (1 U/μL), and 2 μL *E. coli* DNA ligase (5 U/μL); prepare just before use.

3. T4 DNA polymerase (1 U/μL).

4. Stop solution: 0.5 *M* EDTA (pH 8.0), sterile stock.

5. Incubation chamber: 22°C, 12°C, and 16°C.

6. Purification spin column: 100 basepair cutoff filter, such as a Microcon-50 centrifugal filter (Amicon, Beverly, MA), for example.

2.3. Generation of Antisense RNA Using In Vitro Transcription with Promoter-Driven RNA Polymerase

1. 10X In vitro transcription (IVT) buffer: 400 m*M* Tris-HCl (pH 8.0) at 25°C, 100 m*M* $MgCl_2$, 50 m*M* DTT, and 5 mg/mL nuclease-free bovine serum albumin (BSA).

2. T7 RNA polymerase (80 U/μL).

3. IVT reaction mix: 8 μL DEPC-treated ddH_2O, 4 μL 10X IVT buffer, 4 μL 10 m*M* dNTP mix (10 m*M* each of ATP, GTP, CTP, and UTP), 2 μL RNasin (25 U/μL), and 2 μL T7 RNA polymerase; prepare just before use.

4. Incubation mixer: 37°C; 100*g* vortex for 30 s between every 30-min interval.

5. Incubation chamber: 65°C and 37°C.

6. Purification spin column: 100 base-pair cutoff filter.

2.4. Generation of Sense-Strand cDNA Using Reverse Transcription with Random Primers

1. Random deca-oligonucleotide primer mix: dephosphorylated 5′-dNNNNNNN NNN-3′ (N = dATP, dGTP, dCTP or dTTP in a completely random order; total 100 pmol/μL).

2. AMV reverse transcriptase (50 U/μL) and 10X first-strand buffer (500 m*M* Tris-HCl [pH 8.5] at 25°C, 80 m*M* MgCl$_2$, 300 m*M* KCl, and 10 m*M* DTT).
3. Second reverse transcriptase mix: 4 μL DEPC-treated ddH$_2$O, 4 μL 10X first-strand buffer, 4 μL of 10 m*M* dNTP mix (10 m*M* each for dATP, dGTP, dCTP, and dTTP), 2 μL RNasin (25 U/μL), and 4 μL AMV reverse transcriptase; prepare just before use.
4. Incubation chamber: 65°C, 42°C, and 50°C.
5. Purification spin column: 100 bp cutoff filter.

2.5. cDNA Double-Stranding Using High-Fidelity Taq DNA Polymerization with Oligo-(dT)-Promoter Primers

1. Oligo-(dT)$_{24}$-T7 promoter primer: dephosphorylated 5′-dAAGCTTAGAT ATCT AATACG ACTCACTATA GGGAATTTTT TTTTTTTTTT TTTTTTTTT-3′ (100 pmol/μL).
2. 10X cDNA double-stranding buffer: 500 m*M* Tris-HCl (pH 9.2) at 25°C, 160 m*M* (NH$_4$)$_2$SO$_4$, 20 m*M* MgCl$_2$; prepare fresh.
3. cDNA double-stranding reaction mix: 12 μL DEPC-treated ddH$_2$O, 4 μL 10X cDNA double-stranding buffer, 1.5 μL 10 m*M* dNTP mix (10 m*M* each of dATP, dGTP, dCTP, and dTTP), 0.7 μL *Taq* DNA polymerase (5 U/μL), and 0.3 μL Pwo DNA polymerase (5 U/μL); prepare just before use.
4. Incubation chamber: 94°C, 50°C, and 68°C.

3. Methods

3.1. Generation of First-Strand cDNA Using Reverse Transcription with Oligo-(dT)-Promoter Primers

The starting material is 2–20 μg total RNA *(1,9)* or 0.1–3 μg mRNA *(6)*. Poly(A$^+$) RNA is selected using promoter-linked poly(dT) primers, which contains about 20–26 deoxythymidylate oligonucleotides. The first-strand cDNA is synthesized by reverse transcription from the poly(A+) RNA with promoter-linked poly(dT) primers. As shown in **Fig. 1**, the promoter used here is a T7 bacteriophage RNA promoter element.

1. Primer annealing: Suspend RNA in 5 μL of DEPC-treated water, mix well with 1 μL oligo-(dT)$_{24}$-T7 primer, heat to 65°C for 5 min for minimizing secondary structure, cool to 50°C for 10 min for primer hybridization, and then cool on ice.
2. First-strand cDNA synthesis: Add 14 μL of first reverse transcriptase mix and heat to 42°C for 50 min. Add another 1 μL of reverse transcriptase and mix. Continue to incubate the reaction at 42°C for 30 min, heat to 50°C for 10 min, and then cool on ice. The RNA is still attached noncovalently to the cDNA.

3.2. cDNA Double-Stranding Using a Replacement Reaction with DNA Polymerase–RNase H–Ligase Cocktail

In this method *(4)*, which was developed by Okayama and Berg in 1982 and modified by Gubler and Hoffman in 1983, the product of first-strand cDNA–mRNA hybrid is used as a template for a nick-translation reaction. The use of RNase H activity produces a nicked and gapped mRNA sequence, serving as a series of RNA primers that are used by Klenow fragment of *E. coli* DNA polymerase I during the synthesis of the second strand of cDNA (*see* **Note 4**).

1. Second-strand cDNA synthesis: Add 20 μL of second-strand reaction mix to the first-strand cDNA. Incubate the reaction at 22°C for 80 min and then 12°C for another 30 min.
2. Blunt-end the cDNA: Add 10 μL of T4 DNA polymerase to the reaction, incubate at 16°C for 10 min.
3. Reaction stop: Add 5 μL of stop solution to the reaction and mix well.
4. Buffer exchange and sample concentration: Load the reaction into a purification spin column, spin 10 min at 14,000*g* (*see* **Note 3**) and discard the flowthrough. Add 200 μL of DEPC-treated ddH$_2$O into the spin column to wash the cDNA, spin 10 min at 14,000*g*, and discard the flowthrough. Add 20 μL of DEPC-treated ddH$_2$O into the spin column to dissolve the cDNA, place the spin column upside down in a new collecting microtube, and spin 3 min at 3000*g*. Store the 20 μL of the purified cDNA in a –20°C freezer or perform the next step immediately.

3.3. Generation of Anti-Sense RNA Using In Vitro Transcription with Promoter-Driven RNA Polymerase

The promoter region of the double-stranded cDNA is now serving as a recognition site for RNA polymerase during an in vitro transcription reaction. The in vitro transcription provides linear amplification up to 2000-fold of the amount of starting materials *(1,9)*. The proofreading capability of the RNA polymerase ensures the fidelity of the resulting RNA products. Because the promoter is incorporated in the opposite orientation of sense mRNA, the resulting product is aRNA rather than mRNA.

1. In vitro transcription reaction: Add 20 μL of the IVT reaction mix to the purified cDNA and mix well. Incubate the reaction at 37°C for 4–5 h and occasionally mix the reaction every 30 min for better RNA elongation (*see* **Note 5**).
2. Buffer exchange and sample concentration: Load the reaction into a purification spin column, spin 10 min at 14,000*g*, and discard the flowthrough. Add 200 μL of DEPC-treated ddH$_2$O into the spin column to wash the aRNA, spin 10 min at 14,000*g*, and discard the flowthrough. Add 20 μL of DEPC-treated ddH$_2$O into the spin column to dissolve the aRNA, place the spin column upside down in a new collecting microtube, and spin 3 min at 3000*g*. Store the 20 μL of the purified aRNA in a –80°C freezer or perform next step immediately.

3.4. Generation of Second-Strand cDNA Using Reverse Transcription with Random Primers

The aRNA is further reverse-transcribed to second-strand cDNA using random primers. Because the aRNA carries poly(T) sequences in its 5′ end, the resulting second-strand cDNA is tailed by poly(dA) oligonucleotides, which are useful for generating promoter-linked double-stranded cDNA.

1. Primer annealing: Add 2 μL of random deca-oligonucleotide primer mix to the purified aRNA, mix well and heat to 65°C for 5 min for minimizing secondary structure, cool to 50°C for 1 min for primer hybridization, and then cool on ice.
2. Second-strand cDNA synthesis: Add 18 μL of second reverse transcriptase mix and heat to 42°C for 50 min. Add another 1 μL of reverse transcriptase and mix. Continue to incubate the reaction at 42°C for 30 min, heat to 50°C for 10 min, and then cool on ice. The aRNA is still attached noncovalently to the second-strand cDNA.
3. Denaturation: Heat the reaction at 94°C for 3 min and then cool on ice immediately.
4. Buffer exchange and sample concentration: Load the reaction into a purification spin column, spin 10 min at 14,000g, and discard the flowthrough. Add 200 μL of DEPC-treated ddH$_2$O into the spin column to wash the second-strand cDNA, spin 10 min at 14,000g, and discard the flowthrough. Add 20 μL of DEPC-treated ddH$_2$O into the spin column to dissolve the second-strand cDNA, place the spin column upside down in a new collecting microtube, and spin 3 min at 3000g. Store the 20 μL of the purified second-strand cDNA in a –20°C freezer or perform next step immediately.

3.5. cDNA Double-Stranding Using High-Fidelity Taq DNA Polymerization with Oligo-(dT)-Promoter Primers

The aRNA of the aRNA–cDNA hybrid is denatured and the cDNA is used to form a double-stranded cDNA library using one cycle PCR-like reaction with the promoter-linked poly(dT) primers. Then, another round of aRNA amplification can be achieved by in vitro transcription using the promoter-linked double-stranded cDNA as a template following the steps from **Subheadings 3.3.–3.5.**

1. Primer annealing: Add 2 μL of oligo-(dT)$_{24}$-T7 primer to the purified second-strand cDNA, heat to 94°C for 3 min for aRNA removal, cool to 50°C for 10 min for primer hybridization, and then cool on ice.
2. cDNA double-stranding: Add 18.5 μL of cDNA double-stranding reaction mix to the reaction, mix well, and then incubate the reaction at 68°C for 10 min.
3. cDNA library assessment: *See* Chapter 13.

4. Notes

1. As mentioned and proven in Eberwine, and Van Gelder, and their works *(1,6,9)*, the linear amplification of transcription means that the amplified RNA population should reflect its abundance in the original RNA population. Because the transcriptional amplification of nucleic acids only requires the recognition of promoter regions by RNA polymerases and all promoters possess the same affinity to their respective polymerases, the transcription-based nucleic acid amplification technology will provide less preferential amplification than the PCR-based technology. It is noted that the preferential amplification of PCR is caused by the different affinities between different primers (forward and reverse primers) to their templates as well as between the primers and different templates. It is clear that the transcriptional amplification can bypass the use of multiple primers and provide a better amplification rate (up to 2000 fold; cycle) to mirror the mRNA population.

2. Moloney murine leukemia virus reverse transcriptase usually creates a 2- to 4-base nontemplate nucleotide overhang (mostly dC) to the 3' end of a newly reverse-transcribed cDNA upon reaching the 5' end of the mRNA template. It is not known how much efficiency of the template-switching effect can be generated by such a short overhang.

3. Relative Centrifugal Force (RCF) $(g) = (1.12 \times 10^{-5})(\text{rpm})^2 r$, where r is the radius in centimeters measured from the center of the rotor to the middle of the spin column, and rpm is the speed of the rotor in revolutions per minute.

4. In practice, the second-strand cDNA synthesized by Gubler and Hoffman's method is very efficient. Reaction temperatures between 12°C and 22°C usually produces relatively full-length cDNA, lacking only a few nucleotides corresponding to the 5' end of mRNA. Higher temperatures can partially denature the termini of the cDNA and induce the 3'→5' exonuclease activity of DNA polymerase to remove more than terminal two bases. However, some nicks and gaps may remain in the second-strand cDNA because of the incomplete action of DNA ligase and/or polymerase. For larger-sized mRNAs, such nicks and gaps occasionally truncate the full-length sequences and skew the library toward smaller sizes.

5. The most stable and efficient IVT reaction occurs during the first 2 h of incubation at 37°C. The rate of RNA synthesis decreases considerably (40–50%) after 3 h of incubation or below 37°C incubation. A longer reaction may increase yield, but the possibility of degradation by RNase increases. Occasionally gentle mix can prevent the stall of crowded RNA polymerases on a template and enhance full-length synthesis. The overall rate of RNA polymerization is maximal between pH 7.7 and 8.3, but it remains about 70% of maximum at pH 7.0 or 9.0. High concentrations of NaCl, KCl, or NH_4Cl above 75 mM will inhibit the reaction.

References

1. Eberwine, J., Yeh, H., Miyashiro, K., Cao, Y., Nair, S., Finnell, R., et al. (1992) Analysis of gene expression in single live neurons. *Proc. Natl. Acad. Sci. USA* **89,** 3010–3014.

2. Crino, P. B., Trajanowski, J. Q., Dichter, M. A., and Eberwine, J. (1996) Embryonic neuronal markers in tumerous sclerosis: single-cell molecular pathology. *Proc. Natl. Acad. Sci. USA* **93,** 14,152–14,157.

3. O'Dell, D. M., Raghupathi, P., Crino, P. B., Morrison, B., Eberwine, J. H., and McIntosh, T. K. (1998) Amplification of mRNAs from single, fixed, TUNEL-positive cells. *BioTechniques* **25,** 566–570.

4. Gubler, U. and Hoffman, B. J. (1983) *Gene* **25,** 263–269.

5. Matz, M., Shagin, D., Bogdanova, E., Britanova, O., Lukyanov, S., Diatchenko, L., et al. (1999) Amplification of cDNA ends based on template-switching effect and step-out PCR. *Nucleic Acids Res.* **27,** 1558–1560.

6. Phillips, J. and Eberwine, J. H. (1996) Antisense RNA amplification: a linear amplification method for analyzing the mRNA population from single living cells. *Methods* **10,** 283–288.

7. Wang, E., Miller, L. D., Ohnmacht, G. A., Liu, E. T., and Marincola, F. M. (2000) High-fidelity mRNA amplification for gene profiling. *Nat. Biotech.* **18,** 457–459.

8. Ying, S.-Y., Liu, H. M., Lin, S. L., and Chuong, C.-M. (1999) Generation of full-length cDNA library from single human prostate cancer cells. *BioTechniques* **27,** 410–414.

9. Van Gelder, R. N., von Zastrow, M. E., Yool, A., Dement, W. C., Barchas, J. D., and Eberwine, J. H. (1990) Amplified RNA synthesized from limited quantities of heterogeneous cDNA. *Proc. Natl. Acad. Sci. USA* **87,** 1663–1667.

10

Amplification of Representative cDNA Pools from Microscopic Amounts of Animal Tissue

Mikhail V. Matz

1. Introduction

The possibility of amplifying total complementary DNA (cDNA) obtained from small amounts of biological material is not yet routinely considered, despite the fact that obtaining amounts of material suitable for direct processing by standard methods is often time-consuming and expensive and may be even impossible. Perhaps the most significant obstacle to the full appreciation of the technique is the widespread belief that polymerase chain reaction (PCR) amplification severely distorts the original cDNA profile, so that some cDNA species dramatically rise in abundance while others diminish and may even become completely lost. However, we found that there are just a few simple rules that should be followed to ensure that the amplified sample is minimally distorted and fully representative (i.e., contains all types of message originally present in RNA, even the least abundant ones). This was demonstrated in our own experiments on differential display (1) and elsewhere in application of amplified cDNA as a probe for gene profiling by array technology (2–5). According to our experience in gene hunting in various biological models, amplified cDNA can substitute for normal, nonamplified cDNA in virtually all tasks. Moreover, in PCR-based gene hunting techniques such as rapid amplification of cDNA ends (RACE) (6,7; see also Chapter 5 of this volume), subtraction (8), or differential display (9), the amplified cDNA usually outperforms the normal one, because all backgrounds are predictable and can be easily kept under control.

From: *Methods in Molecular Biology, vol. 221: Generation of cDNA Libraries: Methods and Protocols*
Edited by: S.-Y. Ying © Humana Press Inc., Totowa, NJ

1.1. Total RNA Isolation

We use the procedure presented here rather than commercial kits, because this technique is suitable for virtually all animals. It is based on the well-known protocol of Chomczynski and Sacchi *(10)*, with one difference: All of the procedures are performed at neutral pH instead of acidic, as it was originally suggested. Also, the step of RNA precipitation with lithium chloride (LiCl) is added, because it results in very stable RNA preparations and considerably improves the consecutive procedures of cDNA synthesis. We have successfully applied the protocol to RNA isolation from representatives of 13 phyla of multicellular animals. As an alternative, a popular Trizol method (Gibco/Life Technologies) may be used in many cases, although it may not perform well on some nonstandard objects, such as jellyfish. Kits for RNA isolation that utilize columns (such as Qiagen's RNeasy kit) are generally not recommended for nonstandard samples.

The protocol is designed for rather large tissue samples (tissue volume, 10–100 µL), which normally yield about 10–100 µg of total RNA. The protocol for really microscopic amounts of starting material (expected to yield about 1 µg RNA or less) is the same but does not include second phenol–chloroform extraction (**Subheading 3.1., step 4**) and LiCl precipitation (**Subheading 3.1., step 6**). Additionally, the final "pellet" should be dissolved in 5 µL instead of 40 µL of water and transferred directly to cDNA synthesis, omitting the agarose gel analysis.

1.2. cDNA Synthesis

We provide two alternatives for preparing amplified total cDNA from the isolated RNA: method A and method B (**Fig. 1**). Both methods provide a possibility to amplify a cDNA fraction corresponding to messenger (polyA[+]), RNA, starting from total RNA. The fraction of ribosomal RNA in the amplified sample, as it was determined in EST sequencing project based on amplified cDNA, is 15–20%, represented mostly by small subunit ribosomal RNA. This is the same figure that is normally obtained with standard methods of cDNA synthesis *(11)*.

Method A ("classical") is to synthesize a double-stranded cDNA by a conventional means (employing DNA polymerase I/RNAseH/DNA ligase enzyme cocktail for second-strand synthesis), then ligate adaptors and amplify the sample using adaptor-specific primers. The structure of the adaptors evokes a PCR-suppression effect *(12)* and provides a method for selective amplification of only those cDNA molecules that contain both adaptor sequence and T-primer sequence [i.e., corresponding to the polyA(+) fraction of RNA]. The principles behind this method are described in **ref. *13***. The obvious advantage of this

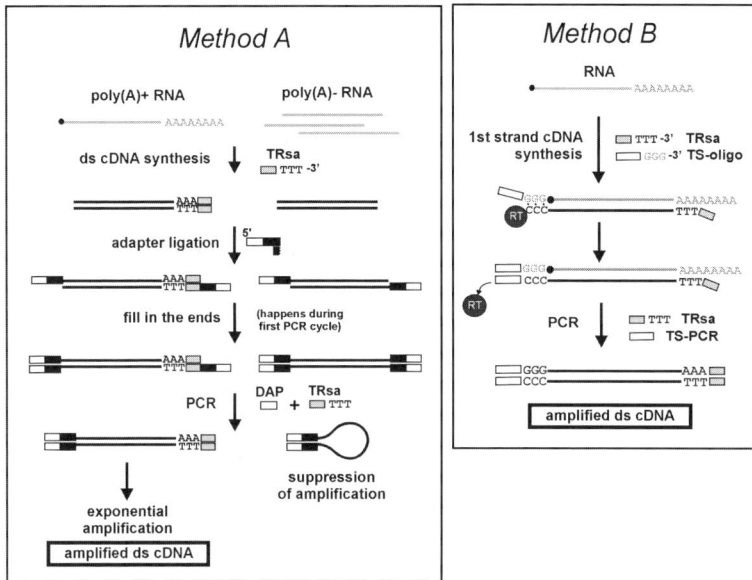

Fig. 1. Schematic outlines of cDNA amplification methods.

method is its high efficiency. A representative cDNA sample (with representation of 10^7 and higher) can be prepared from as little as 20–30 ng of total RNA. However, the method is rather laborious.

Method B is implemented in the SMART cDNA synthesis kit available from Clontech. It utilizes one surprising feature of Moloney murine leukemia virus reverse transcriptase (MMLV RT)—its ability to add a few nontemplate deoxynucleotides (mostly C) to the 3′ end of a newly synthesized cDNA strand upon reaching the 5′ end of the RNA template. Oligonucleotide-containing oligo(rG) sequence on the 3′ end, which is called "template-switch oligo" (TS-oligo), will base pair with the deoxycytidine stretch produced by MMLV RT when added to the RT reaction. Reverse transcriptase then switches templates and continues replicating using the TS-oligo as a template. Thus, the sequence complementary to the TS-oligo can be attached to the 3′ terminus of the first strand of cDNA synthesized and may serve as a universal 5′ terminal site for primer annealing during total cDNA amplification *(14)*. Recently, an improvement to the original procedure was reported *(15)*. The addition of MnCl$_2$ to the reaction mixture after first-strand synthesis, followed by a short incubation, increases the efficiency of nontemplate C addition to the cDNA and thus results in higher overall yield following cDNA amplification.

Although method B is simpler and faster than method A, its somewhat reduced efficiency means that a cDNA sample of suitable representation (more than 10^6) requires a minimum of 1 μg of total RNA. It should be noted that both techniques (as they are described here) provide material not only for total cDNA amplification, but also for RACE, a procedure for obtaining unknown flanks of a fragment. This procedure is indispensable for cloning complete coding regions of proteins. Different RACE techniques are available for each of the methods of cDNA amplification described here (**refs. 6** and **7** for methods A and B, respectively, *see* Chapter 5), both based on a PCR-suppression effect *(12)*.

2. Materials

2.1. Total RNA Isolation (see Note 1)

1. Dispersion buffer (buffer D): 4 *M* guanidine thiocyanate, 30 m*M* disodium citrate, 30 m*M* β-mercaptoethanol (pH 7.0–7.5) (*see* **Note 2**).
2. Buffer-saturated phenol, pH 7.0–8.0 (Gibco/Life Technologies).
3. Chloroform–isoamyl alcohol mix (24 : 1).
4. 96% Ethanol.
5. 80% Ethanol.
6. 12 *M* Lithium chloride.
7. Coprecipitant: SeeDNA reagent (Amersham) or glycogen.
8. Fresh Milli-Q water.
9. Agarose gel (1%) containing ethidium bromide.

2.2. cDNA Synthesis

2.2.1. Method A Using Conventional Second-Strand Synthesis (see Note 3)

1. SuperScript II reverse transcriptase, 200 U/μL (Life Technologies) or 20X PowerScript reverse transcriptase (Clontech) with provided buffer.
2. 0.1 *M* dithiothreitol (DTT).
3. dNTP mix, 10 m*M* each.
4. 5X Second-strand buffer: 500 m*M* KCl, 50 m*M* ammonium sulfate, 25 m*M* MgCl$_2$, 0.75 m*M* β-NAD, 100 m*M* Tris-HCl (pH 7.5), and 0.25 mg/mL bovine serum albumin (BSA).
5. 20X Second-strand enzyme cocktail: 6 U/μL DNA polymerase I, 0.2 U/μL RNase H, 1.2 U/μL *E. coli* DNA ligase.
6. T4 DNA polymerase (1–3 U/μL).
7. T4 DNA ligase 2–4 U/μL with provided buffer (New England Biolabs or equivalent).
8. T/M buffer: 10 m*M* Tris-HCl (pH 8.0) and 1 m*M* MgCl$_2$.
9. Buffer-saturated phenol (pH 7.0–8.0) (Gibco/Life Technologies).
10. Chloroform–isoamyl alcohol mix (24 : 1).
11. Long-and-Accurate PCR enzyme mix (Advantage2 polymerase mix by Clontech, LA-PCR by Takara, Expand Taq by Boehringer, or equivalent; *see* **Note 4**).
12. 10X PCR buffer: provided with the enzyme mix or, if KlenTaq-based homemade mix is used: 300 m*M* tricine–KOH (pH 9.1), 160 m*M* ammonium sulfate, 30 m*M* MgCl$_2$, 0.2 mg/mL BSA.

Invertebrate-optimized sets of oligonucleotides for total cDNA amplification

For Method A ("classical"):

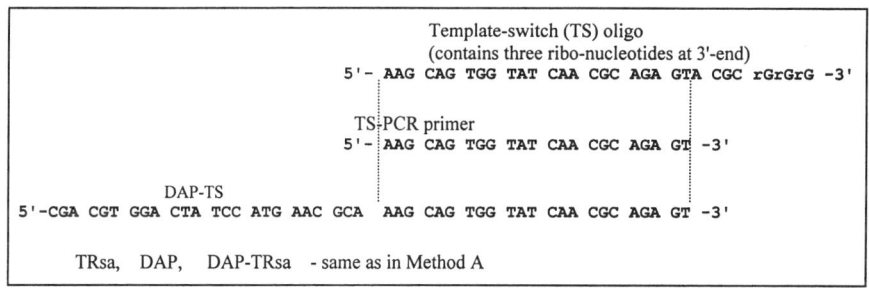

Use of the oligos:
TRsa - for 1st strand cDNA synthesis
adapter - to ligate to ds cDNA
DAP, TRsa - to amplify adapter-ligated cDNA

For Method B ("SMART"):

```
                                    Template-switch (TS) oligo
                                    (contains three ribo-nucleotides at 3'-end)
                         5'- AAG CAG TGG TAT CAA CGC AGA GTA CGC rGrGrG -3'

                    TS-PCR primer
                    5'- AAG CAG TGG TAT CAA CGC AGA GT -3'

         DAP-TS
5'-CGA CGT GGA CTA TCC ATG AAC GCA  AAG CAG TGG TAT CAA CGC AGA GT -3'

    TRsa,   DAP,   DAP-TRsa  - same as in Method A
```

Use of the oligos:
TRsa, TS-oligo - for 1st strand cDNA synthesis
TRsa, TS-PCR - to amplify 1st strand cDNA

Fig. 2. Invertebrate-optimized sets of oligonucleotides for cDNA amplication.

13. Yeast tRNA, 10 µg/µL.
14. 3 *M* sodium acetate (pH 5).
15. Fresh Milli-Q water.
16. Agarose gel (1%) containing ethidium bromide.
17. Oligonucleotides (*see* **Fig. 2** and **Note 5**).

2.2.2. Method B Using Template-Switching Effect

1. SuperScript II reverse transcriptase, 200 U/µL (Life Technologies) or 20X PowerScript reverse transcriptase (Clontech) with provided buffer.
2. 20 m*M* MnCl$_2$.
3. 0.1 *M* DTT.

4. dNTP mix, 10 m*M* each.
5. Long-and-Accurate PCR enzyme mix with buffer (*see* **Note 4**).
6. 10X PCR buffer: provided with the enzyme mix or, if KlenTaq-based homemade mix is used: 300 m*M* tricine–KOH (pH 9.1), 160 m*M* ammonium sulfate, 30 m*M* MgCl$_2$, and 0.2 mg/mL BSA.
7. Agarose gel (1%) containing ethidium bromide.
8. Fresh Milli-Q water.
9. Oligonucleotides (*see* **Fig. 2** and **Note 5**).

3. Methods
3.1. Total RNA Isolation

1. Dissolve the tissue sample in buffer D (*see* **Note 6**).
2. Spin the sample at maxumum speed on table microcentrifuge for 5 min at room temperature to remove debris. Transfer the supernatant to a new tube.
3. Put the tube on ice; add equal volumes of buffer-saturated phenol and mix. There will be no phase separation at this moment. Add 1/5 volume of chloroform–isoamyl alcohol (24:1) and vortex the sample. Two distinct phases will separate. Vortex three to four more times with about 1-min intervals between steps. Incubate the tube on ice between steps. Spin at maximum speed on a table microcentrifuge for 30 min at 4°C. Remove and save the upper, aqueous phase. Take care to avoid warming the tube with your fingers because the interphase may become invisible.
4. Repeat **step 3**.
5. Add 1 µL of coprecipitant and then add an equal volume of 96% ethanol and mix. Spin immediately at maxumum speed on a table microcentrifuge at room temperature for 10 min. The precipitate may not form a pellet, being instead spread over the back wall of the tube and thus being almost invisible even with coprecipitant added. Wash the pellet once with 0.5 mL of 80% ethanol. Dry the pellet briefly until no liquid is seen in the tube (do not overdry).
6. Dissolve the pellet in 100 µL of fresh Milli-Q water. If the pellet cannot be dissolved completely, remove the debris by spinning the sample at maxumum speed on a table microcentrifuge for 3 min at room temperature. Transfer the supernatant to a new tube, then add an equal volume of 12 *M* LiCl and chill the solution at –20°C for 30 min. Spin at maxumum speed on a table microcentrifuge for 15 min at room temperature. Wash the pellet once with 0.5 mL of 80% ethanol and dry as previously done. The precipitated RNA is usually invisible because the coprecipitant does not precipitate in LiCl.
7. Dissolve the pellet in 40 µL of fresh Milli-Q water.
8. Load 2 µL of the solution onto a standard (nondenaturing) 1% agarose gel to check the amount and integrity of the RNA. Add ethidium bromide (EtBr) to the gel to avoid the additional (potentially RNAse prone) step of gel staining. Load a known amount of some DNA on a neighboring lane to use as a standard for determining the RNA concentration. Intact RNA should exhibit sharp band(s) of ribosomal RNA (*see* **Fig. 3A** and **Notes 7–10**).

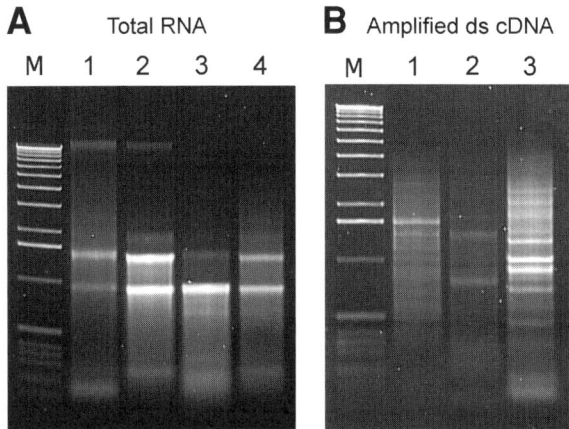

Fig. 3. **(A)** Nondenaturing agarose electrophoresis of total RNA from various invertebrate sources. Lane 1, unidentified sponge; lane 2, comb jelly *Bolinopsis infundibulum* (phylum Ctenophora); lane 3, planarian *Girardia tigrina* (phylum Platyhelminthes); lane 4, stony coral *Montastraea cavernosa* (phylum Cnidaria). M: 50 ng of 1-kb DNA ladder (Gibco/Life Technologies). **(B)** Amplified total cDNA from various sources. Lane 1, comb jelly; lane 2, planarian; lane 3, mollusk *Tridacna* sp. M: 50 ng of 1-kb DNA ladder (Gibco/Life Technologies). The product on lane 2 needs one more PCR cycle, whereas the product on lane 3 is already slightly overcycled (by one to two cycles), but is still very suitable for further manipulations.

3.2. cDNA Synthesis

3.2.1. Method A ("Classical")

3.2.1.1. METHOD A FIRST-STRAND cDNA SYNTHESIS

1. To 5 µL RNA solution in water (0.03–3 µg of total RNA), add 1 µL of 10 µ*M* primer TRsa and cover with mineral oil. Incubate at 65°C for 3 min and then put the tube on ice.
2. Add 2 µL 5X first-strand buffer (provided with reverse transcriptase), 1 µL of 0.1 *M* DTT, 1 µL reverse transcriptase, and 0.5 µL of dNTP mix (10 m*M* each), incubate at 42°C for 1 h, and then put the tube on ice.

3.2.1.2. METHOD A SECOND-STRAND cDNA SYNTHESIS

3. To the first-strand cDNA solution, add 49 µL of Milli-Q H_2O, 1.6 µL of dNTP mix (10 m*M* each), 16 µL of 5X second strand reaction buffer, and 4 µL of 20X second-strand enzyme cocktail. (The total volume of the reaction mix is about 80 µL.) Incubate at 16°C for 1.5 h and then put the tube on ice.
4. Add 1 µL T4 DNA polymerase; incubate 0.5 h at 16°C to polish ends.
5. Stop the reaction by heating at 65°C for 5 min.

6. Take the reaction mix from under the oil, put in new tube, and add 0.5 volume phenol and then 0.5 volume chloroform–isoamyl alcohol (24:1). Vortex the solution and spin at maxumum speed on a table microcentrifuge for 10 min. Transfer the upper, aqueous phase into new tube.

7. Add carrier (DNA [Amersham] or glycogen) and precipitate DNA by adding 0.1 volume (8 µL) 3 M sodium acetate (pH 5.0) and 2.5 volume (200 µL) of 95% ethanol at room temperature. Spin immediately for 15 min at maximum speed on a table microcentrifuge at room temperature.

8. Wash the pellet with 80% ethanol; air-dry the pellet for about 5 min at room temperature. Dissolve pellet in 6 µL H_2O.

3.2.1.3. METHOD A ADAPTOR LIGATION

9. To the 6 µL of double-stranded cDNA (ds-cDNA), add 2 µL of adaptor (10 µM), 1 µL of 10X ligation buffer, and 1 µL T4 DNA ligase and incubate overnight at 16°C.

10. To the ligation mixture, add 90 µL Milli-Q water and 10 µg yeast transfer RNA (tRNA). Purify by QiaQuick PCR purification kit (Qiagen, follow manufacturer's instructions) and elute with 40 µL of T/M buffer. Alternatively, dilute the ligation mixture fivefold by adding 40 µL of Milli-Q water to it (*see* **Note 11**).

3.2.1.4. METHOD A cDNA AMPLIFICATION

11. Prepare the PCR mixture (note that final concentration of primers is 0.1 µM) as follows: add 3 µL of 10X PCR buffer, 1 µL of dNTP mix (10 mM of each), 1.5 µL of 2 µM TRsa primer, 1.5 µL of 2 µM DAP primer, 1 µL of fivefold dilution of ligation mixture or 20 µL of Qia-Quick purified sample of adapter-ligated cDNA, H_2O to 30 µL, and KlenTaq/Pfu homemade polymerase mixture corresponding to 8 U of KlenTaq (*see* **Note 3**). When using commercial polymerase mixtures, follow the manufacturer's recommendations.

12. Perform cycling: 94°C for 30 s, –65°C for 1 min, –72°C for 2 min 30 s (block control), 95°C for 10 s, –65°C for 30 s, –72°C for 2 min 30 s (tube control or simulated tube control). Check 2 µL of the product on a 1% agarose gel after 12 cycles, keeping the PCR tube at room temperature while the electrophoresis runs. If nothing is seen, put the tube back into the thermal cycler and do five more cycles. If the product is barely visible, do only three more cycles. It is very important to determine the minimal number of cycles required to amplify the product until it is readily detectable on an agarose gel with EtBr staining (*see* **Notes 12–15**).

3.2.2. Method B for cDNA Synthesis Using Template-Switching Effect

3.2.2.1. METHOD B FIRST-STRAND cDNA SYNTHESIS

1. To 4 µL RNA solution in water (1–3 µg of total RNA) add 1 µL of 10-µM primer TRsa and cover with mineral oil. Incubate at 65°C for 3 min; put on ice.

2. Add 2 µL 5X first-strand buffer provided with reverse transcriptase, 1 µL of 0.1 M DTT, 1 µL of 5 µM TS-oligo, 1 µL reverse transcriptase, and 0.5 µL dNTP mix (10 mM each). Incubate at 42°C for 1 h, then add 1 µL of 20 mM $MnCl_2$ and

incubate for an additional 15 min at 42°C. Heat to 65°C and incubate for 3 min to stop the reaction. The product can be stored at –20°C for several months.

3.2.2.2. cDNA Amplification

3. Prepare the PCR mixture (final concentration of primers is 0.1 µM) as follows: 3 µL of 10X PCR buffer, 1 µL of dNTP mix (10 mM each), 1.5 µL of 2 µM TRsa primer, 1.5 µL of 2 µM TS-PCR primer, 1.5 µL of fivefold dilution of first-strand cDNA (from **step 2**), Milli-Q H$_2$O to 30 µL, and KlenTaq/Pfu homemade polymerase mixture corresponding to 8 U of KlenTaq (*see* **Note 4**). When using commercial polymerase mixtures, follow the manufacturer's recommendations.

4. Perform cycling: 94°C for 30 s, –65°C for 1 min, –72°C for 2 min 30 s (block control); 95°C for 10 s, –65°C for 30 s, –72°C for 2 min 30 s (tube control or simulated tube control). To determine the exact number of PCR cycles required to amplify cDNA, use the same strategy as described for Method A, but do 17 cycles before the first check on agarose gel. Typically, it takes about 17 cycles if there was 1 µg of total RNA at the start. If the number of cycles is 22–24, a 10^6-representation sample still can be accumulated by making amplification in 10-fold larger volume (i.e., making it in ten 30 µL tubes instead of one) and then pooling them. In this way, 10 times more product of first-strand synthesis may be put into PCR while avoiding the background problems because of nonincorporated cDNA synthesis oligomers. However, this approach leads to slightly more distorted cDNA sample in comparison to direct "less than 20" amplification (*see* **Notes 12–15**).

4. Notes

1. There is widespread belief that RNA is very unstable and, therefore, all of the reagents and materials for its handling should be specially treated to remove possible RNAse activity. We have found that purified RNA is rather stable and, ironically, too much anti-RNAse treatment can become a source of problems. This especially applies to diethyl pyrocarbonate (DEPC)-treating of aqueous solutions, which often leads to RNA preparations that are very stable but completely unsuitable for cDNA synthesis. Simple precautions such as wearing gloves, avoiding speaking over open tubes, using aerosol-barrier tips, and using fresh Milli-Q water for all solutions are sufficient to obtain stable RNA preparations. All organic liquids (phenol, chloroform, and ethanol) can be considered essentially RNAse-free by definition, as well as the dispersion buffer containing 4 M guanidine thiocyanate.

2. Normally, the dispersion buffer does not require titration. If the pH comes out significantly lower than 7.0, try another batch of guanindine or disodium citrate. The buffer may be stored for years at 4°C in the dark.

3. For cDNA synthesis, it is recommended to use reagents provided in the Marathon kit (Clontech), with the exception of oligonucleotides.

4. For Long-and-Accurate (LA)-PCR enzyme mixtures I strongly recommend enzyme the ones based on KlenTaq polymerase (Ab Peptides) or its analogs

(such as AdvanTaq polymerase, Clontech) instead of nontruncated *Taq* variants. In our experience, this enzyme produces the least distortion to the cDNA sample during amplification. The LA mixture can be prepared by adding 1 U of cloned Pfu polymerase (Stratagene) for every 30 U of KlenTaq polymerase. Calculate the required amount of mix assuming that 25 U of KlenTaq are required for 100 µL PCR.

5. The set of oligonucleotides presented in **Fig. 2** has been extensively tested on a number of various invertebrates and consistently produced good results. It is primarily designed for cDNA amplification and RACE, but it can be also successfully applied to preparation of samples for suppression subtractive hybridization *(7)*, because the potentially interfering oligo-derived flanking sequences are removed by *Rsa*I digestion.

6. The volume of tissue should be not more than one-fifth of the buffer D volume. To avoid RNA degradation, tissue dispersion should be done as quickly and completely as possible, ensuring that cells do not die slowly on their own. To adequately disperse a piece of tissue usually takes 2–3 min of triturating using a pipet, taking all or nearly all volume of buffer into the tip each time. The piece being dissolved must go up and down the tip, so it is sometimes helpful to cut the tip to increase the diameter of the opening for larger tissue pieces. Tissue dispersion can be done at room temperature. The dispersed samples can be stored at 4°C for several days (exceptions, such as *Balanoglossus*—acorn worm, phylum *Hemichordata*—that contain high concentrations of highly reactive iodine in its tissues, are rare).

The tissue dispersed in buffer D produces a highly viscous solution. The viscosity is usually the result of genomic DNA. This normally has no effect on the RNA isolation (except for dictating longer periods of spinning at the phenol–chloroform extraction steps), unless the amount of dissolved tissue was, indeed, too large. However, in some cases (e.g., freshwater planarians or mushroom anemones) mucus produced by the animal contributes to viscosity. This substance tends to copurify with RNA, making it very difficult to collect the aqueous phase at the phenol–chloroform extraction step. It likewise lowers the efficiency of cDNA synthesis. The RNA sample contaminated with such mucus, although completely dissolved in water, does not enter the agarose gel during electrophoresis. The EtBr-stained material stays in the well, probably because the mucus adsorbs RNA. Including cysteine in buffer D can diminish the mucus problem. To buffer D, add 0.1 vol of solution containing 20% l-cysteine hydrochloride hydrate and 50 mM tricine–KOH (pH 7) (takes a lot of titration!). The cysteine solution should be freshly prepared. After dissolving the tissue, incubate the sample for 2 h at 4°C and then proceed with the above protocol.

7. RNA degradation can be assessed using nondenaturing electrophoresis. The first sign of RNA degradation on the nondenaturing gel is a slight smear starting from the rRNA bands and extending to the area of shorter fragments, such as seen in **Fig. 3**, lanes 3 and 4. The RNA showing this extent of degradation is still good for further procedures. However, if the downward smearing is so pronounced that

the ribosomal RNA (rRNA) bands do not have a discernible lower edge, the RNA preparation should be discarded. The amount of RNA can be roughly estimated from the intensity of the rRNA staining by ethidium bromide in the gel, assuming that the dye incorporation efficiency is the same as for DNA (the ribosomal RNA may be considered a double-stranded molecule because of its extensive secondary structure).

8. The rule for vertebrate rRNA—that in intact total RNA, the upper (28s) rRNA band should be twice as intense as the lower (18s) band—does not apply to invertebrates. The overwhelming majority have 28s rRNA with a so-called "hidden break" *(16)*. It is actually a true break right in the middle of the 28s rRNA molecule, which is called hidden because under nondenaturing conditions the rRNA molecule is being held in one piece by the hydrogen-bonding between its secondary structure elements. The two halves, should they separate, are each equivalent in electrophoretic mobility to 18s rRNA. In some organisms, the interaction between the halves is rather weak, so the total RNA preparation exhibits a single 18s-like rRNA band even on nondenaturing gel (**Fig. 3A**, lane 3). In others, the 28s rRNA is more robust, so it is still visible as a second band, but it rarely has twice the intensity of the lower one (**Fig. 3A**, lanes 1, 2, and 4).

9. Curiously, genomic DNA contamination is reproducible for a particular species but varies between species. However, it never exceeds the amount seen in **Fig. 3A**, lanes 1 and 2—a weak band of high molecular weight. Such an extent of contamination does not affect further procedures. In fact, the methods of cDNA amplification described here tolerate genomic DNA up to 50% of the total sample mass, without losing specificity or efficiency.

10. To store the isolated RNA, add 0.1 vol of 3 M sodium acetate and 2.5 volumes of 96% ethanol to the RNA in water and mix thoroughly. The sample may be stored for several years at –20°C.

11. Using the Qia-Quick purified ligation mixture removes the excess of nonincorporated adapter oligomers and prevents them from interfering in the subsequent PCR. The step is necessary when the starting amount of RNA was lower than 0.3–0.5 µg, to make it possible to take all of the generated cDNA into subsequent PCR. For higher initial amounts, the purification step may be replaced by fivefold dilution of the mixture, followed by PCR starting with 1 µL of the dilution. Thus, the nonincorporated oligos are simply diluted to a non-interfering concentration. In this case, only one-fiftieth part of the available adapter-ligated cDNA goes into PCR, but because of the excess of RNA at the start, this is usually enough to generate a representative cDNA sample. If you are not quite sure which variant to choose, start with dilution. If you find that cDNA amplification requires too many cycles (more than 20; *see* **Note 11**), purify the remaining ligation mixture by Qia-Quick and take it all into PCR. It is important to use T/M buffer, which contains 10 mM Tris-HCl and 1 mM MgCl$_2$, for elution. Elution with plain water leads to denaturation of DNA because of electrostatic repulsion of strands in low-salt conditions, which decreases the specificity of amplification and promotes background stemming from genomic DNA.

Amplification guidelines

$$N = 2^{(40-n)} \text{ (Lukyanov-Matz equation)}$$

N - number of DNA molecules at the start of amplification, **n** - number of PCR cycles required to amplify the product to the concentration 5-10 ng/μl.

1 DNA molecule 1 kb long weights about 10^{-18} g

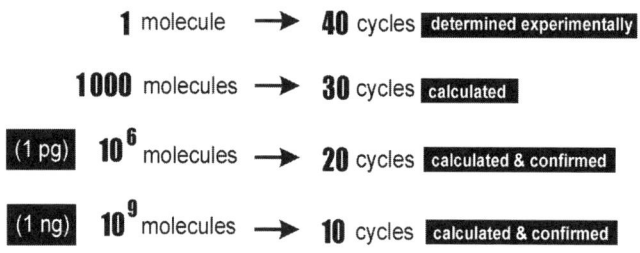

1 molecule ➝ **40** cycles `determined experimentally`

1000 molecules ➝ **30** cycles `calculated`

(1 pg) 10^{6} molecules ➝ **20** cycles `calculated & confirmed`

(1 ng) 10^{9} molecules ➝ **10** cycles `calculated & confirmed`

Fig. 4. Amplification guidelines.

12. The number of PCR cycles required to amplify a visible amount of cDNA (i.e., about 5–10 ng/μL) is a key parameter for assessing the representation of an amplified sample. There is a simple link between the initial number of target DNA molecules and number of PCR cycles required to amplify the sample (**Fig. 4**), as it was empirically determined during the work on in vitro cloning *(17,18)*. Using these guidelines, it can be calculated that a sample consisting of 10^6 molecules (a representation sufficient for most cDNA tasks) or more would require 20 or less PCR cycles to be amplified. In other words, if it took less than 20 PCR cycles to amplify your cDNA, this is a very representative sample. In our practice, we prefer to achieve at least one order of magnitude higher representation (i.e., get robust cDNA product in 16–17 cycles) to ensure that we have even the rarest messages.

13. The amplified cDNA at agarose gel should look like a smear (which may contain some bands, corresponding to the most abundant cDNA species) with the average length about 1 kb (*see* **Fig. 3B**). If it comes out much less, this may be a sign of pronounced RNA degradation during cDNA synthesis (if the total RNA was confirmed to be intact), which is usually the result of the poor quality of reverse transcriptase. Try another batch of it. Alternatively, something may be wrong with the PCR system. Most likely, the polymerase mixture is bad, but it is better to replace all of the reaction components.

14. It is recommended to store the product of amplification as a master sample. The unpurified PCR product produced by a KlenTaq-based enzyme mixture can be stored at –20°C for several years. If large amount of cDNA is required for further

procedures (e.g., cloning), use aliquots of the master sample to amplify more material. Dilute the aliquot of the master sample 50-fold in deionized water and add 1 µL of the dilution per each 20 µL of the PCR mixture prepared as at **step 10**. Do exactly 10 PCR cycles, this will generate a product in a concentration equal to the master sample. Do not apply more cycles attempting to generate more material, as overcycling produces the most pronounced distortions of cDNA profile. Instead, prepare a large volume of PCR mixture, distribute it into several tubes (30 µL per tube), and pool them after amplification is complete. If pure DNA is required for further procedures, the amplified cDNA may be cleaned using the Qia-Quick PCR purification kit (Qiagen) according to the provided protocol, but using T/M buffer for elution instead of the provided buffer.

15. If you intend to clone the product of cDNA amplification, it is necessary to perform a "chase" step after the product is amplified. The conditions for amplification recommended here include using a low working concentration of primers (0.1 µM), which greatly enhance the specificity of poly(A)$^+$-fraction amplification. However, there is a very good chance that a substantial fraction of the sample will be denatured at the end of PCR, because there will be already no primers available to initiate the synthesis of the complementary strand (especially if slight overcycling occurred). Obviously, for cloning, it is highly desirable to have the entire PCR product double-stranded. To ensure this without sacrificing the specificity of amplification, do the following. Run the PCR with a low primer concentration, as recommended, until the product is amplified. Then, keeping the completed PCR reaction in the thermocycler at 72°C, inject an additional amount of primers there (up to 0.2 µM more of each), and perform two nondenaturing "chase" cycles: 77°C for 1 min, –65°C for 1 min, 72°C for 3 min. Purify the product by Qia-Quick PCR purification kit (Qiagen) before cloning (use T/M buffer for elution).

References

1. Matz, M., Usman, N., Shagin, D., Bogdanova, E., and Lukyanov, S. (1997) Ordered differential display: a simple method for systematic comparison of gene expression profiles. *Nucleic Acids Res.* **25,** 2541–2542.
2. Gonzalez, P., Zigler, J. S., Jr, Epstein, D. L., and Borras, T. (1999) Identification and isolation of differentially expressed genes from very small tissue samples. *Biotechniques* **26,** 884–886, 888–892.
3. Spirin, K. S., Ljubimov, A. V., Castellon, R., Wiedoeft, O., Marano, M., Sheppard, D., et al. (1999) Analysis of gene expression in human bullous keratopathy corneas containing limiting amounts of RNA. *Invest. Opthalmol. Vis. Sci.* **40,** 3108–3115.
4. Wang, E., Miller L. D., Ohnmacht, G. A., Liu, E. T., and Marincola, M. (2000) High-fidelity mRNA amplification for gene profiling. *Nat. Biotechnol.* **18,** 457–459.
5. Livesey, F. J., Furukawa, T., Steffen, M. A., Church, G. M., and Cepko, C. L. (2000) Microarray analysis of the transcriptional network controlled by the photoreceptor homeobox gene Crx. *Curr. Biol.* **10,** 301–310.

6. Chenchik, A., Diachenko, L., Moqadam, F., Tarabykin, V., Lukyanov, S., and Siebert, P. D. (1996) Full-length cDNA cloning and determination of mRNA 5′ and 3′ ends by amplification of adaptor-ligated cDNA. *Biotechniques* **21,** 526–534.

7. Matz, M., Shagin, D., Bogdanova, E., Britanova, O., Lukyanov, S., Diatchenko, L., et al. (1999) Amplification of cDNA ends based on template-switching effect and step-out PCR. *Nucleic Acids Res.* **27,** 1558–1560.

8. Diatchenko, L., Lukyanov, S., Lau, Y. F., and Siebert, P. D. (1999) Suppression subtractive hybridization: a versatile method for identifying differentially expressed genes. *Methods Enzymol.* **303,** 349–380.

9. Matz, M. V. and Lukyanov, S. A. (1998) Different strategies of differential display: areas of application. *Nucleic Acids Res.* **26,** 5537–5543.

10. Chomczynski, P. and Sacchi, N. (1987) Single-step method of RNA isolation by acid guanidinium thiocyanate–phenol–chloroform extraction. *Anal. Biochem.* **162,** 156–159.

11. Lee, Y. H., Huang, G. M., Cameron, R. A., Graham, G., Davidson, E. H., Hood, L., et al. (1999) EST analysis of gene expression in early cleavage-stage sea urchin embryos. *Development* **126,** 3857–3867.

12. Siebert, P. D., Chenchik, A., Kellogg, D. E., Lukyanov, K. A., and Lukyanov, S. A. (1995) An improved PCR method for walking in uncloned genomic DNA. *Nucleic Acids Res.* **23,** 1087–1088.

13. Lukyanov, K. A., Diachenko, L., Chenchik, A., Nanisetti, A., Siebert, P. D., Usman, N. Y., et al. (1997) Construction of cDNA libraries from small amounts of total RNA using the suppression PCR effect. *Biophys. Biochem. Res. Comm.* **230,** 285–288.

14. Schmidt, W. M. and Mueller, M. W. (1999) CapSelect: a highly sensitive method for 5′ CAP-dependent enrichment of full length cDNA in PCR-mediated analyses of mRNAs. *Nucleic Acid Res.* **27,** e31.

15. Chenchik, A., Zhu, Y. Y., Diatchenko, L., Li, R., Hill, J., and Siebert, P. D. (1998) *Gene Cloning and Analysis by RT-PCR* (Siebert, P. and Larrick, J., eds.), BioTechniques Books, Natick, MA, pp. 305–319.

16. Ishikawa, H. (1977) Evolution of ribosomal RNA. *Comp. Biochem. Physiol. B* **58,** 1–7.

17. Lukyanov, K. A., Matz, M. V., Bogdanova, E. A., Gurskaya, N. G., and Lukyanov, S. A. (1996) Molecule by molecule PCR amplification of complex DNA mixtures for direct sequencing: an approach to in vitro cloning. *Nucleic Acids Res.* **24,** 2194–2195.

18. Fradkov, A. F., Lukyanov, K. A., Matz, M. V., Diatchenko, L. B., Siebert, P. D., and Lukyanov, S. A. (1998) Sequence-independent method for in vitro generation of nested deletions for sequencing large DNA fragments. *Anal. Biochem.* **258,** 138–141.

11

Single-Cell cDNA Library Construction Using Cycling aRNA Amplification

Shi-Lung Lin

1. Introduction

Several existing methods for constructing complementary DNA (cDNA) libraries require several thousand cells and involve the lengthy procedure of reverse transcription (RT), restriction, adaptor ligation, and vector cloning, which usually fail to maintain the completeness of a cDNA library, resulting in a loss of rare cDNAs. However, gene expression analysis of specific cell populations within a heterogeneous tissue is essential for research in vivo, requiring a method to generate cDNA libraries from a very small number of specific cells. The generation of amplified antisense RNA (aRNA) by incorporating an oligo-(dT) primer coupled to a T7 RNA polymerase promoter sequence during RT has been developed to linearly increase transcriptional copies of mRNAs from a limited amount of promoter-linked cDNAs *(1,2)* (*see* **Note 1**). However, the aRNAs prepared from a single live neuron has been reported to cover 50–75% of the total mRNA population *(2,3)*, indicating that rare mRNAs were lost during the amplification procedure.

One possibility for the loss of rare mRNA species in the previous methods may result from the use of random primers and diluted RNA amount. We present a novel method for generating full-length cDNA libraries that combines *in situ* reverse transcription and the terminal transferase tailing reaction *(4)*. To prevent mRNAs from degradation during extraction, the initial steps of *in situ* reverse transcription, first-strand cDNA tailing, and aRNA amplification are preferably performed using fixed and permeabilized cells (*see* **Note 2**). The intracellular poly(A⁺) RNA is amplified up to 2000-fold prior to be extracted from cells. Subsequent RT-PCR polymerase chain reaction generates a full-

From: *Methods in Molecular Biology, vol. 221: Generation of cDNA Libraries: Methods and Protocols*
Edited by: S.-Y. Ying © Humana Press Inc., Totowa, NJ

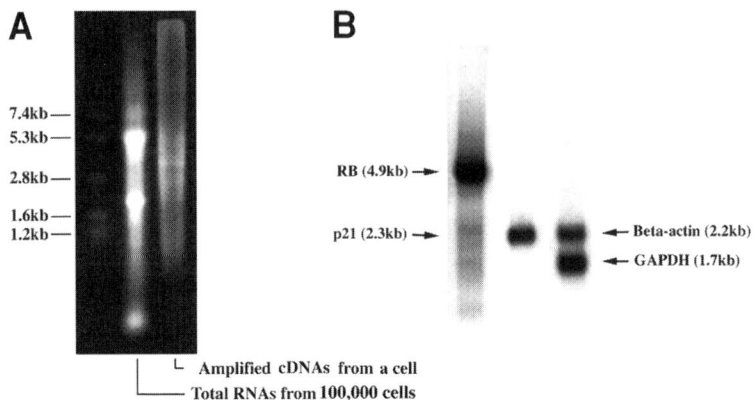

Fig. 1. **(A)** One percent formaldehyde–agarose gel analysis of the cDNA library from different numbers of cultured LNCaP cells. Lane 1, RNA markers from 1.2 to 7.4 kb; lane 2, total RNA repertoire from 100,000 cells; lane 3, cDNA library from a few single cells. The cDNA library obtained from single cells was sufficient to produce a smear visualizable upon staining with ethidium bromide. **(B)** Southern blot analysis of RB, p21, β-actin, and GAPDH in the cDNA library generated from 10 LNCaP cells, showing corrected sizes of all tested gene transcripts from 1.7 (GAPDH) to 4.9 kb (RB).

length cDNA library in a quantity easily manipulatable for gene expression analysis; the resultant cDNA are amplified to total more than 10^9 the amount of original mRNA. Moreover, the tailing reaction protects the 3' end of first-strand cDNA, which codes the 5' end of a mRNA sequence. A more full-length library construct, therefore, can be assessed for gene cloning. It is noted that the maintenance of the populationary ratio of individual RNA species has also been improved, indicating the necessity of preserving intact mRNA termini for high-quality cDNA library generation. Our previous report *(4)* has shown that the cDNA library so generated is suitable for Southern analysis of correctly sized genes up to 5 kb (*see* **Fig. 1**).

The method can be broken down into six steps (A–F, **Fig. 2**): (A) cell fixation and permeabilization, (B) first reverse transcription with oligo-(dT)-promoter primers, (C) first-strand cDNA tailing and double-stranding, (D) in vitro or in-cell transcription, (E) second reverse transcription using the externally added tail, and (F) optional PCR amplification. Although the starting material can be either from 0.1–2 µg total RNAs or from as few as a single cell, we usually use 20–200 cells for more consistent amplification results. It has also been shown that too large an amount of starting material can cause lower efficiency of tailing reaction. The chance to generate a complete cDNA library from more than 200 cells will depend on the relative amount of terminal transferase (TdT)

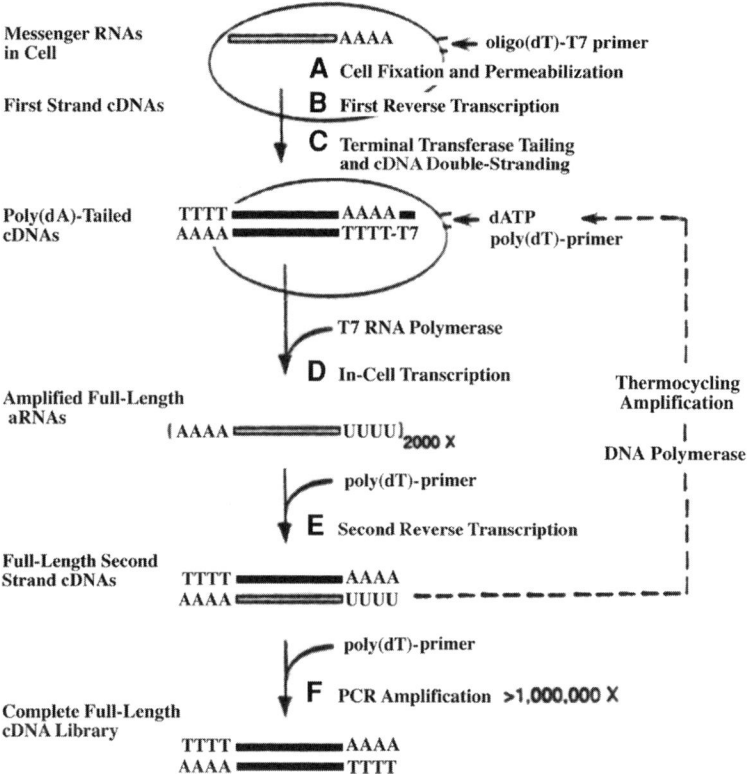

Fig. 2. Flowchart of current method for generating a full-length cDNA library from a few single cells. Cells fixation and permeabilization protect intracellular mRNA from degradation and allow enzymes to penetrate cell membranes. Upon RT using an oligo-(dT)-T7 promoter and tailing with poly(dA) oligonucleotide, the cDNA library was amplified by in-cell transcription up to 2000-fold and then PCR for another 10^6-fold. The final PCR products, therefore, yield more than 10^9 times the amount of original mRNAs.

activity to the first-strand cDNA molecules. The coverage of TdT activity is less effective when outnumbered cDNA molecules interact with limited TdT enzyme in a short tailing period of time. A longer incubation time does not improve the coverage efficiency of a tailing reaction and usually causes longer tails to bias the corrected size of a cDNA sequence. Therefore, we suggest at least 50 U of TdT for maximal 1 μg total RNAs or 20 ng mRNAs in a short tailing reaction.

Two methods for the reamplification of the final cDNA library are presented. In one method, because the amplified aRNA contains a poly(A) tail that is similar to its originated mRNA, the aRNA amplification can be reiterated following the same procedure using aRNA as the starting material. The second procedure uses PCR to amplify the cDNA based on its poly(dA) tail in both termini. It is clear that PCR provides a much simpler approach for high-quantity cDNA generation, but three major disadvantages are also created. First, preferential amplification of some abundant genes is a well-known problem for PCR-based methodology. Second, low fidelity of the resultant DNA product is another generic problem caused by most thermostable DNA polymerases. In addition to these two problems, we also found that PCR using poly(dT) primers tends to increase the mean size of a cDNA library after multiple cycles of amplification. Such size increase is a result from the concatenate linkage between a poly(dA) tail and a poly(dT) termini of different cDNA templates during PCR. Therefore, although the PCR product is useful for Southern or reverse Northern blotting, we recommend to perform only the cycling amplification procedure of this method without final PCR for cDNA preparation and labeling for high-resolution gene-chip analysis.

2. Materials

2.1. Generation of First-Strand cDNA Using Reverse Transcription with Oligo-(dT)-Promoter Primers

1. Diethyl pyrocarbonate (DEPC) H_2O: Stir double-distilled water (ddH$_2$O) with 0.1% DEPC for more than 12 h and then autoclave at 120°C under about 1.2 kgf/cm^2 for 20 min, twice.
2. Oligo-(dT)24-T7 promoter primer: dephosphorylated 5'-dAAGCTTAGAT ATC TAATACG ACTCACTATA GGGAATTTTT TTTTTTTTT TTTTTTTTT-3' (100 pmol/µL).
3. AMV reverse transcriptase (50 U/µL) and 10X reverse transcription buffer (500 mM Tris-HCl (pH 8.5) at 25°C, 80 mM MgCl$_2$, 300 mM KCl, 10 mM dithiothreitol [DTT]).
4. First reverse transcriptase mix: 7 µL DEPC-treated ddH$_2$O, 2 µL 10X reverse transcription buffer, 2 µL of 10 mM dNTP mix (10 mM each for dATP, dGTP, dCTP, and dTTP), 1 µL RNasin (25 U/µL), and 2 µL AMV reverse transcriptase: prepare just before use.
5. Incubation chamber: 65°C, 42°C, 50°C, and 94°C.
6. Purification spin column: 100 bp cutoff filter (e.g., a Microcon-50 centrifugal filter [Amicon, Beverly, MA], as shown here).

2.2. cDNA Amplification Using Terminal Transferase Tailing with Poly(dA) Oligonucleotides

1. 10X Terminal transferase tailing buffer: 500 mM Tris-HCl (pH 8.0) at 25°C, 400 mM KCl, 80 mM MgCl$_2$, and 100 mM DTT; prepare fresh.
2. Terminal deoxynucleotidyl transferase (TdT, 25 U/µL).

3. Terminal transferase reaction mix: 3 µL DEPC-treated ddH$_2$O, 1 µL of 10 mM dNTP, 3 µL of 10X terminal transferase tailing buffer, 1 µL of 10 mM dATP, and 2 µL terminal deoxynucleotidyl transferase; prepare just before use.
4. Incubation mixer: 37°C, 100g vortex for 30 s between every 5-min interval.
5. Incubation chamber: 94°C.

2.3. cDNA Double-Stranding Using DNA Polymerization with Poly(dT) Primers

1. Poly(dT)$_{24}$ primer: dephosphorylated 5′-dTTTTTTTTTT TTTTTTTTTT TTTT-3′ (100 pmol/µL).
2. 10X cDNA double-stranding buffer: 500 mM Tris-HCl (pH 9.2) at 25°C, 160 mM (NH$_4$)$_2$SO$_4$, and 20 mM MgCl$_2$; prepare fresh.
3. cDNA double-stranding reaction mix: 10 µL DEPC-treated ddH$_2$O, 5 µL of 10X first-strand buffer, 2 µL of 10 mM dNTP mix (10 mM each for dATP, dGTP, dCTP, and dTTP), 0.7 µL *Taq* DNA polymerase (5 U/µL), and 0.3 µL Pwo DNA polymerase (5 U/µL); prepare just before use.
4. Incubation chamber: 94°C, 50°C, and 68°C.
5. Nonionic detergent (octylphenoxy)polyethanol (2%).
6. Purification spin column: 100 bp cutoff filter.

2.4. Generation of Full-Length Antisense RNA Using In Vitro Transcription with Promoter-Driven RNA Polymerase

1. 10X In-vitro transcription (IVT) buffer: 400 mM Tris-HCl (pH 8.0) at 25°C, 100 mM MgCl$_2$, 50 mM DTT, and 5 mg/mL nuclease-free bovine serum albumin (BSA).
2. T7 RNA polymerase (80 U/µL).
3. IVT reaction mix: 8 µL DEPC-treated ddH$_2$O, 4 µL of 10X IVT buffer, 4 µL of 10 mM dNTP mix (10 mM each of ATP, GTP, CTP, and UTP), and 2 µL RNasin (25 U/µL); prepare just before use.
4. Incubation mixer: 37°C, 100g vortex for 30 s between every 30-min interval.

2.5. Generation of Sense-Strand cDNA Using Reverse Transcription with Poly(dT) Primers

1. Poly(dT)$_{24}$ primer: dephosphorylated 5′-dTTTTTTTTTT TTTTTTTTTT TTTT-3′ (100 pmol/µL).
2. AMV reverse transcriptase (50 U/µL) and 10X reverse transcription buffer (500 mM Tris-HCl [pH 8.5] at 25°C, 80 mM MgCl$_2$, 300 mM KCl, and 10 mM DTT).
3. Second reverse transcriptase mix: 9 µL DEPC-treated ddH$_2$O, 2 µL of 10X reverse transcription buffer, 3 µL of 10 mM dNTP mix (10 mM each of dATP, dGTP, dCTP, and dTTP), 1 µL RNasin (25 U/µL), and 3 µL AMV reverse transcriptase; prepare just before use.

4. Incubation chamber: 65°C, 42°C, and 50°C.
5. Purification spin column: 100 bp cutoff filter.

2.6. cDNA Double-Stranding Using DNA Polymerization with Oligo-(dT)-Promoter Primers

1. Oligo(dT)$_{24}$-T7 promoter primer: dephosphorylated 5′-dAAGCTTAGAT ATC TAATACG ACTCACTATA GGGAATTTTT TTTTTTTTTT TTTTTTTTT-3′ (100 pmol/µL).
2. 10X cDNA double-stranding buffer: 500 mM Tris-HCl (pH 9.2) at 25°C, 160 mM $(NH_4)_2SO_4$, and 20 mM MgCl$_2$; prepare fresh.
3. cDNA double-stranding reaction mix: 12 µL DEPC-treated ddH$_2$O, 4 µL of 10X cDNA double-stranding buffer, 1.5 µL of 10 mM dNTP mix (10 mM each of dATP, dGTP, dCTP, and dTTP), 0.7 µL *Taq* DNA polymerase (5 U/µL), and 0.3 µL Pwo DNA polymerase (5 U/µL); prepare just before use.
4. Incubation chamber: 94°C, 50°C, and 68°C.

2.7. Optional cDNA Amplification Using PCR Reaction with Poly(dT) Primers

1. Poly(dT)$_{24}$ primer: dephosphorylated 5′-dTTTTTTTTTT TTTTTTTTTT TTTT-3′ (100 pmol/µL).
2. 10X cDNA double-stranding buffer: 500 mM Tris-HCl (pH 9.2) at 25°C, 160 mM $(NH_4)_2SO_4$, and 20 mM MgCl$_2$; prepare fresh.
3. cDNA PCR reaction mix: 12 µL DEPC-treated ddH$_2$O, 5 µL of 10X cDNA double-stranding buffer, 2 µL of 10 mM dNTP mix (10 mM each of dATP, dGTP, dCTP, and dTTP), 0.7 µL *Taq* DNA polymerase (5 U/µL), and 0.3 µL Pwo DNA polymerase (5 U/µL); prepare just before use.
4. Incubation chamber or thermocycler machine: 94°C, 50°C, and 68°C.

3. Methods

3.1. Generation of First-Strand cDNA Using Reverse Transcription with Oligo-(dT)-Promoter Primers

The starting material is 0.1–2 µg total RNA *(1–3)* or a few permeabilized cells *(4)* (*see* **Note 3**). Poly(A$^+$) RNA is selected using promoter-linked oligo-(dT) primers, which contains about 20–26 deoxythymidylate oligonucleotides. The first-strand cDNA is synthesized by reverse transcription from the poly(A$^+$) RNA with promoter-linked oligo-(dT) primers. As shown in **Fig. 1**, the promoter used here is a T7 bacteriophage RNA promoter element.

1. Primer annealing: Suspend RNA in 5 µL of DEPC-treated water, mix well with 1 µL oligo-(dT)$_{24}$-T7 primer, heat to 65°C for 5 min for minimizing secondary structure, cool to 50°C for 10 min for primer hybridization, and then cool on ice.
2. First-strand cDNA synthesis: Add 14 µL of first reverse transcriptase mix and heat to 42°C for 50 min. Add another 1 µL of reverse transcriptase and mix.

Continue to incubate the reaction at 42°C for 30 min, heat to 50°C for 10 min, and then cool on ice. The RNA is still attached noncovalently to the cDNA.

3. Denaturation: Heat the reaction at 94°C for 3 min and then cool on ice immediately.

4. Primer removal and buffer exchange: Load the reaction into a purification spin column, spin for 10 min at 14,000g, and discard the flowthrough (*see* **Note 4**). Add 200 μL of DEPC-treated ddH$_2$O into the spin column to wash the cDNA, spin for 10 min at 14,000g, and discard the flowthrough. Add 20 μL of DEPC-treated ddH$_2$O into the spin column to dissolve the cDNA, place the spin column upside down in a new collecting microtube, and spin 3 min at 3000g. Store the 20 μL of the purified cDNA in a –20°C freezer or perform the next step immediately.

3.2. cDNA Amplification Using Terminal Transferase Tailing with Poly(dA) Oligonucleotides

In this method, which was reported by Ying and Chuong in 1999, the first-strand cDNA is dA-tailed using terminal deoxyribonucleotidyl transferase (TdT), and a poly(dT) primer is applied to initiate the second-strand cDNA synthesis by a two-cycle PCR-like reaction. Although this external priming procedure preserves better full-length conformation of the mRNA, the efficiency of the TdT tailing reaction seems to depend on the particular 3′ termini of different first-strand cDNA species, resulting in uneven coverage. Such a problem, however, can be improved by adequate TdT activity rate and constant reaction vortex (*see* **Note 5**). Practically, 1 U of TdT is required for tailing every picomole of cDNA in a mild shaking incubator (100g), if the average size of cDNA is 3 kb.

1. TdT tailing reaction: Add 10 μL of terminal transferase reaction mix to the purified cDNA and mix well. Incubate the reaction at 37°C for 30 min and occasionally mix the reaction every 5 min for better tailing coverage.

2. Reaction stop: Heat the reaction at 94°C for 3 min and cool on ice immediately.

3.3. cDNA Double-Stranding Using DNA Polymerization with Poly(dT) Primers

The mRNA portion of the resulting messenger RNA (mRNA)–cDNA hybrid is denatured, and double-stranded cDNA was formed using a one-cycle PCR-like reaction with poly(dT) primers. Then, the amplification of cDNA representative can be achieved by in vitro transcription using the promoter-linked double-stranded cDNA as template.

1. Primer annealing: Add 2 μL of poly(dT)$_{24}$ primer to the reaction, heat to 94°C for 3 min for mRNA removal, cool to 50°C for 1 min for primer hybridization, and then cool on ice.

2. cDNA double-stranding: Add 20 µL of cDNA double-stranding reaction mix to the reaction, mix well, and then incubate the reaction at 68°C for 10 min.

3. Breaking cell membrane: For using permeabilized cells as starting material, add 200 µL of 2% nonionic detergent (octylphenoxy)polyethanol to the reaction and vortex at 100*g* for 10 min.

4. Primer removal and buffer exchange: Load the reaction into a purification spin column, spin for 10 min at 14,000*g*, and discard the flowthrough. Add 200 µL of DEPC-treated ddH$_2$O into the spin column to wash the double-stranded cDNA, spin for 10 min at 14,000*g*, and discard the flowthrough. Add 20 µL of DEPC-treated ddH$_2$O into the spin column to dissolve the double-stranded cDNA, place the spin column upside down in a new collecting microtube, and spin for 3 min at 3000*g*. Store the 20 µL of the purified cDNA in a –20°C freezer or perform the next step immediately.

3.4. Generation of Full-Length Antisense RNA Using In Vitro Transcription with Promoter-Driven RNA Polymerase

The promoter of the double-stranded cDNA is now served as a recognition site for RNA polymerase during an in vitro transcription reaction. The in vitro transcription provides linear amplification up to 2000-fold of the amount of starting materials *(1,2)*. The proofreading capability of the RNA polymerase ensures the fidelity of the resulting nucleic acid products. Because the promoter is incorporated in the opposite orientation of mRNA, the resulting product is antisense RNA (aRNA) rather than sense mRNA.

1. In vitro transcription reaction: Add 20 µL of the IVT reaction mix to the purified cDNA and mix well. Incubate the reaction at 37°C for 2–3 h and occasionally mix the reaction every 30 min for better RNA elongation (*see* **Note 6**).

2. Buffer exchange and sample concentration: Load the reaction into a purification spin column, spin for 10 min at 14,000*g*, and discard the flowthrough. Add 200 µL of DEPC-treated ddH$_2$O into the spin column to wash the aRNA, spin for 10 min at 14,000*g*, and discard the flowthrough. Add 20 µL of DEPC-treated ddH$_2$O into the spin column to dissolve the aRNA, place the spin column upside down in a new collecting microtube, and spin for 3 min at 3000*g*. Store the 20 µL of the purified aRNA in a –80°C freezer or perform the next step immediately.

3.5. Generation of Second-Strand cDNA Using Reverse Transcription with Poly(dT) Primers

Because the full-length aRNA is flanked by poly(T)- and poly(A)-oligonucleotide in its 5′ and 3′ termini, respectively, the second-strand cDNA can be formed by reverse transcription of the aRNA using poly(dT) primers.

1. Primer annealing: Add 2 µL of poly(dT)$_{24}$ primer to 10 µL of the purified aRNA, mix well and heat to 65°C for 5 min for minimizing secondary structure, cool to 50°C for 1 min for primer hybridization, and then cool on ice.

2. Second-strand cDNA synthesis: Add 18 µL of second reverse transcriptase mix and heat to 42°C for 50 min. Add another 1 µL of reverse transcriptase and mix. Continue to incubate the reaction at 42°C for 30 min, heat to 50°C for 10 min, and then cool on ice. The aRNA is still attached noncovalently to the second-strand cDNA.

3. Denaturation: Heat the reaction at 94°C for 3 min and then cool on ice immediately.

4. Buffer exchange and sample concentration: Load the reaction into a purification spin column, spin for 10 min at 14,000g, and discard the flowthrough. Add 200 µL of DEPC-treated ddH$_2$O into the spin column to wash the second-strand cDNA, spin for 10 min at 14,000g, and discard the flowthrough. Add 20 µL of DEPC-treated ddH$_2$O into the spin column to dissolve the second-strand cDNA, place the spin column upside down in a new collecting microtube, and spin for 3 min at 3000g. Store the 20 µL of the purified second-strand cDNA in a –20°C freezer or perform the next step immediately.

3.6. cDNA Double-Stranding Using DNA Polymerization with Oligo-(dT)-Promoter Primers

The aRNA of the aRNA–cDNA hybrid is denatured and the cDNA is used to form a double-stranded cDNA library using a one-cycle PCR-like reaction with the promoter-linked oligo-(dT) primers. Then, another round of aRNA amplification can be achieved by in vitro transcription using the promoter-linked double-stranded cDNA as the template, following the steps from **Subheadings 3.4.–3.6.**

1. Primer annealing: Add 2 µL of oligo-(dT)$_{24}$-T7 primer to the purified second-strand cDNA, heat to 94°C for 3 min for aRNA removal, cool to 50°C for 10 min for primer hybridization, and then cool on ice.

2. cDNA double-stranding: Add 18.5 µL of cDNA double-stranding reaction mix to the reaction, mix well, and then incubate the reaction at 68°C for 10 min.

3.7. Optional cDNA Amplification Using PCR Reaction with Poly(dT) Primers

The resultant double-stranded cDNA contains the poly(dA–T) region in the both termini, which can be used as a primer binding site for poly(dT) primers during PCR. Although the following PCR procedure can easily provide a high quantity full-length cDNA library, some PCR-based disadvantages may bias the quality of a cDNA library. For high-fidelity gene analysis, the cycling steps from **Subheadings 3.4.–3.6** are recommended.

1. Primer annealing: Add 2 µL of poly(dT)$_{24}$ primer and 1 µL of the cDNA to 27 µL of DEPC-treated ddH$_2$O, heat to 94°C for 3 min for cDNA denaturation, cool to 50°C for 1 min for primer hybridization, and then cool on ice.

2. PCR reaction: Add 20 µL of cDNA double-stranding reaction mix to the reaction, mix well, and then incubate the reaction at 68°C for 10 min. Repeatedly incubate the reaction at 94°C for 1 min, 50°C for 1 min, and then 68°C for 7 min up to 25 cycles.

3. Primer removal and buffer exchange: Load the reaction into a purification spin column, spin for 10 min at 14,000*g*, and discard the flowthrough. Add 200 µL of DEPC-treated ddH$_2$O into the spin column to wash the second-strand cDNA, spin for 10 min at 14,000*g*, and discard the flowthrough. Add 20 µL of DEPC-treated ddH$_2$O into the spin column to dissolve the second-strand cDNA, place the spin column upside down in a new collecting microtube, and spin for 3 min at 3000*g*. Store the 20 µL of the purified cDNA in a –20°C freezer until use up to 2 mo.

4. cDNA library assessment (*see* Chapter 13).

4. Notes

1. As described and proven in Eberwine, and Van Gelder, and their colleagues' publications (*1,2*), the linear amplification of transcription means that the amplified RNA population should reflect its abundance in the original RNA population. Because the transcriptional amplification of nucleic acids only requires the recognition of promoter regions by RNA polymerases and all promoters possess the same affinity to their respective polymerases, the transcription-based nucleic acid amplification technology will provide less preferential amplification than the PCR-based technology. It is noted that the preferential amplification of PCR is caused by the different affinity between different primers (forward and reverse primers) to their templates as well as between the primers and different templates. It is clear that the transcriptional amplification can bypass the use of multiple primers and also provide a better amplification rate (up to 2000-fold per cycle) to mirror the mRNA population.

2. Because the penetration of formaldehyde is less effective than acetone, methanol, or ethanol/acetone mixture, the dissected sample tissues should be as small as possible (<2 mm size cubes) to enable rapid fixation.

3. Isolated cells were preserved in 500 µL of ice-cold 10% formaldehyde in suspension buffer (0.15 *M* NaCl (pH 7.0), and 1 m*M* EDTA) for the following fixation and permeabilization procedure (*5*). After 1 h of incubation with occasional agitation, fixed cells were collected with Microcon-50 filters and washed by 350 µL of ice-cold PBS with vigorous pipetting. The collection and wash were repeated at least once. The fixed cells were then permeabilized in 500 µL of 0.5% nonionic detergent (octylphenoxy)polyethanol for 1 h with frequent agitation. After that, three collections and washes were given to the cells as earlier but using 350 µL of ice-cold PBS containing 0.1 *M* glycine instead. The cells were finally mixed with 0.1 µ*M* poly(dT)$_{24}$ primer and resuspended in the same buffer with vigorous pipetting to evenly distribute them into small aliquots (about 50 cells in 10 µL) for storage. They could be stored at –80°C for up to 2 wk.

4. Relative Centrifugal Force (RCF) (g) = $(1.12 \times 10^{-5})(\text{rpm})^2 \cdot r$, where r is the radius in centimeters measured from the center of the rotor to the middle of the spin column and rpm is the speed of the rotor in revolutions per minute.

5. The first-strand cDNA is poly(dA)-tailed by TdT using provided condition that should produces an average 35- to 65-base overhang. The incorporation rate is increased about 80–90% with occasionally gentle mix, but drops to about 50% without mix. The efficiency of TdT tailing seems to be varied among different mRNA species, but can be improved by occasionally gently mixing in a short period of incubation time. The use of multiple poly(dT)$_{22}$N'N'' primers (N' = dA, dC, or dG; and N'' = dA, dT, dC or dG) instead of poly(dT)$_{24}$ primers for the synthesis of second-strand cDNA can improve the final poly(dA) tail to be about 20–26 residues. The use of cobalt-based buffers is not recommended in this protocol.

6. The most stable and efficient IVT reaction occurs during the first 2-h incubation at 37°C. The rate of RNA synthesis decreases considerably (40–50%) after 3 h of incubation or below 37°C incubation. A longer reaction may increase the yield, but the possibility of degradation by RNase increases. Occasionally, gentle mixing can prevent the stall of crowded RNA polymerases on a template and enhance full-length synthesis. The overall rate of RNA polymerization is maximal between pH 7.7 and 8.3, but it remains about 70% of maximum at pH 7.0 or 9.0. High concentrations of NaCl, KCl, or NH$_4$Cl above 75 mM will inhibit the reaction.

References

1. Van Gelder, R. N., von Zastrow, M. E., Yool, A., Dement, W.C., Barchas, J. D., and Eberwine, J. H. (1990) Amplified RNA synthesized from limited quantities of heterogeneous cDNA. *Proc. Natl. Acad. Sci. USA* **87,** 1663–1667.

2. Eberwine, J., Yeh, H., Miyashiro, K., Cao, Y., Nair, S., Finnell, R., et al. (1992) Analysis of gene expression in single live neurons. *Proc. Natl. Acad. Sci. USA* **89,** 3010–3014.

3. Crino, P. B., Trajanowski, J. Q., Dichter, M. A., and Eberwine, J. (1996) Embryonic neuronal markers in tumerous sclerosis: single-cell molecular pathology. *Proc. Natl. Acad. Sci. USA* **93,** 14,152–14,157.

4. Ying, S. Y., Lui, H. M., Lin, S. L., and Chuong, C. M. (1999) Generation of full-length cDNA library from single human prostate cancer cells. *BioTechniques* **27,** 410–414.

5. Embleton, M. J., Gorochov, G., Jones, P. T., and Winter, G. (1992) In-cell PCR from mRNA: amplifying and linking the rearranged immunoglobulin heavy and light chain V-genes within single cells. *Nucleic Acids Res.* **20,** 3831–3837.

12

mRNA/cDNA Library Construction
Using RNA–Polymerase Cycling Reaction

Shi-Lung Lin and Shao-Yao Ying

1. Introduction

Molecular profiling of single-cell gene expression permits the high-definition investigation of intracellular gene activity and physiological status of the cells under certain special conditions, such as pathogenesis *(1,2)*, cancer staging *(3)*, drug treatment, and developmental processes *(4)*. Traditionally, gene transcripts were extracted from lysed cells with phenol–chloroform followed by precipitation, and messenger RNAs (mRNA) were further purified by oligo-(dT)-dextran media *(5)*. However, the tedious procedures of extraction, chromatography, and precipitation could not maintain the completeness of a whole mRNA repertoire, resulting in a significant loss of rare RNA (<10 copies/cell) populations. Such loss could be as much as 30% of the original repertoire. The requirement of bulk tissue samples for a better population coverage was another drawback of the phenol–chloroform extraction methods. A minimum of several thousand cells is needed for an acceptable quality of RNA extraction. Because of tissue heterogeneity, these methods usually provided neither reliable nor reproducible results. Unfortunately, it is impossible to collect adequate amounts of pure or homogeneous samples for these methods because of a tremendous difficulty in sample dissection and RNA preservation, especially the preservation of rare mRNA species.

A breakthrough improvement of mRNA preparation is now based on a in vitro transcription (IVT) reaction, which provides linear amplification of a whole poly(A)$^+$ RNA repertoire up to 2000-fold per cycle from limited numbers of cells *(6,7)*. By incorporating an RNA promoter into cDNA templates, these transcription-based methods amplified nucleotides by RNA polymerization.

From: *Methods in Molecular Biology, vol. 221: Generation of cDNA Libraries: Methods and Protocols*
Edited by: S.-Y. Ying © Humana Press Inc., Totowa, NJ

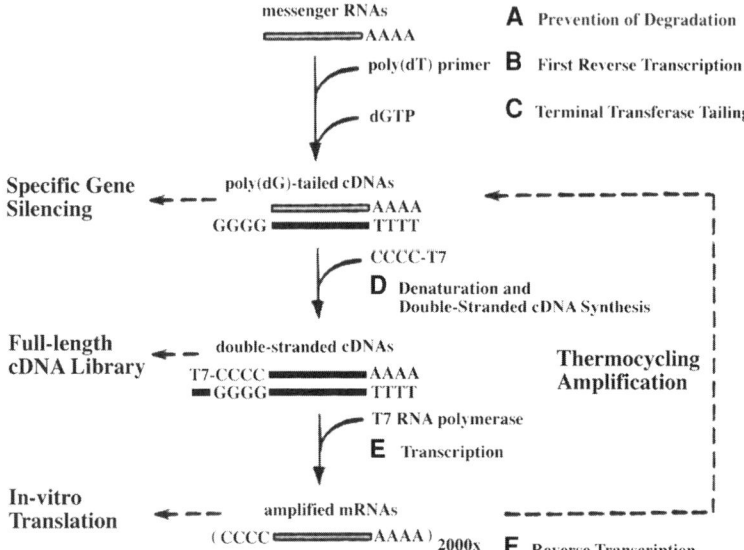

Fig. 1. An illustration of the RNA-PCR thermocycling procedure. The cycling steps **D–F** can be repeated at least one time for the linear amplification of a mRNA library by in vitro transcription. Advantageously, the reactions of steps **A–F** can be continuously performed in a reverse transcription and in vitro transcription (RT&T) buffer. The cycling of reverse and in vitro transcription reactions provides more flexibility for the enzymatic synthesis of single-stranded RNAs, RNA–DNA hybrids, and double-stranded DNAs, which are ready for a variety of biochemical applications such as mRNA library preparation for microarray analysis (*see* **Figs. 3** and **4**), probe preparation for specific gene detection *(7)*, full-length gene cloning, in vitro translation for protein synthesis, and gene knockout analysis through a posttranscriptional gene silencing mechanisms *(8)*.

The identification of some useful mRNA markers for certain disease detection has been reported *(2,3)*. Recently, a novel thermocycling procedure, RNA-polymerase cycling reaction (RNA-PCR), further achieved full-length mRNA amplification and successfully displayed cancer-stage-specific gene expression by Northern blot analysis *(3)*. To the best of our knowledge, this is the first procedure that has been tested to generate a full-length mRNA library from as few as 20 tissue cells (2-pg mRNAs) for profiling cancer stages in vivo. In brief, the RNA-PCR (polymerase chain reaction) procedure (*see* **Fig. 1**) is based on (A) prevention of degradation, (B) reverse transcription of mRNAs with poly-(dT) primers, (C) poly(dC)-tailing of the first-strand cDNAs, (D) denaturation and then double-stranding the DNA templates with oligo-(dG)-

T7-promoter primers, (E) in vitro transcription from the promoter region to generate multiple RNA sequences, and (F) repeating steps A–E without step C to achieve the desired poly(A)$^+$ RNA amount for analysis.

This method is capable of generating cell-type-specific poly(A)$^+$ RNA libraries up to 5 kb in the full-length conformation of most mRNAs, and up to 12 kb in a shorter 5′ truncated form of larger mRNA species. A high G-C content RNA, however, tends to be a little shorter than its original size. In general, the good integrity of total cellular mRNAs should appear as a smear between approx 500 bases and 5 kb on an electrophoresis gel and is composed of a median size of around 2 kb *(5)*. It is noteworthy that the full-length conformation at this range actually covers more than 90% of a whole mRNA population in cells. Based on our electrophoresis data, the quality of an RNA-PCR-derived mRNA library has reached the same quality as a smear between 300 and 7.4 kb without ribosomal RNA and genomic DNA contamination on a 1% formaldehyde–agarose gel (*see* **Fig. 2A**). Northern blot analysis of p16, a rare gene usually not shown in phenol–chloroform extracted RNAs, was clearly detected in a RNAPCR-derived library, whereas the signal of GAPDH (a highly abundant gene) was observed in all tested libraries, indicating a better preservation of rare mRNA species by RNA-PCR amplification. Moreover, RB (4.9 kb), β-actin (2.2 kb), and GAPDH (1.7 kb) gene transcripts were all measured in their corrected full-length sizes (**Fig. 2B**), further confirming a potential full-length conformation up to at least 5 kb. The utilization of thermostable reverse transcriptases in our current protocol has improved the full-length potential up to 9 kb and the 5′-end start codon of the resulting mRNA reading frames can be well preserved for further in vitro translation. In addition to the linear amplification of a IVT-based reaction, the RNA-PCR-derived RNAs could, therefore, proportionally represent most of mRNA populations in their original makeup.

To test high-yield and linear amplification, we have routinely generated 30 µg of amplified mRNAs in a 40-µL reaction mixture after three rounds of RNA-PCR amplification from about 20 single cells (approx 1 ng total RNAs). This represents a 1.5×10^7-fold increase based on a comparison between the amount of synthesized poly(A)$^+$ RNAs and that of theoretically presumed mRNAs within a cell (0.1 pg). Even after 10-fold dilution of current enzymatic activities, a more than 20-fold increase of specific mRNA sequences was measured in each cycle of transcriptional amplification (*see* **Fig. 2C**). Such high-yield amplification has been proven to be a linear amplification process, as shown in **Fig. 2D**. Because of the strict proof-reading feature of RNA polymerases, linear amplification is a natural property of transcriptional amplification methods *(7,9)*. Linear amplification maintains the accurate ratio of each expressed gene transcript in an amplified library for representing the

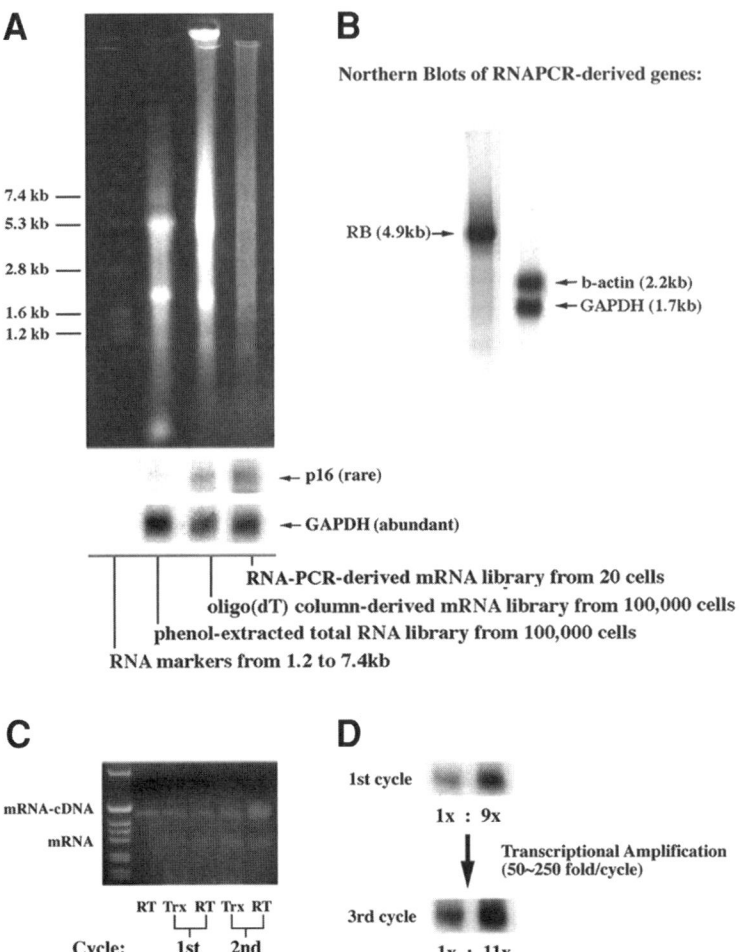

A

7.4 kb —
5.3 kb —

2.8 kb —

1.6 kb —
1.2 kb —

B

Northern Blots of RNAPCR-derived genes:

RB (4.9kb)→

←— b-actin (2.2kb)
←— GAPDH (1.7kb)

←— p16 (rare)

←— GAPDH (abundant)

RNA-PCR-derived mRNA library from 20 cells
oligo(dT) column-derived mRNA library from 100,000 cells
phenol-extracted total RNA library from 100,000 cells
RNA markers from 1.2 to 7.4kb

C

mRNA-cDNA

mRNA

Cycle: RT Trx RT Trx RT
           1st   2nd

D

1st cycle

1x : 9x

Transcriptional Amplification
(50~250 fold/cycle)

3rd cycle

1x : 11x

Fig. 2. Analyses of basic RNA-PCR features. (**A**) Comparison between RNA libraries prepared by phenol–chloroform extraction (lane 2), oligo-(dT) chromatographic column (lane 3), and RNA-PCR (lane 4) fractionated on a 1% formaldehyde–agarose gel, all ranging from 300 to above 7.4 kb based on RNA markers. A uniform smearing pattern of all three products indicates good RNA quality and quantity. p16, a rare and quickly degraded gene transcript, can be clearly identified in the RNA-PCR-derived library but not the others, whereas the abundant GAPDH and β-actin transcript was detected by Northern blots in all three libraries. (**B**) The amplification level of a specific gene transcript (activin) between two cycles of RNA-PCR has shown a significant 10-fold increase after utilization of a 1/10-fold enzymatic activity (20 U of T7 RNA polymerase). A greater than 250-fold amplification rate has been detected when 200 U of T7 RNA polymerase was applied to a RNA-PCR reaction (*3*). However, such a tremendous amplification rate cannot be observed by gel electrophoresis without dilution. (**C**) The ratio of amplified gene products in (**D**) was analyzed by Northern blotting at two predetermined concentrations (1:9) after two cycles of RNA-PCR amplification. The final ratio (1:11) was considered to closely match the original 1:9 ratio, indicating that the transcriptional amplification is a fairly linear amplification procedure.

original RNA composition. Indeed, the correct ratio composition of a whole mRNA population is critical to warrant the reproducibility and representation of its resulting gene analyses. However, previous methods for RNA preparations could not provide any evidence for linear amplification of rare RNAs *(6,7,9,10)*. To this end, the RNA-PCR-derived RNA library was experimentally tested to provide better lineage, coverage, and representation of RNA amplification for high-density microarray hybridizations, as shown in **Fig. 3**.

Reliable reproducibility and representation of a RNA-PCR-derived mRNA library have been confirmed using microarray analysis. When applied to affymetrix U95A2 gene chips ($n = 3$), both phenol–chloroform extracted and RNA-PCR-amplified RNA libraries displayed about 4200 expressed genes on a total about 12,670 gene chips ($33.5 \pm 0.3\%$ and $33.2 \pm 0.4\%$, respectively), representing a very similar size of RNA populations. It should be noted that the RNA-PCR-amplified library is amplified from 20 LNCaP cells, whereas the total RNA library is extracted from about 10^6 cells of the same. Among all expressed genes, 17 of them were completely missing in 1 of the libraries, indicating a 0.4% of representation loss that may have occurred during different handling. Less than 2% of the average population (approx 102 genes) was detected to be differentially displayed more than threefold changes, showing a very good mutual representation capacity. From the computing results of scatterplots (**Fig. 3**, left), a highly linear correlation of gene coverage was found in the abundant and moderate mRNA species of both. A more intense signal for rare RNA population was detected in the RNA-PCR-derived mRNAs, indicating a better preservation of most rare species. Although they were not perfectly matched with each other, the above results have demonstrated much more promising compatibility than those from the comparison of the extracted total RNAs to an aRNA library amplified by Eberwine's conventional aRNA amplification method from 20 LNCaP cells (**Fig. 3**, right). It is known that the aRNA amplification has been widely applied to prepare labeled probes for microarray hybridization *(7,11)*. However, because of the utilization of random primers for cycling amplification, the compared aRNA library ($n = 3$) displayed a more skewed and less abundant population containing an average of 2243 expressed genes (approx 17.8%).

Using 8000 gene DNA microarrays provided by the National Cancer Institute (NCI), we have also performed a similar experimental comparison among a standard reference RNA library from NCI, amplified aRNAs generated from 10 ng of the standard, and RNA-PCR-derived RNAs amplified from 10 ng of the standard ($n = 2$). It showed an average 83% linear correlation between the standard and RNA-PCR-derived RNA libraries (*see* **Fig. 4**, middle), whereas only 49% correlation was detected between the standard RNA and aRNA library (**Fig. 4**, right). Based on 0.1 pg mRNA per cell and each cell containing

A

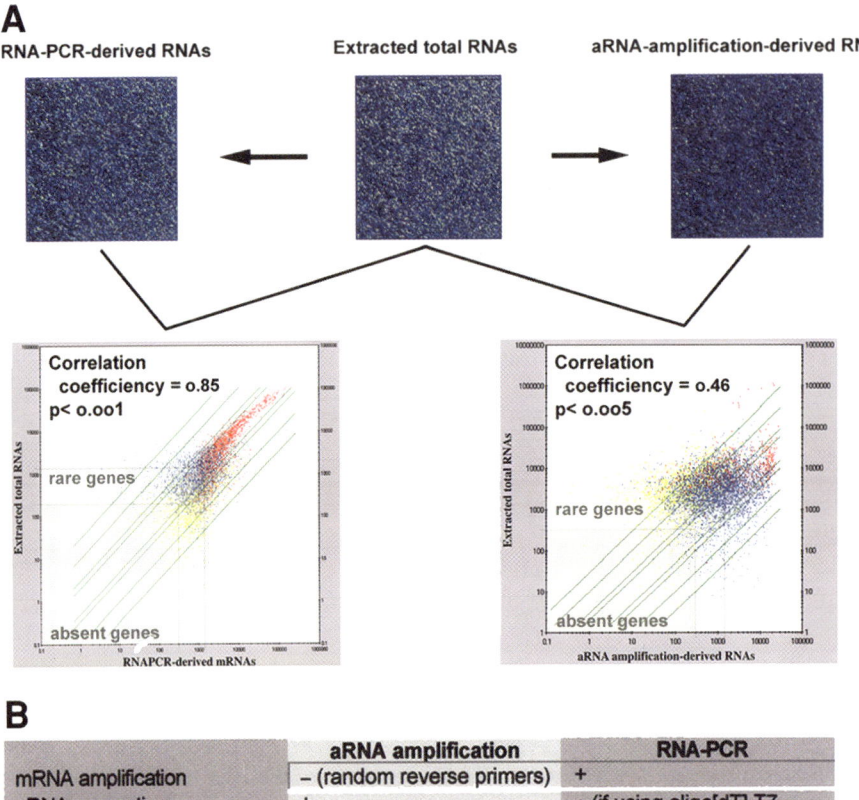

B

	aRNA amplification	RNA-PCR
mRNA amplification	– (random reverse primers)	+
aRNA generation	+	+ (if using oligo[dT]-T7 primers for resulting mRNA amplification)
Full-length cDNA amplification	– (random reverse primers)	+
Single-cell amplification	+ (only for specific genes)	+ (for a whole library)
Linear amplification	+ (theoretical proposal)	+ (experimental proof)

Fig. 3. The comparison of differential gene expression patterns by microarray analyses. (**A**) Among RNA-PCR-derived poly(A)+ RNAs from 20 cells, aRNAs amplified by Eberwine's method (*6,7*), and traditional phenol–chloroform extracted total RNAs from 10^6 cells, the results showed a highly compatible population in the abundant and moderate mRNA species of the extracted and the RNA-PCR-derived RNAs (left) but not the aRNAs (right). The rare mRNAs (white zone between two squares) were, however, markedly different among all three groups. The gray shadow area indicated a marginal reading zone of signal detection, showing genes absent in the compared libraries. The 45° green lines parallel to the *x/y* slope marked the differential changes from one (near the center) to eight (to both axes) folds. (**B**) A list of distinct features between RNA-PCR and aRNA amplification methods, which may contribute to the different results of microarray analysis.

2% mRNAs, the 10-ng standard of referenced total RNAs is equivalent to the mRNA amount of 200 cells. It is a reasonable cell number for sample collection by a laser capture machine (LCM). For single-cell microarray analysis, this result suggests that RNA-PCR preserves much better mRNA information using fewer single cells than previous methods. However, the current method is not suitable for analysis of bulk RNA sources because of its limitation in terminal deoxynucleotidyl transferase (TdT) tailing procedure. For RNA samples greater than 5 μg, Eberwine's aRNA amplification method is still the best choice *(11)*.

2. Materials

2.1. Generation of First-Strand cDNA Using Reverse Transcription with Poly(dT) Primers

1. Diethyl pyrocarbonate (DEPC) H_2O: Stir double-distilled water with 0.1% DEPC for more than 12 h and then autoclave at 120°C under about 1.2 kgf/cm^2 for 20 min, twice.
2. Poly(dT)$_{24}$ primer: dephosphorylated 5'-dTTTTTTTTTTT TTTTTTTTTT TTTT-3' (100 pmol/μL) (*see* **Note 1**).
3. AMV reverse transcriptase (50 U/μL) and 10X reverse transcription buffer (500 mM Tris-HCl [pH 8.5] at 25°C, 80 mM MgCl$_2$, 300 mM KCl, and 10 mM dithiothreitol [DTT]).
4. First reverse transcriptase mix: 7 μL DEPC-treated ddH$_2$O, 2 μL 10X reverse transcription buffer, 2 μL of 10 mM dNTP mix (10 mM each of dATP, dGTP, dCTP, and dTTP), 1 μL RNasin (25 U/μL), and 2 μL AMV reverse transcriptase; prepare just before use.
5. Incubation chamber: 65°C, 42°C, and 50°C.
6. Purification spin column: 100 bp cutoff filter (e.g., a Microcon-50 centrifugal filter [Amicon, Beverly, MA], as shown here).

2.2. cDNA Amplification Using Terminal Transferase Tailing with Poly(dC) Oligonucleotides

1. 10X terminal transferase tailing buffer: 500 mM Tris-HCl (pH 8.0) at 25°C, 400 mM KCl, 80 mM MgCl$_2$, and 100 mM DTT; prepare fresh.
2. Terminal deoxynucleotidyl transferase (25 U/μL).
3. Terminal transferase reaction mix: 3 μL DEPC-treated ddH$_2$O, 1 μL of 10 mM dNTP, 3 μL of 10X terminal transferase tailing buffer, 1 μL of 10 mM dCTP, and 2 μL terminal deoxynucleotidyl transferase; prepare just before use.
4. Incubation mixer: 37°C, 100g vortex for 30 s between every 5-min interval.
5. Incubation chamber: 37°C and 94°C.

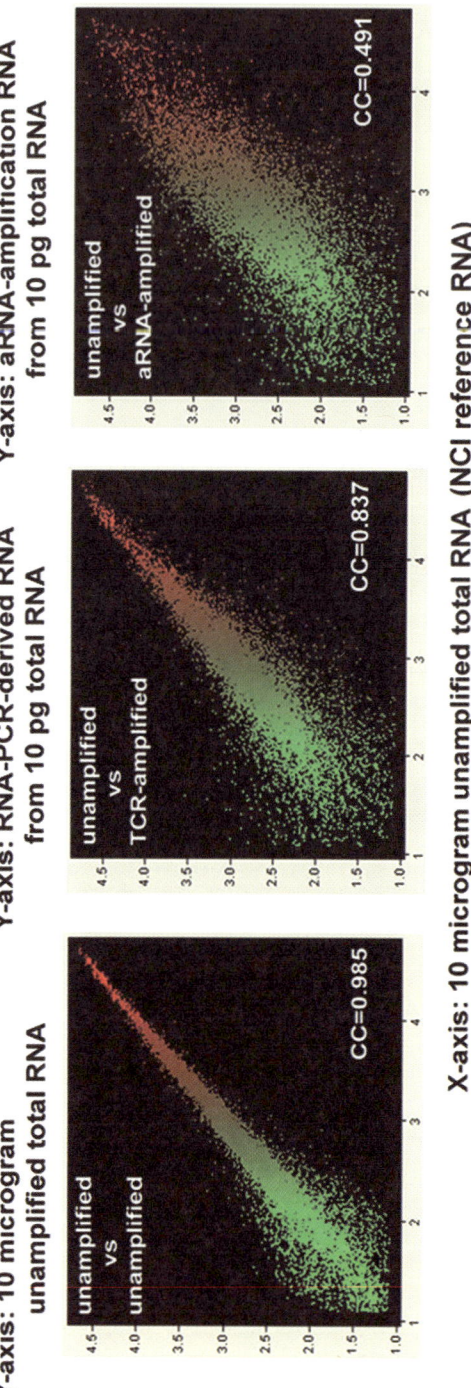

Y-axis: 10 microgram unamplified total RNA

unamplified
vs
unamplified

CC=0.985

Y-axis: RNA-PCR-derived RNA from 10 pg total RNA

unamplified
vs
TCR-amplified

CC=0.837

Y-axis: aRNA-amplification RNA from 10 pg total RNA

unamplified
vs
aRNA-amplified

CC=0.491

X-axis: 10 microgram unamplified total RNA (NCI reference RNA)

Fig. 4. Microarray analysis using human DNA gene-chips (n = 2 for each group), two-cycle amplification products of RNA-PCR-derived RNA from 10 ng total RNA referenced by National Cancer Institute (NCI) display an average 83.7% correlation coefficiency (CC) compared to 10 µg of the original reference RNA. Because our preset threshold for acceptable variation is onefold change, such high CC rate indicates that >83% of the original mRNA population has been well preserved in almost the same composition and ratio. Traditional aRNA amplification products from 10 ng reference RNA, however, display a lower 49% CC rate, which may result from the use of random primers and, therefore, loss of full-length RNA composition during cycling amplification.

2.3. cDNA Double-Stranding Using DNA Polymerization with Oligo(dG)-Promoter Primers

1. Oligo-(dG)$_{10}$N-T7 RNA promoter primer mix: dephosphorylated 5'-GGCAGT-GAAT TGTAATACGA CTCACTCACT ATAGGGAAGG CGGGGGGGGN-3' (N = A, T, or C; total 100 pmol/μL including 35 pmol/μL 5'-GGCAGTGAAT TGTA ATACGA CTCACTCACT ATAGGGAAGG CGGGGGGGA-3', 35 pmol/μL 5'-GGCAGTGAAT TGTAATACGA CTCACTCACT ATAGGGAAGG CGGG GGGGT-3', and 30 pmol/μL 5'-GGCAGTGAAT TGTAATACGA CTCACT CACT ATAGGGAAGG CGGGGGGGC-3').
2. 10X cDNA double-stranding buffer: 500 mM Tris-HCl (pH 9.2) at 25°C, 160 mM (NH$_4$)$_2$SO$_4$, 20 mM MgCl$_2$; prepare fresh.
3. cDNA double-stranding reaction mix: 10 μL DEPC-treated ddH$_2$O, 5 μL of 10X cDNA double-stranding buffer, 2 μL of 10 mM dNTP mix (10 mM each of dATP, dGTP, dCTP, and dTTP), 0.7 μL *Taq* DNA polymerase (5 U/μL), and 0.3 μL Pwo DNA polymerase (5 U/μL); prepare just before use.
4. Incubation chamber: 94°C, 50°C, and 68°C.
5. Purification spin column: 100 bp cutoff filter.

2.4. Generation of Full-Length Sense RNA Using In Vitro Transcription with Promoter-Driven RNA Polymerase

1. 10X In vitro transcription (IVT) buffer: 400 mM Tris-HCl (pH 8.0) at 25°C, 100 mM MgCl$_2$, 50 mM DTT, and 5 mg/mL nuclease-free bovine serum albumin (BSA).
2. T7 RNA polymerase (80 U/μL).
3. IVT reaction mix: 8 μL DEPC-treated ddH$_2$O, 4 μL 10X IVT buffer, 4 μL of 10 mM dNTP mix (10 mM each of ATP, GTP, CTP, and UTP), 2 μL RNasin (25 U/μL), and 2 μL T7 RNA polymerase; prepare just before use.
4. Incubation mixer: 37°C, 100g vortex for 30 s between every 30-min interval.

2.5. Another Round of Thermocycling Amplification Using RNA-PCR with Sense RNA

1. Poly(dT)$_{24}$ primer: dephosphorylated 5'-dTTTTTTTTTT TTTTTTTTTT TT TT-3' (100 pmol/μL).
2. 10X RT&T buffer: 600 mM Tris-HCl (pH 8.3) at 25°C, 300 mM KCl, 80 mM MgCl$_2$, 100 mM DTT, and 5M betaine.
3. RNA-PCR reaction mix: 2 μL DEPC-treated ddH$_2$O, 2 μL of 10X RT&T buffer, 2 μL of 10 mM dNTP mix (10 mM each for dATP, dGTP, dCTP, and dTTP), 1 μL RNasin (25 U/μL), and 2 μL AMV reverse transcriptase (50 U/μL); prepare just before use.
4. DEPC-treated ddH$_2$O.
5. Oligo-(dG)$_{10}$N-T7 RNA promoter primer mix: dephosphorylated 5'-GGCAGT GAAT TGTAATACGA CTCACTCACT ATAGGGAAGG CGGGGGGGGN-3' (N = A, T or C; total 100 pmol/μL).
6. *Taq* DNA polymerase (5 U/μL) and Pwo DNA polymerase (5 U/μL).

7. Incubation chamber: 65°C, 42°C, 50°C, 94°C, and 68°C.
8. Purification spin column: 100 bp cutoff filter.

3. Methods

3.1. Generation of First-Strand cDNA Using Reverse Transcription with Poly(dT) Primers

The starting material can be either 0.1 ng to 1 μg total RNA *(3)* or permeabilized cell preparation *(3,12)* (*see* **Notes 2** and **3**). Poly(A⁺) RNA is selected using poly(dT) primers, which contains about 20–26 deoxythymidylate oligonucleotides. The first-strand cDNA is synthesized by reverse transcription from the poly(A⁺) RNA with the poly(dT) primers. As shown in **Fig. 1**, the promoter used here is a T7 bacteriophage RNA promoter element.

1. Primer annealing: Suspend RNA in 5 μL of DEPC-treated water, mix well with 1 μL poly(dT)$_{24}$ primer, heat to 65°C for 5 min for minimizing secondary structure, cool to 50°C for 1 min for primer hybridization, and then cool on ice.
2. First-strand cDNA synthesis: Add 14 μL of first reverse transcriptase mix and heat to 42°C for 50 min. Add another 1 μL of reverse transcriptase and mix. Continue to incubate the reaction at 42°C for 30 min, heat to 50°C for 10 min, and then cool on ice. The RNA is still attached noncovalently to the cDNA.
3. Denaturation: Heat the reaction at 94°C for 3 min and then cool on ice immediately.
4. Primer removal and buffer exchange: Load the reaction into a purification spin column, spin for 10 min at 14,000g, and discard the flowthrough (*see* **Note 4**). Add 200 μL of DEPC-treated ddH$_2$O into the spin column to wash the cDNA, spin for 10 min at 14,000g, and discard the flowthrough. Add 20 μL of DEPC-treated ddH$_2$O into the spin column to dissolve the cDNA, place the spin column upside down in a new collecting microtube, and spin 3 min at 3000g. Store the 20 μL of the purified cDNA in a –20°C freezer or perform the next step immediately.

3.2. cDNA Amplification Using Terminal Transferase Tailing with Poly(dC) Oligonucleotides

In this method, which was reported by Lin and Ying in 2000, the first-strand cDNA is dC-tailed using TdT and a promoter-linked oligo(dG) primer is applied to initiate the second-strand cDNA synthesis. As shown in **Fig. 1**, the promoter used here is a T7 bacteriophage RNA promoter element. Although this external priming procedure preserves better full-length conformation of the mRNA, the efficiency of the TdT tailing reaction seems to depend on the particular 3′ termini of different first-strand cDNA species, resulting in uneven coverage. Such a problem, however, can be improved by adequate TdT activity rate and constant reaction vortex (*see* **Note 5**). Practically, 1 U of TdT is required for tailing every picomole of cDNA in a mild shaking incubator (100g), if the average size of cDNA is 3 kb.

1. TdT tailing reaction: Add 10 μL of terminal transferase reaction mix to the purified cDNA and mix well. Incubate the reaction at 37°C for 30 min and occasionally mix the reaction every 5 min for better tailing coverage.
2. Reaction stop: Heat the reaction at 94°C for 3 min and cool on ice immediately.

3.3. cDNA Double-Stranding Using DNA Polymerization with Oligo-(dG)-Promoter Primers

The mRNA of the resulting mRNA–cDNA hybrid is denatured, and double-stranded cDNA was formed using a modified two-cycle PCR-like reaction with promoter-linked oligo-(dG) primers. Then, the amplification of cDNA representative can be achieved by in vitro transcription using the promoter-linked double-stranded cDNA as the template. The full-length construct of the cDNA template is protected and flanked by poly(dC)- and poly(dA)-oligonucleotide in its 5′ and 3′ termini, respectively.

1. Primer annealing: Add 2 μL of oligo-(dG)$_{10}$N-T7 RNA promoter primer mix to the reaction, heat to 94°C for 3 min for mRNA removal, cool to 50°C for 10 min for primer hybridization, and then cool on ice.
2. cDNA double-stranding: Add 20 μL of cDNA double-stranding reaction mix to the reaction, mix well, and then incubate the reaction at 68°C for 10 min. Repeat the thermocycling incubation from 94°C for 3 min, 50°C for 10 min, and then 68°C for 10 min, one more time.
3. Breaking cell membrane: For using permeabilized cells as starting material, add 200 μL of 2% nonionic detergent (octylphenoxy)polyethanol to the reaction and vortex at 100g for 10 min.
4. Primer removal and buffer exchange: Load the reaction into a purification spin column, spin for 10 min at 14,000g, and discard the flowthrough. Add 200 μL of DEPC-treated ddH$_2$O into the spin column to wash the double-stranded cDNA, spin for 10 min at 14,000g, and discard the flowthrough. Add 20 μL of DEPC-treated ddH$_2$O into the spin column to dissolve the double-stranded cDNA, place the spin column upside down in a new collecting microtube, and spin for 3 min at 3000g. Store the 20 μL of the purified cDNA in a –20°C freezer or perform the next step immediately.

3.4. Generation of Full-Length Sense RNA Using In Vitro Transcription with Promoter-Driven RNA Polymerase

The promoter of the double-stranded cDNA is now served as a recognition site for RNA polymerase during an in vitro transcription reaction. The in vitro transcription provides linear amplification up to 2000-fold of the amount of starting materials *(6,7)*. The proofreading capability of the RNA polymerase ensures the fidelity of the resulting nucleic acid products. Because the promoter is incorporated in the same orientation of mRNA, the resulting product is sense RNA (mRNA) rather than antisense RNA. A cap structure can be added to the sense RNA for further peptide synthesis (*see* **Note 6**).

1. In vitro transcription reaction: Add 20 μL of IVT reaction mix to the purified cDNA and mix well. Incubate the reaction at 37°C for 2–3 h and occasionally mix the reaction every 30 min for better RNA elongation (*see* **Note 7**).
2. Buffer exchange and sample concentration: Load the reaction into a purification spin column, spin for 10 min at 14,000g, and discard the flowthrough. Add 200 μL of DEPC-treated ddH$_2$O into the spin column to wash the poly(A$^+$) RNA, spin for 10 min at 14,000g, and discard the flowthrough. Add 20 μL of DEPC-treated ddH$_2$O into the spin column to dissolve the poly(A$^+$) RNA, place the spin column upside down in a new collecting microtube, and spin for 3 min at 3000g. Store the 20 μL of the purified poly(A$^+$) RNA in a –80°C freezer or perform the next step immediately.

3.5. Another Round of Thermocycling Amplification Using RNA-PCR with Sense RNA

The sense RNA so generated is flanked by poly(dC)- and poly(A)-oligonucleotide in its 5′ and 3′ termini, respectively. These homopolymeric tails not only maintain the full-length conformation of the mRNA but also serve as templates for the poly(dT) and promoter-linked oligo-(dG) primers. The cycling of the above transcriptional amplification can be reiterated using the sense RNA directly in the next round of RNA-PCR reaction following the cycling steps of **Subheading 3.4.** and **3.5.** (*see* **Note 8**).

1. Primer annealing: Add 1 μL poly(dT)$_{24}$ primer to 10 μL of the purified poly(A$^+$) RNA, heat to 65°C for 5 min for minimizing secondary structure, cool to 50°C for 1 min for primer hybridization, and then cool on ice.
2. First-strand cDNA synthesis: Add 9 μL of RNA-PCR reaction mix to the reaction, and heat to 42°C for 50 min. Add another 1 μL of reverse transcriptase and mix. Continue to incubate the reaction at 42°C for 30 min, heat to 50°C for 10 min, and then cool on ice. The RNA is still attached noncovalently to the cDNA.
3. Denaturation: Add 16 μL of DEPC-treated ddH$_2$O and 2 μL of oligo-(dG)$_{10}$N-T7 RNA promoter primer mix to the reaction and incubate the reaction at 94°C for 3 min and then 50°C for 10 min.
4. cDNA double-stranding: Add 0.7 μL *Taq* DNA polymerase and 0.3 μL Pwo DNA polymerase to the reaction and incubate at 68°C for 10 min.
5. Primer removal and buffer exchange: Load the reaction into a purification spin column, spin for 10 min at 14,000g, and discard the flowthrough. Add 200 μL of

DEPC-treated ddH$_2$O into the spin column to wash the double-stranded cDNA, spin for 10 min at 14,000g, and discard the flowthrough. Add 20 μL of DEPC-treated ddH$_2$O into the spin column to dissolve the double-stranded cDNA, place the spin column upside down in a new collecting microtube, and spin for 3 min at 3000g. Store the 20 μL of the purified cDNA in a –20°C freezer or perform the next step immediately.

6. cDNA library assessment (*see* Chapter 13).

4. Notes

1. When performed with a specific primer complementary to the 3'-targeted sequence of a desired mRNA in conjunction with the oligo-(dG)-T7 primer, the 5' end of a specific mRNA can be generated by RNA-PCR for 5'-UTR (untranslated region) analysis. The design of these sequence-specific primers is based on the same principle used by PCR (50–55% G-C rich). On the other hand, in addition to the 3'-end sequence-specific primer, we can also use another sequence-specific promoter–primer to amplify certain domain within the code-reading frame of the mRNA for further research. The design of these promoter–primers, however, requires a higher G-C content (60–65%) working at the same annealing temperature as the sequence-specific primers because of their unmatched promoter regions. For example, the new annealing temperature for the sequence-matched region of a promoter-primer is [2(dA + dT) + 3(dC + dG)] (5/6), not including the promoter region. Please remember that all primers were purified by polyacrylamide gel electrophoresis before used in an RNA-PCR reaction.

2. Isolated cells were preserved in 500 μL of ice-cold 10% formaldehyde in suspension buffer (0.15 M NaCl [pH 7.0], 1 mM EDTA) for the following fixation and permeabilization procedure *(12)*. After 1-h incubation with occasional agitation, fixed cells were collected with a Microcon-50 filter and washed by 350 μL of ice-cold PBS with vigorous pipetting. The collection and wash were repeated at least once. The fixed cells were then permeabilized in 500 μL of 0.5% nonionic detergent (octylphenoxy)polyethanol for 1 h with frequent agitation. After that, three collections and washes were given to cells, as earlier, but using 350 μL of ice-cold phosphate-buffered saline (PBS) containing 0.1 M glycine instead. The cells were finally mixed with 0.1 μM poly(dT)$_{24}$ primer and resuspended in the same buffer with vigorous pipetting to evenly distribute them into small aliquots (about 50 cells in 10 μL) for RNA-PCR. They could be stored at –80°C for up to 2 wk.

3. The adequate amount of fixed cells for RNA-PCR ranged from more than 20 but less than 200 cells (if each cell contains 0.1 pg mRNAs) because of the completeness of TdT tailing reactions. The chance to generate a good mRNA library (300 bp~ to 5 kb) from <20 cells is less than 50% based on our tests. The chance to generate a complete mRNA library from more than 200 cells will depend on the relative amount of terminal transferase (TdT) activity to the first-strand cDNA molecules. The TdT activity is less effective when too many cDNA molecules interact with TdT in a limited tailing reaction. In brief, we

currently know that the concentration of TdT determines the completeness of an RNA-PCR-derived library, whereas that of RNA polymerases and reverse transcriptases determines the amplification rate of a RNA-PCR reaction. Therefore, we suggest that please use at least 50 U of TdT for every 0.1 ng mRNAs in a tailing reaction and more than 60 U each of RNA polymerases and reverse transcriptases in a 20-µL transcription reaction.

4. Relative Centrifugal Force (RCF) (g) = $(1.12 \times 10^{-5}) \cdot (rpm)^2 \cdot r$, where r is the radius in centimeters measured from the center of the rotor to the middle of the spin column and rpm is the speed of the rotor in revolutions per minute.

5. The first-strand cDNA is poly(dC)-tailed by TdT using the provided condition that should produces an average 8–15-base overhang. The incorporation rate is increased about 75–85% with occasionally gentle mix, but drops to about 50% without mixing. The efficiency of TdT tailing seems to be varied among different mRNA species but can be improved by occasionally gently mixing in a short period of incubation time. The length of homopolymeric tails should be limited by the special designs of returning primers or promoter–primers, as mentioned in **ref.** *(3)* (e.g., an equal mixture of T7-oligo-(dG)$_{10-12}$N primers; N= dA, dT, or dC). The homopolymeric region of a returning primer should range from 7 to 16 bases, most preferably from 10 to 12 bases. The use of cobalt-based buffers is not recommended in this protocol.

6. The RNA-PCR has been tested to provide amplified full-length mRNAs for in vitro translation (*see* Chapter 25). A cap nucleotide can be added to the 5′ end of the amplified mRNAs during the transcriptional amplification. Unlike normalized RNAs, the capped mRNAs can be directly used in protein synthesis and may help to isolate such protein activity if its folding is correct. The preferred cap nucleotides include P1-5′-(7-methyl)-guanosine-P3-5′-adenosine-triphosphate and P1-5′-(7-methyl)-guanosine-P3-5′-guanosine-triphosphate. Such protein products are useful for protein differential display on a two-dimensional gel.

7. The most stable and efficient IVT reaction occurs during the first 2-h of incubation at 37°C. The rate of RNA synthesis decreases considerably (40–50%) after a 3-h incubation or below 37°C incubation. A longer reaction may increase yield, but the possibility of degradation by RNase increases. Occasionally, gentle mixing can prevent the stall of crowded RNA polymerases on a template and enhance full-length synthesis. The overall rate of RNA polymerization is maximal between pH 7.7 and 8.3, but it remains about 70% of maximum at pH 7.0 or 9.0. High concentrations of NaCl, KCl, or NH$_4$Cl above 75 mM will inhibit the reaction.

8. To reach about 1.5×10^7-fold amplification of mRNAs, 3–4 cycles of RNA-PCR were needed to perform for about 20–50 cells and 2–3 cycles for about 125–200 cells. The optical density (OD) (A$_{260}$/A$_{280}$) value ranged from about 1.7 to 2.0 for mRNA products and mRNA–cDNA hybrids and from about 1.6 to 1.9 for double-stranded cDNA products, depending at which cycling step you stop the RNA-PCR reaction. Remove enzymes with protein-remover filters (Microcon) before OD detection. A lower OD value may indicate an insufficient

amplification rate, enzyme-related variation, or RNase contamination. We preferred to run a 1% formaldehyde–agarose gel to exam the quality of RNA-PCR products. Identification of some rare mRNAs (<six copies/cell) by RT-PCR from the RNAPCR-derived library is another way to observe its quality.

References

1. Thykjaer, T., Workman, C., Kruhoffer, M., Demtroder, K., Wolf, H., Andersen, L. D., et al. (2001) Identification of gene expression patterns in superficial and invasive human bladder cancer. *Cancer Res.* **61,** 2492–2499.
2. Crino, P. B., Trojanowski, J. Q., Dichter, M. A., and Eberwine, J. (2001) Embryonic neuronal markers in tuberous sclerosis: single-cell molecular pathology. *Proc. Natl. Acad. Sci. USA* **93,** 14,152–14,157.
3. Lin, S. L., Chuong, C. M., Widelitz, R. B., and Ying, S. Y. (1999) In vivo analysis of cancerous gene expression by RNA-polymerase chain reaction. *Nucleic Acid Res.* **27,** 4585–4589.
4. Goh, S. H., Park, J. H., Lee, Y. J., Lee, H. G., Yoo, H. S., Lee, I. C., et al. (2000) Gene expression profile and identification of differentially expressed transcripts during human intrathymic T-cell development by cDNA sequencing analysis. *Genomics* **70,** 1–18.
5. Sambrook, J., Fritsch, E. F., and Maniatis, T. (1989) Construction and analysis of cDNA libraries, in *Molecular Cloning*, 2nd ed. Cold Spring Harbor Laboratory, Cold Spring Harbor, NY, pp. 8.11–8.19.
6. Ying, S. Y., Liu, H. M., Lin, S. L., and Chuong, C. M. (1990) Amplified RNA synthesized from limited quantities of heterogeneous cDNA. *Proc. Natl. Acad. Sci. USA* **87,** 1663–1667.
7. Eberwine, J., Yeh, H., Miyashrio, K., Cao, Y., Nair, S., Finnell, R., et al. (1992) Analysis of gene expression in single live neurons. *Proc. Natl. Acad. Sci. USA* **89,** 3010–3014.
8. Lin, S. L., Chuong, C. M., and Ying, S. Y. (2001) A novel mRNA–cDNA interference phenomenon for silencing bcl-2 expression in human LNCaP cells. *Biochem. Biophys. Res. Commun.* **281,** 639–644.
9. Compton, J. (1991) Nucleic acid sequence-based amplification. *Nature* **350,** 91, 92.
10. O'Dell, D. M., Raghupathi, R., Crino, P. B., Morrison, B., 3rd, Eberwine, J. H., and McIntosh, T. K. (1998) Amplification of mRNAs from single, fixed, TUNEL-positive cells. *BioTechniques* **25,** 566–570.
11. Wang, E., Miller, L. D., Ohnmacht, G. A., Liu, E. T., and Marincola, F. M. (2000) High-fidelity mRNA amplification for gene profiling. *Nat. Biotech.* **18,** 457–459.
12. Embleton, M. J., Gorochov, G., Jones, P. T., and Winter, G. (1992). In-cell PCR from mRNA: amplifying and linking the rearranged immunoglobulin heavy and light chain V-genes within single cells. *Nucleic Acids Res.* **20,** 3831–3837.

13

Quality Assessment of cDNA Libraries

Hans-Jürgen Fülle

1. Introduction

A perfect complementary DNA (cDNA) library would provide an accurate and complete representation of all messenger RNA (mRNA) sequences expressed in a particular source, whether a cell, tissue, or organism. Of primary importance during cDNA library construction is the use of a high-quality preparation of RNA, poly (A)⁺ RNA, or other specific subsets of mRNA as starting material for generating a double-stranded DNA copy. Monitoring quality during cDNA synthesis to ensure generation of full-length cDNA is another crucial step in the process. Because only a high-quality cDNA library will yield high-quality data, including valid sequence information, rigorous quality assessment is also needed after library construction (i.e., prior to and during the use of a library for downstream applications). Applying a set of common quality criteria to newly generated cDNA libraries will become increasingly important *(1)*.

The quality of a cDNA library can be assessed by a number of criteria. Strict quality control testing prior to its use should include (1) the determination of complexity and titer, (2) the determination of frequency of recombinants, and (3) an analysis of the average insert size and range. In addition, (4) a number of other tests can be performed to ensure, for example, normal representation of sequences of housekeeping genes, absence of any contaminating sequences, and intactness of cDNA on their 5′ ends. Last but not least, DNA sequencing of selected recombinants can be a useful approach for determining the overall fidelity of the cloning process, depending on the techniques used for cDNA synthesis.

Taken together, there is not one single approach to assess whether a library is indeed perfect or less than perfect. Ultimately, of course, the best criterion

From: *Methods in Molecular Biology, vol. 221: Generation of cDNA Libraries: Methods and Protocols*
Edited by: S.-Y. Ying © Humana Press Inc., Totowa, NJ

of the quality of a cDNA library remains the successful identification and isolation of those sequences originally sought.

2. Materials

1. cDNA library, unamplified or amplified (*see* **Note 1**).
2. Plating bacteria from a suitable *Escherichia coli* host strain: Use only strains of *E. coli* that are suitable for supporting the growth of the selected bacteriophage λ vector and recommended by the manufacturer or distributor of the library. Dilute an overnight bacterial culture to a final concentration (optical density [OD_{600}]) of 1.0 in 10 mM MgSO$_4$ and store at 4°C for up to 1 wk (*see* **Note 2**).
3. Bacterial culture media, agar plates, and 0.7% top agarose: To optimize adsorption of λ phage to host bacteria include 10 mM MgSO$_4$ in all Luria-Bertani (LB) media. When growing overnight bacterial cultures for λ phage transduction, also include 0.2% maltose. Plates used for phage titering should be 1–2 d old and free of surface moisture. Prewarm plates at 37°C for 30 min before use.
4. SM buffer plus gelatin, X-Gal, and IPTG: The suspension medium (SM) buffer plus gelatin is 100 mM NaCl, 8 mM MgSO$_4$·7H$_2$O, 50 mM Tris-HCl (pH 7.5), and 0.01% (w/v) gelatin; sterilize by autoclaving and store in aliquots at room temperature. For 100 mM IPTG (isopropyl-thio-β-galactoside), add 0.238 g IPTG to 10 mL of H$_2$O, filter sterilize, and store at –20°C. For 100 mM X-gal (5-bromo-4-chloro-3-indoyl-β-D-galactopyranoside), add 0.409 g X-gal to 10 mL of dimethyl formamide (DMF), store at –20°C. *Caution:* DMF vapor is irritating to skin and mucous membranes; handle under a fume hood.
5. Reagents for polymerase chain reaction (PCR), including specific oligonucleotide primers.
6. Reagents for agarose gel electrophoresis.

3. Methods

3.1. Complexity of a cDNA Library and Determination of Titer

Obtaining information about the complexity (i.e., the total number of independent recombinants) and the titer of a cDNA library is critical to confirm with an adequate probability that a specific recombinant is represented and can be isolated. Complexity—together with titer and frequency of recombinants—is one of the main determinants for the minimal number of recombinants that must be screened (*see* **Note 3**).

The library titer is usually expressed as plaque-forming units per milliliter (pfu/mL) for cDNA libraries constructed in bacteriophage λ and its derivatives and is determined by serial dilution. A library should be titered or retitered before each plating and screening, as the titer can drop considerably during long-term storage *(2,3)*.

In a primary, unamplified cDNA library, determining the titer will give an estimate of the number of independent recombinants (*see* **Note 4**). The complexity of a primary library with good quality should be 1×10^6 or larger. This usually corresponds to a primary titer of 1×10^6 pfu/mL or higher. A library of this complexity should be representative of the complexity of the starting messenger RNA (mRNA) population and readily yield "low abundance" recombinants (*see* **Note 5**). However, screening of 1×10^7 independent recombinants or more may be required if the desired clone is present at very low levels. This could be the case, for example, if the library is derived from a mixed population of mRNAs from several cell or tissue types (*see* **Note 6**).

Although it is widely recommended that unamplified libraries be used for screening, most commercial cDNA libraries have been amplified at least once to generate an almost limitless source of cDNA clones to screen. Titers for amplified λ-based libraries are typically between 1×10^9 and 1×10^{11} pfu/mL, sometimes even higher. However, because an amplified library is comprised of sibling pfu sets rather than independent recombinants, its titer alone is usually not a good indicator of its complexity. Amplification involves resampling of the primary library, invariably reduces complexity, and tends to distort the overall representation of mRNAs. A biased representation of the starting mRNA population is often the result *(2,3)*.

1. Prepare plating bacteria from a suitable *E. coli* bacterial host strain (*see* **Note 2**). Prepare LB agar plates and 0.7% LB top agarose.
2. Make a series of 10-fold dilutions (e.g., 10^{-1}–10^{-6}) of the cDNA library, using SM buffer plus gelatin as a diluent or an equivalent buffer recommended by the manufacturer or distributor of the library.
3. Titer the library by infecting 0.2-mL aliquots of plating bacteria in sterile 17×100-mm polypropylene tubes with 10-µL aliquots of phage dilutions. Also include, as a control, SM buffer that does not contain bacteriophage λ.
4. Mix by gently shaking and incubate tubes for 15 min at 37°C to allow adsorption of phage to cells and start of infection.
5. Add 3 mL of melted 0.7% LB top agarose at approx 48°C to each aliquot of infected bacteria.
6. Quickly mix by gently shaking and immediately pour entire mixture evenly onto the surface of a prewarmed 100-mm LB agar plate.
7. Allow top agarose to harden at room temperature, then invert and incubate plates at 37°C for 6–12 h or until distinct plaques developed just large enough to be counted.
8. Count the number of plaques on a readily countable plate (30–300 plaques); that is, count the small circular areas of clearing that indicate phage-infected and lysed bacteria on the confluent bacterial lawn. Compute the titer considering the corresponding dilution factor using the following formula:

$$\text{Titer (in pfu/mL)} = \frac{100 \times (\text{no. of plaques})}{\text{Library dilution}}$$

3.2. Determination of Frequency of Recombinants

The proportion of recombinants vs nonrecombinants in a cDNA library should be larger than 90% to keep to a minimum the dilution of rare recombinants and thus the number of plaques to be screened. Two methods, blue/white and PCR screening, are available to determine the frequency of recombinants.

3.2.1. Blue/White Screening

The percentage of recombinants can often be conveniently verified during titer determination if blue/white color screening for recombinants is available through α-complementation with the chosen vector (*see* **Note 7**).

1. Titer the library as described (*see* **Subheading 3.1.**), but use 3 mL LB top agarose supplemented with 50 µL each of 100 m*M* X-gal and IPTG.
2. Score an appropriate plate for the number of colorless plaques vs total number of plaques (colorless *plus* blue). This ratio is a measure of the proportion of the library that consists of recombinant bacteriophages (*see* **Note 8**).

3.2.2. PCR Screening

Polymerase chain reaction screening can be performed directly on bacteriophage plaque lysates to determine the presence or absence of inserts (*see also* **Subheading 3.3.**). Universal oligonucleotide primers are employed that are complementary to vector-specific sequences flanking the cloning site (e.g., λgt10- or λgt11-specific sequences or RNA polymerase promoter sequences like T3, T7, or SP6) *(3)*.

1. Prepare a PCR master mix *minus* thermostable DNA polymerase according to standard methods *(2,3)* and dispense in an appropriate number of PCR amplification tubes.
2. Lightly touch a bacteriophage plaque with the end of a small plastic pipet tip. Transfer the tip to one of the PCR amplification tubes and quickly rinse with master mix. Repeat for a total of 10 or more randomly selected plaques.
3. Liberate phage DNAs by boiling for 2 min. Cool tubes on ice and spin briefly to collect contents.
4. Add thermostable DNA polymerase (conventional or capable of amplifying long templates) and perform 30 cycles of standard PCR. Adapt conditions to specific equipment, reagent, reaction volume, and primer requirements *(2,3)*.
5. Controls include tubes that contain either known recombinant or known nonrecombinant bacteriophage or no template DNA.
6. Analyze resulting PCR products by agarose gel electrophoresis and ethidium bromide staining according to standard methods *(2,3)*.
7. Score for the presence of small DNA fragments that correspond to the distance between primer sites in parental nonrecombinant bacteriophage vs total number

of plaques picked. This ratio is a measure of the proportion of the library that consists of nonrecombinant bacteriophages.

3.3. Analysis of Size Range and Average Insert Size

An analysis of the average insert size and its range should be performed to control for an appropriate and complete representation of full-length cDNA clones, particularly those derived from long mRNAs. This may be less of an issue when size selection of newly synthesized double-stranded cDNA (usually >500 bp) has been performed during construction of the library. Nevertheless, the distribution of the insert sizes is an important parameter to consider when assessing the quality of a cDNA library.

The average insert size in cDNA libraries is often 2 kb or less, although large insert libraries have become available. However, even libraries with a smaller average insert size can be sufficiently useful, especially when the approximate length of the desired cDNA is known to be within the insert size range of the library (*see* **Note 9**).

The insert sizes of a small number of randomly selected recombinants can be determined by isolation of bacteriophage DNA and digest with an appropriate restriction enzyme. However, the use of the PCR can dramatically reduce the time and effort required to characterize cDNAs cloned in vectors. When two universal oligonucleotide primers flanking the cloning sites are available, PCR can be used to quickly screen a cDNA library for inserts (*see also* **Subheading 3.2.2.**) and to determine their size. With the introduction of long-distance (LD) PCR, it is possible to characterize a library by performing insert screening of all cDNA clones regardless of size *(4,5)*.

1. Prepare a LD PCR master mix *minus* thermostable DNA polymerase according to standard methods *(2,3)* and dispense in an appropriate number of PCR amplification tubes.
2. Lightly touch a bacteriophage plaque with the end of a small plastic pipet tip. Transfer the tip to one of the PCR amplification tubes and quickly rinse with master mix. Repeat for a total of 10 or more randomly selected plaques.
3. Liberate phage DNAs by boiling for 2 min. Cool tubes on ice and spin briefly to collect contents.
4. Add thermostable DNA polymerase capable of amplifying long templates *(4,5)* and perform 30 cycles of standard PCR. Adapt conditions to specific equipment, reagent, reaction volume, and primer requirements *(2,3)*.
5. Controls include tubes that contain either known recombinant or known nonrecombinant bacteriophage or no template DNA.
6. Analyze resulting PCR products by agarose gel electrophoresis and ethidium bromide staining according to standard methods *(2,3)*.
7. Estimate average size and range of inserts.

3.4. Additional Approaches for Quality Assessment of cDNA Libraries

3.4.1. Tests for Normal Representation of Housekeeping Genes

Under certain circumstances, additional quality control tests need to be performed, for example, to ensure normal representation of sequences of housekeeping genes that are expected to be highly represented in a given library. These may include PCR analyses for recombinants encoding glyceraldehyde-3-phosphate dehydrogenase, β-actin, tubulin, clathrin, or ribosomal proteins (e.g., L3 and S6). The general protocol for direct analysis of bacteriophage plaque lysates by LD PCR should be followed (*see* **Subheading 3.3.**), using oligonucleotide primers specific for one or more of these genes (*see* **Note 10**).

3.4.2. Detection of Contamination by Sequence Similarity Searches

Complementary DNA libraries may contain sequence elements of foreign origin and, therefore, may not faithfully represent the genetic information from the biological source (*6*). Such contaminating sequences are most commonly derived from accessory DNAs, including vectors, adapters, linkers, or PCR primers that were used in the process of constructing the library. Occasionally, unintended contamination from other sources occurs that can stem from transposable elements from the cloning host organism becoming integrated during propagation of the library. Contamination may also be the result of impurities in RNA preparations or reagents used in the isolation, purification, or cloning procedures (e.g., nuclear RNA or DNA present in total cellular RNA, yeast genomic DNA present in transfer RNA used as a carrier in the preparation of cDNAs, or cross-contamination with other recombinant DNA used in the laboratory).

Obviously, such contamination has its greatest impact on nucleotide sequence analyses. If not properly identified, it may result in wasted time and effort on meaningless analyses and erroneous conclusions drawn about the biological significance of sequences (*6*).

An on-line sequence similarity search against databases of vector or similar sequences is an efficient and convenient approach to screen sequences derived from cDNA libraries for contamination from accessory DNAs. One of the many noncommercial tools and free databases available for conducting such a search is VecScreen, provided by the National Center for Biotechnology Information (NCBI). VecScreen is a BLASTn-based search engine for the detection of vector and other sequences (*see* **Note 11**).

1. To screen a sequence for contaminating sequences, log onto the VecScreen web page at http://www.ncbi.nlm.nih.gov/VecScreen/VecScreen.html.

2. Type in or cut and paste the nucleotide sequence to be queried in FASTA format and select the "Run VecScreen" button.
3. On the following page, press "FORMAT!" to check the results.
4. The VecScreen output lists all segments of the query sequence that closely match any of the sequences in the UniVec database using a BLAST search (*see* **Note 12**). A link is available with on-line help for the interpretation of results.

In addition, the Basic Local Alignment Search Tool (BLAST) family of programs or similar software can be employed to search for specific foreign sequences known to have been used during library construction or to check for recombinant DNA concurrently used elsewhere in the laboratory. Databases of sequences from potential contaminants should include the BLAST databases of *E. coli* or yeast sequences.

3.4.3. Test for Intactness of cDNA 5' Ends

Rarely performed but considered to provide a good indication for the quality of a cDNA library is testing for the intactness of cDNA 5' ends (*1*).

1. Compare 5' end sequence information derived from randomly selected cDNA clones to known protein sequences (*see* **Note 13**).

3.4.4. DNA Sequencing to Determine Cloning Fidelity

Quality control efforts scrutinizing the fidelity of cloned recombinants may be of particular concern when total cellular RNA rather than cytoplasmic RNA has been isolated as a source to clone cDNAs or when PCR-based cloning techniques have been used. In these instances, a greater potential of generating artifacts during cDNA synthesis may exist, e.g., by introducing randomly ligated, intron-containing, or jumped cDNAs. Direct PCR analyses of selected recombinants can be used to check for the presence of introns in ubiquitously expressed genes (*see* **Subheadings 3.2.2.** and **3.3.**). DNA sequencing or genomic Southern blotting according to standard methods can be other useful approaches to determine the fidelity of cDNAs (*2,3*).

4. Notes

1. In this chapter, only the quality assessment of bacteriophage λ-based cDNA libraries is discussed in detail, but most of the general considerations are readily applicable to plasmid-based libraries as well. Many of the techniques used to characterize the quality of a cDNA library are similar or identical to methods that are described elsewhere in this volume.
2. Numerous protocols are available on how to prepare host bacteria for plating bacteriophage libraries (*2,3*).

3. The complexity of a cDNA library should be sufficiently high to contain full-length representatives of all sequences of interest, including those derived from "low-abundance" or "rare" mRNAs, to ensure that a clone of interest has a good chance of being represented. The abundance of a specific mRNA depends on the cell type. It has been estimated that 10,000–30,000 different mRNA species are present in varying abundance in a typical mammalian cell *(3)*. Statistically, the number of independent recombinants (N) required to achieve a given probability that a low-abundance mRNA will be present at least once in a cDNA library is $N = [\ln (1–P)/\ln (1–n)]$, where P is the probability desired (often 0.99, or 99%) and n is the fractional proportion of total mRNA population represented by a single mRNA species. According to Williams et al., the value of n is about 2.7×10^{-5} or 1/37,000 for a typical "low-abundance" mRNA species (10 copies/cell) in a transformed human fibroblast *(7)*. Therefore, the number of independent recombinants that need to be screened to isolate this mRNA with a probability of 99% is about $N = 170,000$. To ensure that a "rare" recombinant (1 copy/cell) is contained in a library, 1×10^6 or more independent recombinants may be required. Empirically, a library is desired to contain at least five times more individual recombinants (fivefold coverage) than the total expected by the lowest-abundance estimate *(2,3)*.

4. The titer reflects the number of plaque-forming bacteriophages—recombinant *plus* nonrecombinant—rather than the actual number of recombinants (*see* **Subheading 3.2.**).

5. Limited complexity is often less of a problem in microdissected compared to traditional cDNA libraries.

6. Other reasons for a very low abundance of recombinants include sampling variation and preferential cloning of certain sequences. For mRNAs that represent less than 0.1–0.01% of the total mRNA population, the richest cell or tissue source should be used for isolating cDNA clones of interest. Construction of a subtracted cDNA library for enrichment of "rare" mRNAs can be an other option *(2,3)*.

7. In suitable λ vectors, cDNA insertion inactivates the product of the *E. coli lacZ* gene (β-galactosidase), and colorless, clear plaques are formed. Vectors not containing a cDNA insert are *lacZ*-positive nonrecombinants. When plated in an appropriate host on medium containing IPTG and X-gal, parental nonrecombinant vectors produce blue plaques because they can metabolize X-gal (a chromogenic substrate for β-galactosidase activity) or similar substrates in the presence of IPTG (a synthetic inducer of the *lac* repressor).

8. *Caveat:* False positive (nonrecombinant) colorless plaques can sometimes appear because of false inserts or suboptimal levels of IPTG.

9. Complementary DNA libraries derived from microdissected tissues also tend to have smaller insert sizes, mostly as a result of the recovery of fragmented RNA. However, this is usually not considered a problem, as these libraries are intended for gene profiling and not as templates for full-length gene cloning.

10. Alternatively, comparative hybridization with oligonucleotide probes from the 5' and 3' ends of housekeeping genes can be performed to ensure normal representation.

11. VecScreen detects contamination by running a BLAST sequence similarity search against the UniVec vector sequence database. VecScreen then categorizes the matches, eliminates redundant hits, and shows the location of contaminating and suspect segments on a simple graphical display. Screens for vector contamination may also be conducted by running a sequence similarity search, such as BLAST, against other vector sequence databases (e.g., NCBI's vector database) *(6)*.

12. In addition to unique sequence segments from a large number of vectors, the UniVec database also contains sequences for adapters, linkers, and primers commonly used in the process of cloning cDNA *(6)*.

13. Often, the presence of translation initiation codons followed by intact open reading frames is taken as an indication for the presence of full-length cDNA clones in a library, although, ideally, the complete sequence from the 5' cap to the poly(A) addition site should be present.

Acknowledgment

Research performed in the author's laboratory is supported in part by grants DC04281 (to H.-J. F.) and EY03042 (Core Vision Research Center, Doheny Eye Institute) from the National Institutes of Health.

References

1. Anonymous (1998) cDNA Cloning workshop identifies critical issues. *Hum. Genome News* **9,** 17–18.

2. Ausubel, F. M., Brent, R., Kingston, R. E., Moore, D. D., Seidman, J. G., Smith, J. A., et al., eds. (1999) *Short Protocols in Molecular Biology*, 4th ed., Wiley, New York.

3. Sambrook, J. and Russell, D. W., eds. (2001) *Molecular Cloning: A Laboratory Manual*, 3rd ed., Cold Spring Harbor Laboratory, Cold Spring Harbor, NY.

4. Barnes, W. M. (1994) PCR amplification of up to 35-kb DNA with high fidelity and high yield from lambda bacteriophage templates. *Proc. Natl. Acad. Sci. USA* **91,** 2216–2220.

5. Cheng, S., Fockler, C., Barnes, W. M., and Higuchi, R. (1994) Effective amplification of long targets from cloned inserts and human genomic DNA. *Proc. Natl. Acad. Sci. USA* **91,** 5695–5699.

6. NCBI VecScreen, Contamination In Sequence Databases, http://www.ncbi.nlm.nih. gov/VecScreen/contam.html (Accessed Nov. 11, 2002).

7. Williams, J. G. (1981) The preparation and screening of a cDNA clone bank, in *Genetic Engineering* (Williamson, R., ed.), Academic, London, pp. 1–59.

14

Assessment of the Quality of mRNA Libraries by Agarose Gel Electrophoresis

Tsen-Yin Lin and Shao-Yao Ying

1. Introduction

Gel electrophoresis has been widely used in separating and purifying macromolecules such as proteins and nucleic acids that differ in size, shape, charge, and conformations. The gel basically acts as a tube in which the long-chain macromolecules such as RNAs contract and migrate in an extended configuration *(1)*. When charged molecules are placed in an electric field, they migrate toward either the positive (anode) or negative (cathode) pole according to their charge; polymers of a small organic molecule such as agarose and polyacrylamide are matrix which serves as a molecular sieve. Because of their negatively charged phosphate backbone, nucleic acid molecules migrate toward the anode.

Agarose gel electrophoresis is the most common method of analyzing RNA species, which has a relatively low resolving power but a large range of separation from 50 bp to 500 kb in concentrations from 3.0% to 0.1%, respectively *(2)*. By using gels with different concentrations of agarose, one can resolve different sizes of RNA fragments. Higher concentrations of agarose facilitate separation of small RNAs, whereas low agarose concentrations allow the resolution of larger RNAs. Agarose gels are nontoxic and extremely easy to prepare. Just by completely melting the agarose in the appropriate buffer and then pouring into the model, the gels are formed. Although RNA has a high degree of secondary structure, it is necessary to use a denaturing agents such as formaldehyde or methylmercuric hydroxide in the gel to disrupt the secondary structure and leave RNA in its linear form *(3,4)*.

From: *Methods in Molecular Biology, vol. 221: Generation of cDNA Libraries: Methods and Protocols*
Edited by: S.-Y. Ying © Humana Press Inc., Totowa, NJ

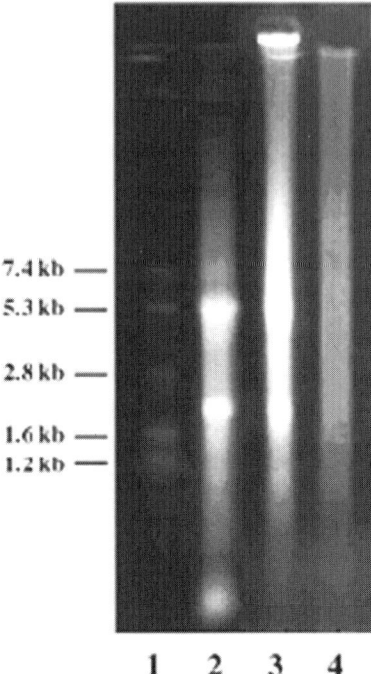

7.4 kb —
5.3 kb —

2.8 kb —

1.6 kb —
1.2 kb —

1   2   3   4

Fig. 1. Comparison between RNA repertoires prepared from phenol–chloroform extraction (lane 2), oligo-(dT) chromatographic column (lane 3), and RNA-PCR (polymerase chain reaction) (lane 4) fractionated on a 1% formaldehyde–agarose gel, all ranging from 300 bases to above 7.4 kb based on RNA markers. A homogenous and uniform smearing pattern of the RNA-PCR products indicates better mRNA quality than produced by the other methods, containing no contamination from nonpoly(A)$^+$ nucleotide species. (From **ref. 6** with permission.)

To detect the separation of various sizes of nucleic acids, the gel is stained with a low concentration of the fluorescent interacting dye, ethidium bromide. The bands of DNAs/RNAs containing as little as 1–10 ng can be viewed under ultraviolet light *(5)*. For agrose gel separation, a homogenous and uniform smearing pattern is considered to be a criterion for good quality of mRNA generated (*see* **Fig. 1**), which shows no contamination with either rRNAs or genomic DNAs *(6)*. As discussed earlier, agarose electrophoresis provides the most common and easiest way of identifying the quality of mRNA before subsequent analysis such as Northern blotting.

2. Materials
2.1. Preparation of Agarose Gel

1. Diethyl pyrocarbonate (DEPC)-treated ddH_2O.
2. 10X MOPS buffer: 200 mM MOPS [3-(N-morpholino)propanesulfonic acid] (pH 7.0), 50 mM NaOAc, and 10 mM EDTA (pH 8.0). Treat with DEPC and autoclave. Adjust volume to 1 L with DEPC double-distilled (dd) H_2O (*see* **Note 1**).
3. RNase-free agarose.
4. 37% Formaldehyde.
5. Ethidium bromide stock: 10 mg/mL EtBr in DEPC ddH_2O (*see* **Note 2**).

2.2. Preparation of RNA Sample

1. RNA loading buffer: 50% (v/v) glycerol, 1 mM EDTA (pH 8.0), and 0.4% bromophenol blue.
2. RNA sample (up to 20 μg).
3. 10X MOPS buffer.
4. 37% Formaldehyde.
5. Formamide (deionized).

2.3. Electrophoresis

1. 1X MOPS buffer: 20 mM MOPS (pH 7.0), 5 mM NaOAc, and 1 mM EDTA (pH 8.0).
2. Horizontal gel electrophoresis apparatus.

2.4. Visualization of RNA by UV Illumination

1. 300-nm-Wavelength ultraviolet (UV) light source.

3. Methods
3.1. Preparation of Agarose Gel

1. Seal the edges of the gel tray with tape and set the mold horizontally on the bench.
2. Add 1.2 g of agarose to 100 mL DEPC ddH_2O in an Erlenmeyer flask and heat it in a microwave oven.
3. Cool the agarose to 60°C in a water bath. Equilibrate for 15 min.
4. Add 10 mL of 10X MOPS buffer, 2 mL of 37% formaldehyde, and 5 μL of EtBr stock to the agarose solution in the hood (*see* **Note 3**).
5. Pour the agarose solution into the gel tray to form a gel of 3–5 mm thickness in a fume hood (*see* **Note 4**).

6. Insert the gel comb after the pouring and allow the gel to solidify for at least 30 min.
7. After the gel has set, place the gel in the electrophoresis tank. Fill the tank with 1X MOPS buffer to cover the gel.

3.2. Preparation of RNA Sample

1. Prepare samples in a sterile 1.5 mL microfuge tube:
   RNA (up to 20 μg)   4.5 μL
   DEPC ddH$_2$O   4.5 μL
   2X MOPS buffer   2.0 μL
   37% Formaldehyde   3.5 μL
   Formamide   10 μL
2. Vortex the sample briefly and centrifuge for a few seconds at maximum speed in a microcentrifuge.
3. Incubate samples at 68°C for 10 min to denature the RNA and immediately place on ice.
4. Add 2 μL of loading buffer to each sample and mix well.

3.3. Electrophoresis

1. Load the denatured RNA samples into the wells of the gel (*see* **Note 5**).
2. Connect the electrophoresis tank to a constant-voltage power supply.
3. Run the gel at a maximum of 5 V/cm gel distance until the bromophenol blue dye has migrated approximately two-thirds of the way through the gel (*see* **Note 6**).
4. Rinse the gel in DEPC ddH$_2$O with three changes of water and apply gentle shaking lasting 15–30 min to remove the formaldehyde.

3.4. Visualization of RNA by UV Illumination

1. Ethidium bromide intercalated with RNA induces increased fluorescence under UV light compared to the uncomplexed dye in solution.
2. Place the gel on UV transparent Perpex and illuminate it with UV light of 300 nm from below (*see* **Note 7**).
3. Observe the stained RNA for analysis (*see* **Note 8**).
4. Record the gel image by taking photography or using gel documentation system.

4. Notes

1. The buffer will turn yellow during autoclaving. This has no influence on gel electrophoresis.
2. Ethidium bromide is toxic and a powerful mutagen. Take appropriate safety precautions when handling. The stock should be stored at 2–8°C in a dark bottle.
3. Formaldehyde is toxic and volatile. Use a fume hood to avoid inhalation.

4. Avoid air bubbles in the gel or trapped between the wells. Air bubbles can be removed with a pipet before gel sets.
5. Do not make sample volumes close to the well's capacity because samples may spill over to adjacent wells. Make sure to include one lane for appropriate molecular-weight markers.
6. Avoid use of high voltages, which may cause trailing and smearing of RNA bands.
7. Ultraviolet light damages RNA. Use a lower-intensity UV light if RNA fragments are to be extracted from the gel *(7)*.
8. The respective ribosomal bands should appear as sharp bands on the gel. If the ribosomal bands are not sharp but appears as a smear toward smaller-sized RNAs, it is possible that the RNA sample degrades before electrophoresis. The 28S ribosomal RNA band should be twice as intense as the 18S rRNA band. If not, this generally indicates that some degradation occurred.

References

1. Deutsch, J. M. (1998) Theoretical studies of DNA during gel electrophoresis. *Science* **240,** 922–924.
2. Sambrook, J., Fritsch, E. F., and Maniatis, T. (1989) *Molecular Cloning: A Laboratory Manual,* Cold Spring Harbor Laboratory, Cold Spring Harbor, NY.
3. Lehrach, H., Diamond, D., Wozney, J. M., and Boedtker, H. (1977) RNA molecular weight determinations by gel electrophoresis under denaturing conditions, a critical reexamination. *Biochemistry* **16,** 4743–4751
4. Ogden, R. C. and Adams, D. A. (1987) Electrophoresis in agarose and acrylamide gels. *Methods Enzymol.* **152,** 61–87.
5. Sharp, P. A., Sugden, B., and Sambrook, J. (1973) Detection of two restriction endonuclease activities in Haemophilus parainfluenzae using analytical agarose–ethidium bromide electrophoresis. *Biochemistry* **12,** 3055–3063
6. Lin, S. L., Chuong, C. M., Widelitz, R. B., and Ying, S. Y. (1999) In vivo analysis of cancerous gene expression by RNA-polymerase chain reaction. *Nucleic Acids Res.* **27,** 4585–4589.
7. Brunk, C. F. and Simpson, L. (1977) Comparison of various ultraviolet sources for fluorescent detection of ethidium bromide-DNA complexes in polyacrylamide gels. *Anal. Biochem.* **82,** 455–462.

15

PACS RT-PCR

*A Method for the Generation and Measurement
of any Poly(A)-Containing mRNA Not Affected
by Contaminating Genomic DNA*

Igor Nepluev and Rodney J. Folz

1. Introduction

The generation and measurement of low-abundance messenger RNA (mRNA) transcripts is possible using techniques such as reverse transcriptase–polymerase chain reaction (RT-PCR) *(1,2)*. There are a number of variables that must be controlled for if accurate and reproducible results are to be obtained *(1)*. A significant limitation to the accurate quantitation of many mRNA species is genomic DNA contamination during the RNA purification step *(3)*. This contaminating genomic DNA often results in artifactual genomic-derived DNA-PCR amplification products. In most cases, exhaustive DNase digestion of the purified RNA sample is critical, especially in those where exon-specific primers cannot be separated by intronic sequences *(3)*. Although DNase digestion appears to work well for poly(A^+) mRNA preparations, DNase pretreatment fails to remove all trace contaminating genomic DNA from total RNA preparations *(3)*. In this method, we will demonstrate a quantitative method for the measurement of any poly(A)-containing mRNA that is not affected by contaminating genomic DNA, which we have termed poly(A) cDNA-specific RT-PCR (PACS).

2. Materials

1. T3 RNA polymerase with transcription optimized 5X buffer and 100 m*M* dithiothreitol (DTT) (Promega, Madison, WI). Store the phage RNA polymerases at –20°C and avoid exposure to frequent temperature changes.

From: *Methods in Molecular Biology, vol. 221: Generation of cDNA Libraries: Methods and Protocols*
Edited by: S.-Y. Ying © Humana Press Inc., Totowa, NJ

2. DNA template, linearized plasmid with T3 promoter sequences as a start site for transcription, and cDNA with 126-bp deletion located in the 3′ untranslated region close to the poly(A) tract.

3. TRIzol reagent (Life Technologies, Rockville, MD). When working with the TRIzol use gloves and avoid breathing vapor.

4. Chloroform, isopropanol, and 75% ethanol.

5. QIAprep spin mini prep kit (Qiagen, Valencia, CA) with buffers: PB, PE, and QIAprep spin column.

6. Restriction enzymes, DNase I (RNase-free), AMV reverse transcriptase, Deep Vent (exo-) DNA polymerase (all from New England BioLabs, Beverly, MA). Store at –20°C; avoid exposure to frequent temperature changes.

7. Diethyl pyrocarbonate (DEPC) (0.1%)-treated, RNase-free water. Add 0.1 mL DEPC to 100 mL of water and shake vigorously to bring DEPC into solution and incubate overnight. Autoclave for 30 min to remove any trace of DEPC. Wear gloves and use a fume hood when using this chemical. DEPC is a suspected carcinogen and should be handled with great care.

8. rNTP mix (2.5 mM each of ATP, CTP, GTP, and UTP), dNTP mix (25 mM each of dATP, dCTP, dGTP, dTTP).

9. [^{35}S]-dATP, 1000 Ci/mmol (Amersham Pharmacia Biotech, Piscataway, NJ); radiation hazard. Gloves should be worn when working with radiolabeled dATP or PCR products.

10. Oligonucleotide primers, oligo-(dT$_{18}$) primer (Stratagene Co., La Jolla, CA). Store at –20°C in a small aliquot; avoid exposure to frequent temperature changes.

11. Agarose and acrylamide mini-gel electrophoresis equipment, thermal cycler, and Storm 860 phosphorimager.

12. TBE buffer: 0.89 M Tris-borate and 0.002 M EDTA.

13. RNA gel loading solution: 95% formamide, 20 mM EDTA, 0.05% bromophenol blue, and 0.05% xylene cyanol FF; store at –20°C.

3. Methods

3.1. Generation of Internal Standards for Competitive RT-PCR

We will demonstrate this method using a full-length mouse extracellular superoxide dismutase (EC-SOD) cDNA *(4)*. After obtaining the cDNA of interest, a deletion mutation is generated in the 3′ region of the cDNA. This can be accomplished by PCR-directed deletion mutagenesis techniques or by taking advantage of unique restriction sites located within the cDNA. As shown in **Fig. 1**, a 126-bp deletion was made in the 3′ noncoding region of a mouse full-length EC-SOD cDNA. The plasmid containing this deletion was named pSK5/27 SS7. Ideally, a small deletion of about 50–100 bp should be generated in order to keep the internal standard cDNA nucleotide composition as close as possible to the original cDNA nucleotide composition so as to minimize

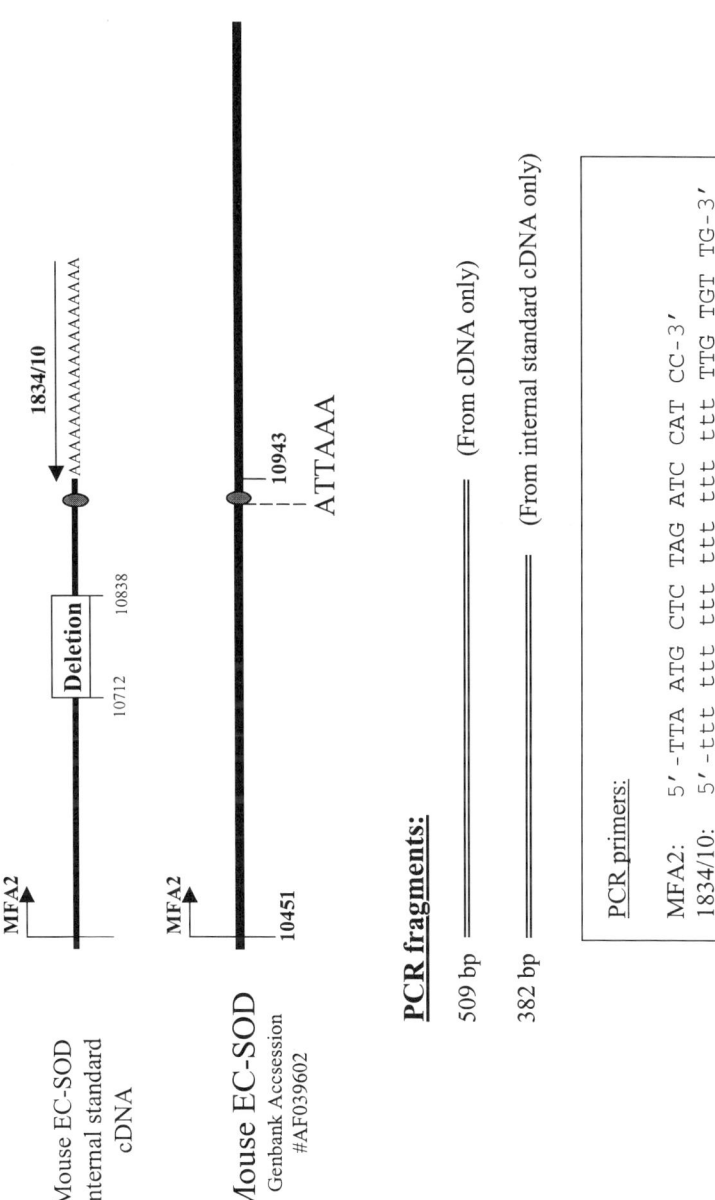

Fig. 1. Localization of the mouse EC-SOD deletion site mutation and the primers used for PCR and RT-PCR analysis. The first (main) poly(A) signal (ATTAAA). Nucleotide numbering is based on GenBank accession no. AF039602. The nucleotide sequence of each of the PCR primers used in this chapter and corresponding PCR size-amplification fragments from DNA or cDNA are shown.

any differential PCR amplification efficiencies between the two. The deletion must be large enough to be easily recognizable on an agarose or acrylamide mini-gel.

3.2. Internal Standard cRNA Preparation

3.2.1. Preparation of Linearized DNA Template

In this example, the plasmid DNA pSK5/27 SS7 (20 µg) is linearized with 40 units of *Xho*I in 0.2 mL of 1X *Xho*I buffer for 2 h at 37°C. This results in the generation of 5′ protruding (overhang) ends. It is important to avoid the use of restriction enzymes that produce 3′ protruding ends. If these enzymes must be used, the linearized template ends must be made blunt with Klenow DNA polymerase prior to transcription. To demonstrate complete linearization and ensure the presence of a clean nondegraded DNA fragment of the expected size, the digested DNA template is analyzed by 1.2% agarose gel electrophoresis. Once completely digested, the linearized DNA is purified using the QIAprep Spin Miniprep Kit protocol using a microcenttrifuge as described next.

3.2.2. QIAprep Miniprep Kit Protocol

1. Mix 5 vol of buffer PB to 1 vol of DNA solution. Next, the solution is centrifuged through the QIAprep membrane at 15,300*g* for 2 min.
2. The QIAprep spin column is washed by adding 0.5 mL of buffer PB and centrifuging for 1 min. Discard the flowthrough.
3. The QIAprep spin column is again washed with 0.75 mL of buffer PE and centrifuged for 1 min at 15,300*g*. After discarding the flowthrough, recentrifuge for an additional 1 min to remove any remaining buffer.
4. To elute the DNA, add 50 µL of nuclease-free water to the center QIAprep column, let stand for 1 min, and centrifuge at 15,300*g* for 1 min in a clean 1.5-mL microfuge tube.

3.2.3. Synthesis of cRNA Internal Standard

The purified linearized plasmid is used as a template to direct in vitro cRNA synthesis as follows:

1. Combine the following reaction components at room temperature and in the order given:

10 µL	Transcription optimized 5X buffer
5 µL	100 m*M* DTT
90 U	Recombinant RNasin ribonuclease inhibitor
20 µL	rNTP mix (2.5 m*M* each of ATP, GTP, UTP, and CTP)
10 µL	DNA template, linearized at 3–4 µg
<u>60 U</u>	<u>Phage T3 RNA Polymerase</u>
50 µL	Final volume

2. Incubate for 2 h at 37°C.
3. Remove the DNA template by adding 2 μL RNase-free DNase I (2 U).
4. Incubate for an additional 30 min at 37°C.
5. Remove unincorporated NTP and purify cRNA:
   a. Add 50 μL nuclease-free water and 1 mL TRIzol reagent.
   b. Shake vigorously for 15 s.
   c. Add 0.2 mL of chloroform, cover the samples tightly, shake vigorously for 15 s, and let stand at room temperature for 2 min.
   d. Centrifuge the homogenate at 15,300g for 15 min at 4°C.
   e. Transfer the aqueous phase to a fresh tube and mix with 0.5 mL isopropanol.
   f. Store samples at room temperature for 5 min.
   g. Centrifuge at 12,000g for 10 min at 4°C.
   h. Remove supernatant and wash RNA pellet with 1 mL of 75% ethanol by vortexing and subsequent centrifugation at 7500g for 5 min at 4°C.
   i. Air-dry the RNA pellet, briefly. However, do not let the RNA pellet dry completely, as it will greatly decrease its solubility.
6. Check the quantity and integrity of the in-vitro synthesized cRNA by gel electrophoresis.
   a. Prepare 1% agarose gel dissolved in 0.5X TBE buffer. Heat in microwave, and after the solution cools, add sodium dodecyl sulfate (SDS) to 0.1%.
   b. Remove a total of 4 μL cRNA and combine with 8 μL RNA gel loading solution. Heat to 85°C for 2 min and load onto the nondenaturing agarose gel.
   c. After electrophoresis, stain the gel with ethidium bromide (0.5 μg/mL) for 20 min. The cRNA should appear as a single band.

3.3. Extraction of Nucleic Acids and Reverse Transcription

Total mouse RNA and DNA were isolated from mouse lungs using TRIzol reagent. Briefly, 250 mg of lung tissue was homogenized for 1 min in 9 vol of TRIzol using a Brinkmann (Westbury, NY) polytron homogenizer setting at 5 (*see* **Note 1**). The size distribution of the RNA can be checked by nondenaturing 1% agarose gel electrophoresis containing 0.1% SDS in 0.5X TBE buffer. Both rRNA and mRNA can be seen as discrete bands ranging from 200 bp to 8 kb and indicates an intact RNA population. RNA extracted by this method can be quantified spectrophotometrically at 260 nm. Typical yields are 10 μg total RNA per 10 mg lung tissue. RNA should be dissolved in RNase-free water and stored at –80°C until further use.

3.4. First-Strand cDNA Synthesis

First-Strand synthesis of cDNA was generated using AMV reverse transcriptase following a protocol outlined from Promega (Madison, WI) with slight modification. One microliter of the oligo-(dT) primer (0.1 μg/μL) was added to 10 μg total RNA (final volume of 10 μL) and heated to 75°C for

5 min and then allowed to cool to 42°C. To this mixture was added the following components: 3 µL AMV RT 5X reaction buffer, 1 µL dNTP mix, and either 1 µL AMV RT (10 U/µL) or 1 µL nuclease-free water. The RT reaction was carried out at 42°C for 40 min, followed by heating at 95°C for 5 min in order to inactivate the reverse transcriptase. The reaction mixture was placed on ice and 85 µL nuclease-free water was added (total volume 100 µL), mixed, and stored at –80°C until further use.

3.5. PCR Reaction

Mouse EC-SOD primers used are shown in **Fig. 1**. Also shown are the expected PCR fragment sizes.

Polymerase chain reactions were performed in a thermal cycler Gene Amp PCR System 2400 (Perkin Elmer, Norwalk, CT) in 20 µL of a solution containing 1X ThermoPol reaction buffer, 0.2 mM each of dATP, dCTP, dGTP, and dTTP, 0.5 U Deep Vent (exo-) DNA polymerase, 1 µCi [^{35}S]-dATP, and a 1-µL template (first-strand cDNA synthesis reaction). Negative control PCR reactions included nonreverse transcribed total mouse lung RNA (800 ng), no DNA or RNA template, internal standard cDNA alone (10^{-4} copies), as well as cDNA prepared from 20 ng mouse total RNA (*see* **Fig. 2**). The amplification program consisted of 35 cycles: 96°C for 30 s, 60°C for 60 s, followed by 72°C for 30 s. After amplification, the reaction was held at 72°C for 7 min.

3.6. Detection and Analysis of RT-PCR Products

Mouse lung homogenates were spiked with increasing amounts of internal EC-SOD cRNA standards (10^6, 10^7, 10^8, and 10^9 copies), subjected to standard RNA isolation methods, reverse-transcribed and PCR amplified using the primer pair MFA2 and 1834/10. This primer pair is predicted to generate a 509-bp fragment if first-strand cDNA is used as the template or 382 bp if cDNA synthesized from mouse EC-SOD internal standard is used. As seen in **Fig. 1**, increasing the amount of competitive internal deletion standard into the mixture results in progressively increased amounts of the 382-bp PCR product (**Fig. 2**, lanes 1–4). When 800 ng of mouse lung total RNA (lane 5) or no RNA or DNA (lane 6) is used as the template, no bands are seen. When the Internal deletion standard alone is used as the template, only the 382-bp PCR fragment is seen (lane 7), whereas when cDNA generated by reverse transcriptase of total mouse lung RNA is used as the template, only the 509-bp PCR fragment is seen (lane 8).

Following completion of PCR cycling, 10 µL of gel loading solution (95% formamide, 20 mM EDTA, 0.05% bromophenol blue, and 0.05% xylene cyanol FF) was added, the tubes were heated to 85°C for 3 min and placed on ice immediately. Samples (5 µL) were loaded onto 5% (w/v) denaturing 8 M

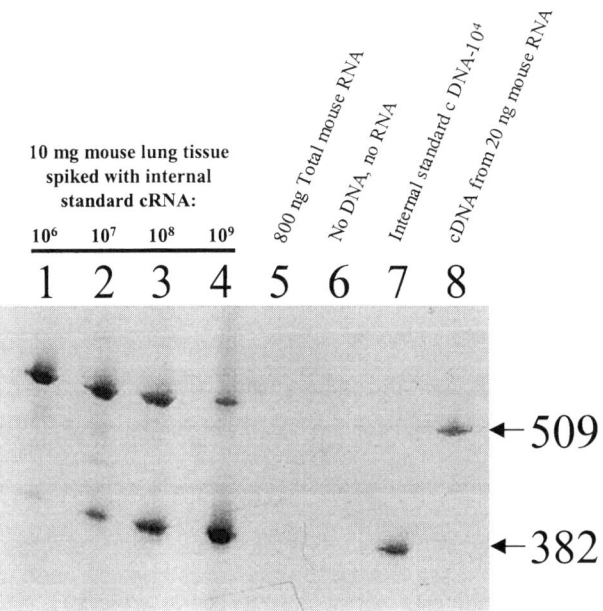

10 mg mouse lung tissue
spiked with internal
standard cRNA:

10^6 10^7 10^8 10^9

800 ng Total mouse RNA
No DNA, no RNA
Internal standard c DNA-10⁻⁴
cDNA from 20 ng mouse RNA

1   2   3   4   5   6   7   8

←509

←382

Fig. 2. PACS quantitative RT-PCR of EC-SOD mRNA obtained from mouse lung. Using the 3′-specific primer set MFA2 and 1834/10, RT-PCR was performed from mouse lung total RNA spiked with internal standard cRNA and analyzed by denaturing polyacrylamide gels. Lanes 1–4 show the results obtained when 10 mg mouse lung tissue total RNA was spiked with increasing amount of EC-SOD internal standard cRNA, reverse-transcribed, and 1% used as the template during the reaction. Lanes 5–8 represent various control reactions. Lane 5, uses nonreverse-transcribed total mouse lung RNA, 800 ng. Lane 6 has no DNA or RNA template. Lane 7 shows the internal deletion standard cDNA, at 10^{-4} copies, used as the template. In lane 8, cDNA generated from 20 ng mouse lung total RNA is shown. The 509-bp PCR fragment is generated by PCR using native EC-SOD mRNA as the template. The 382-bp PCR fragment is generated from the EC-SOD internal standard.

urea polyacrylamide TBE (0.89 M Tris-borate and 0.002 M EDTA) gel and electrophoresed at 10 W for 3 h. The gel was then dried and exposed on a phosphor screen for 18 h. The relative intensity of PCR products was determined by scanning the phosphor screen with a STORM 860 scanner (Molecular Dynamics, Inc., Sunnyvale, CA) and quantifying bands of interest using ImageQuant Analysis Program (version 1.2).

After correcting for the size of the PCR fragment and the A/T content, mouse lung EC-SOD mRNA content was measured to be $4.97 \times 10^7 \pm 0.86 \times 10^7$ copies per 10 mg mouse lung tissue. Under similar conditions, PACS RT-PCR would be predicted to readily determine significant differences in mRNA concentrations of as little as 0.4-fold from controls using small sample sizes.

4. Notes

1. Isolation of RNA from a very small amount of tissue (10 mg) samples can be homogenized in 0.1 mL of TRizol in 1.5-mL Eppendorf tubes by adding 0.5 mL of sand and applying a 1-mL pipet tip to mechanically disrupt the tissue. Following homogenization, add 0.1 mL TRizol and store the homogenate for 5 min at room temperature to permit the complete dissociation of nucleoprotein complexes. Next, add 40 µL of chloroform, cover, shake for 15 s, and let stand for 3 min at room temperature. Extract the RNA as described in **Subheading 3.2.3.**

Acknowledgments

This work was funded, in part, by grants from the National Institutes of Health (HL55166, ES/HL08698) and by a Grant-in-Aid from the American Heart Association to RJF.

References

1. Freeman, W. M., Walker, S. J., and Vrana, K. E. (1999) Quantitative RT-PCR: pitfalls and potential. *BioTechniques* **26(1),** 112–125.
2. Gilliland, G., Perrin, S., Blanchard, K., and Bunn, H. F. (1990) Analysis of cytokine mRNA and DNA: detection and quantitation by competitive polymerase chain reaction. *Proc. Natl. Acad. Sci. USA* **87,** 2725–2729.
3. Mutimer, H., Deacon, N., Crowe, S., and Sonza, S. (1998) Pitfalls of processed pseudogenes in RT-PCR. *BioTechniques* **24(4),** 585–588.
4. Folz, R. J., Guan, J., Seldin, M.F., Oury, T. D., Enghild, J. J., and Crapo, J. D. (1997) Mouse extracellular superoxide dismutase: primary structure, tissue-specific gene expression, chromosomal localization and lung *in situ* hybridization. *Am. J. Respir. Cell. Mol. Biol.* **17,** 393–403.

16

Single-Cell mRNA Library Analysis by Northern Blot Hybridization

Shi-Lung Lin

1. Introduction

The debut of RNA-polymerase cycling reaction (RNA-PCR) has promised to provide linear amplification of a reproducible mRNA library from as few as 20 single cells *(1)*. By incorporating a RNA promoter element during the synthesis of double-stranded complementary DNA (cDNA) templates, a poly(A^+) RNA library can be generated and reamplified from the templates in the same conformation and composition as its mRNA origins (**Fig. 1**). Using microarray analysis, the RNA-PCR-derived poly(A^+) library has been proven to contain above 97% of the original poly(A^+) RNA population and maintain $88 \pm 4\%$ linear correlationship to the populationary ratio of each RNA species (Chapter 12). It has also been tested to generate a full-length mRNA library from as few as 20 homologous tissue cells (2-pg mRNAs) for profiling cancer stages in vivo.

Northern blot analysis of gene expression usually requires abundant mRNA resources (>0.5 µg/lane), which is impossible to acquire from a few homologous tissue cells using traditional RNA extraction methods. However, we have acquired 30 µg of amplified poly(A^+) RNAs in one 50-µL reaction after three rounds of RNA-PCR amplification from about 20 single cells. This represents a 1.5×10^6-fold increase based on the comparison between the amount of the amplified poly(A^+) RNAs and that of theoretically presumed mRNAs within a cell (0.1 pg). It is noted that some rare RNAs can be well preserved by RNA-PCR for further gene analysis. Therefore, RNA-PCR can be a tool for providing unlimited mRNA resources for Northern blot detection at the single-cell scale.

From: *Methods in Molecular Biology, vol. 221: Generation of cDNA Libraries: Methods and Protocols*
Edited by: S.-Y. Ying © Humana Press Inc., Totowa, NJ

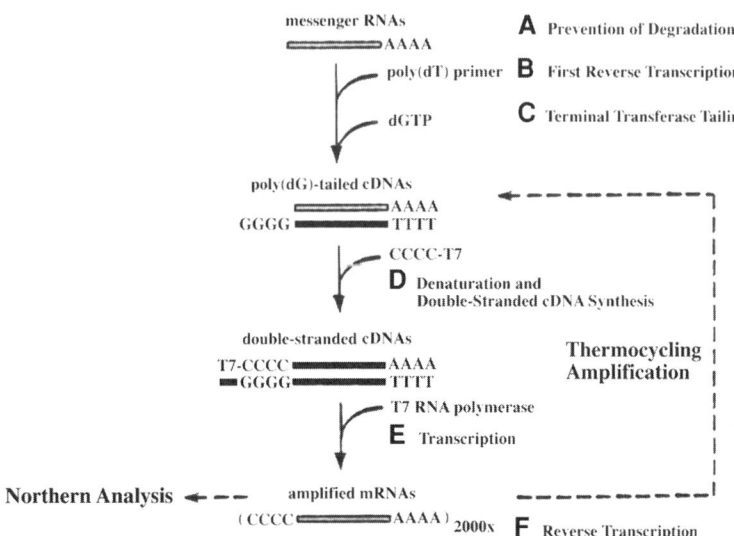

Fig. 1. An illustration of using RNA-PCR-derived poly(A⁺) RNA for Northern blot analysis. The mRNA generated in step D can be fractionated on a formaldehyde–agarose gel for specific gene detection.

One of the most powerful applications of this procedure is to the analyses of pathological sections. In our model, pathological sections were stained by *in situ* hybridization (**Fig. 2A**) with activin probes to identify activin-positive and activin-negative cells in vivo. Following microdissection, poly(A⁺) RNAs were amplified by RNA-PCR from the activin-positive and activin-negative epithelial prostatic cancer cells, respectively. The amplified poly(A⁺) RNAs from malignant, intermediate, and neoplastic patients of prostate cancers were displayed by electrophoresis on a formaldehyde-containing agarose gel (**Fig. 2B**) and then transferred to a nylon membrane as blots.

We has successfully verified the Northern blotting results of this procedure by investigating the behavior of known genes in prostate cancers *(2,3)*. For cultured LNCaP cells, we observed the time-course alterations of p53 and p16 expressions with Northern blotting at times after activin treatment, consistent with previous reports demonstrating by traditional chromatography methods that both genes were slightly upregulated after activin treatment (**Fig. 3A**, left panel). For pathological tissue sections, a similar upregulation pattern was also seen in the patients with intermediate and neoplastic prostate cancers, showing a good correlation (**Fig. 3**, middle panel). However, the expression of p53 was identical in activin-positive and activin-negative cells derived from

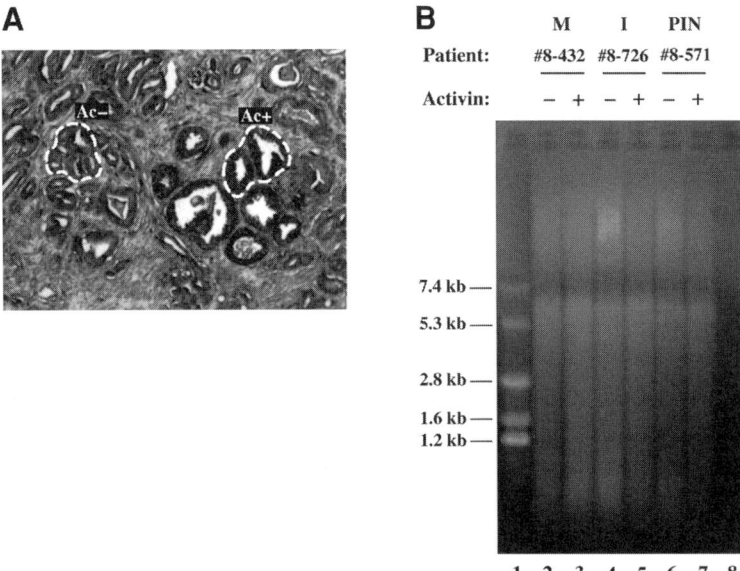

Fig. 2. RNA-PCR using cells microdissected from specific regions of pathological sections. **(A)** Identification of activin-positive and activin-negative prostate cancer epithelial cells by *in situ* hybridization. Dotted circles demarcate the regions for cell isolation with a micromanipulator under a microscope. For clarity, the background of this section was stained by hematoxylin. This staining was not used in the sections used for cell isolation. **(B)** One percent denaturing agarose gel electrophoresis of RNA-PCR products from the above isolated few cells. Three stages of prostate cancers were identified under a microscope and label as malignancy (M), intermediate (I), and prostatic intraepithelial neoplasia (PIN). From left to right: RNA markers (lane 1), mRNAs from activin-negative prostatic cancer cells (lane 2, 4, and 6), mRNAs from activin-positive prostatic cancer cells (lane 3, 5, and 7), and negative control of RNA-PCR without cells (lane 8). (From **ref. *1*** with permission.)

the malignant cancer patient, suggesting a possible loss of apoptotic regulation. This presumption is likely related to the observation that the more progressive the cancer, the fewer activin-positive cells were detected by *in situ* hybridization.

Poly(A^+) RNAs generated from prostate cancer sections were further compared with RNA-PCR-derived poly(A^+) RNAs prepared from LNCaP cells (**Fig. 3B**). Both data showed consistent levels of p53 at the neoplastic stage, apoptosin at the malignant and intermediate stages, and p16 and apoptostatin at all three stages. Some inconsistency was observed; the expression of p21 was

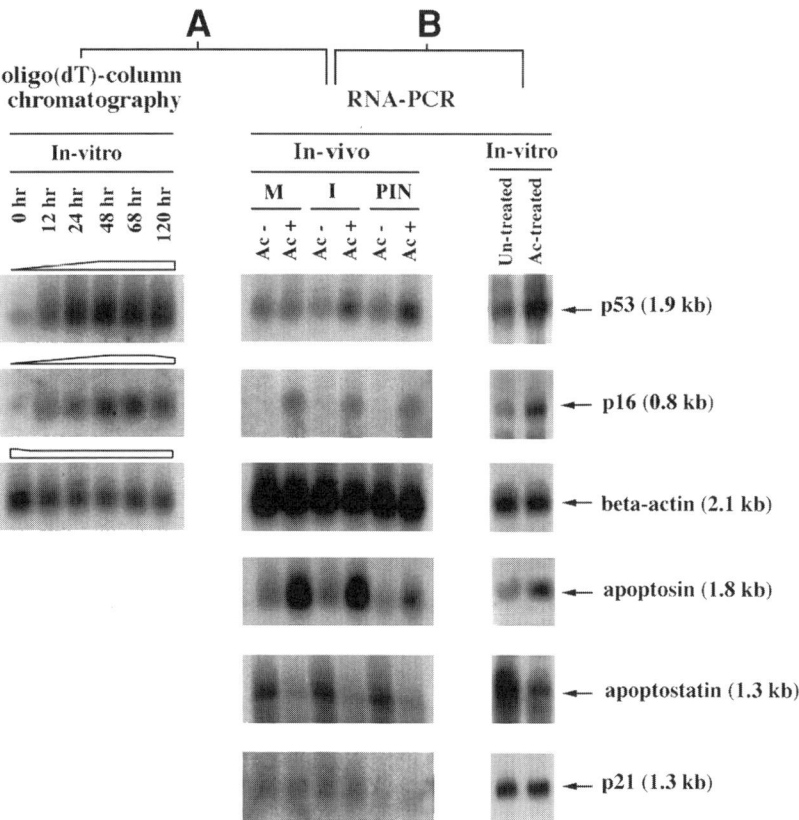

Fig. 3. Northern blot analyses of gene expression changes. **(A)** Comparison between poly(A+) mRNA libraries made by oligo-(dT) column chromatography and RNA-PCR. In vitro mRNAs (left panel) were generated by chromatography from 10⁶ LNCaP cells at different times after activin treatments (after 0, 12, 24, 48, 68, and 120 h). In vivo mRNAs were amplified by RNA-PCR from 20 isolated cancer cells with (Ac+) or without (Ac–) expression of activin. p53, β-actin templates (Ambion, Austin, TX), and synthetic p16 were used as probes in panel A. **(B)** Comparison between in vitro and in vivo mRNAs made by RNA-PCR from 20 cells. The blots of in vitro mRNAs (right panel) displayed differential alterations of gene expressions in LNCaP cells before and after activin treatment (120 h); the blots of in vivo mRNAs (middle panel) showed the actual differential expressions in patients' tissue cells. Probes (about 500–750 bases in length) for panel B were isolated from RNA-PCR-derived cDNA libraries prepared from LNCaP cells and amplified by PCR with sequence-specific primers. The sequences of the probes were confirmed by automated sequencing. All detected gene transcripts matched their original mRNA sizes, indicating that they were of good integrity and most likely full length. (From **ref. 1** with permission.)

weak in vivo but clearly expressed in vitro. This may result from individual and stage-related variations, demonstrating the need to maintain caution in extrapolating cell culture data to tissue sections. With this new, easy, and reliable procedure, re-evaluation of cancer molecular phenotypes at the single cell levels can be performed without artifacts produced in cultured cells.

2. Materials

2.1. Fractionation of Poly(A⁺) RNA Library by Electrophoresis

1. Diethyl pyrocarbonate (DEPC) H_2O: Stir double-distilled water with 0.1% DEPC for more than 12 h and then autoclave at 120°C under about 1.2 kgf/cm² for 20 min, twice.
2. 1% Agarose gel (40 mL): Add 0.4 g agarose powder into 29.4 mL DEPC H_2O and mix it with 4 mL 10X BE buffer (0.2 M sodium borate [pH 8.3] at 25°C and 2 mM EDTA).
3. 37% Formaldehyde (*see* **Note 1**).
4. Running buffer (500 mL): Add 40 mL of 37% formaldehyde into 410 mL DEPC H_2O and mix it with 50 mL 10X BE buffer.
5. Sample preparation (20 µL): Add 2 µg RNA-PCR-derived RNAs in total 5 µL DEPC H_2O and mix it with 2 µL 10X BE buffer, 3.5 µL of 37% formaldehyde, 10 µL formamide, and 0.5 µL ethidium bromide (10 mg/mL) in order. Heat to 55°C for 15 min, then cool on ice and add 2 µL 10X loading dyes (50% glycerol, 0.25% bromophenol blue, and 0.25% xylene cyanol FF in 1 mM EDTA [pH 8.0]).
6. Horizontal gel electrophoresis system.

2.2. Blotting of Fractionated Poly(A⁺) RNA Library

1. 0.05 N NaOH.
2. 20X SSC buffer: 0.3 M trisodium citrate (pH 7.0) and 3 M NaCl.
3. Nylon membrane: 0.2 mm larger than the surface length (L) and width (W) of gel.
4. 3MM blotting papers: One is the same size as the gel; the other one is three times longer in length but with the same width.
5. 3MM blotting thick pads: 0.2 mm smaller than the surface length and width of the gel.
6. Plastic wrap.
7. Blotting platform ($10L \times 8W \times 3H$ [cm³] and container (>750 mL).

2.3. Preparation of Radioactive Probes

1. Random 9-oligonucleotide primer mix: dephosphorylated 5′-dNNNNNNNNN-3′ (N = dATP, dGTP, dCTP, or dTTP in a completely random order; total 100 pmol//µL).
2. 10X Nick-translation buffer: 250 mM Tris-HCl (pH 7.0) at 25°C, 1.35 M KCl, 65 mM MgSO$_4$, 85 mM (NH$_4$)$_2$SO$_4$, 2.25 M β-NAD, 80 mM dithiothreitol (DTT),

and 1 mM each of dCTP, dGTP, and dTTP; prepare fresh.

3. Isotope shielding plate.
4. [α-^{32}P] dATP at 3000 Ci/mmol (*see* **Note 2**).
5. Exo(–) Klenow enzyme (5 U/μL).
6. Stop solution: 0.5 M EDTA (pH 8.0), sterile stock.
7. Incubation chamber: 37°C.
8. Purification spin column: 100 bp cutoff filter (such as Bio-Rad spin column P-30).

2.4. Blot Hybridization with Probes

1. QuikHyb solution (Stratagene, La Jolla, CA).
2. Fragmented salmon sperm DNA (100 μg/mL).
3. Probe denaturing chamber: 94°C.
4. Hybridization chamber with rotors: 65°C.
5. Low-stringent washing buffer: 2X SSC and 0.1% sodium dodecyl sulfate (SDS).
6. High-stringent washing buffer: 0.2X SSC and 0.1% SDS.

2.5. Autoradiography

1. Transparent plastic hybridization bag.
2. X-ray film (such as Kodak BioMax equivalent).
3. Film cassette.
4. –80°C Freezer.

3. Methods

3.1. Fractionation of Poly(A⁺) RNA Library by Electrophoresis

This method is modified from those of Lehrach et al. *(4)* and Goldberg *(5)*. The starting material is preferably 0.5–2 μg RNA-PCR-derived RNAs *(1)*. A larger amount (>10 μg) of the starting material may increase the signal intensity of rare RNA species and nonspecific background.

1. Gel casting ($10L \times 8W \times 0.5H$ [cm^3]): Melt the 1% agarose gel powder in a microwave oven; then, cool to about 50°C and add 6.6 mL of 37% formaldehyde. Gently mix well without causing any bubbles and pour into a gel-casting mold.
2. Electrophoresis: Submerge the precasted gel in an electrophoresis tank containing running buffer and prerun it for 5 min at 3 V/cm. Immediately load the sample preparations into the lanes of the gel and continue to run it at 3 V/cm. The sizes of RNAs can be determined by RNA markers, which are usually loaded in the most outside lanes.
3. Quality analysis: At the end of the electrophoresis (when the xylene cynaol FF has almost migrated to the margin of the gel), align the gel with a transparent ruler and photograph them by ultraviolet illumination.

3.2. Blotting of Fractionated Poly(A⁺) RNA Library

For an occasional use, the capillary elution method is inexpensive and efficient enough to transfer most of RNAs sized less than 5 kb from the gel to a blotting membrane. If heavily routine practice or analysis on much larger-sized RNAs is required, vacuum transfer or electroblotting should be considered.

1. Removal of formaldehyde: Rinse the gel by DEPC-treated water and soak it in 0.05 N NaOH for 15 min. Then, rinse the gel by DEPC-treated water and soak it in 20X SSC buffer for 30 min.
2. Blotting: Place a blotting platform in a RNase-free container containing 750 mL of 20X SSC buffer and make sure that the platform is raised at least 1 cm above the buffer surface. Rinse the longer 3MM blotting paper in 20X SSC buffer and place it on the surface of the platform. Make sure that there is no air bubble between the blotting paper and the platform. Both ends of the 3MM paper should be submerged in the buffer to provide capillary elution force all of the time. Place the gel upside down on the 3MM paper; then, rinse the nylon membrane and use it to cover all of the gel surface. Again, make sure there is no bubble between the gel and the nylon membrane. Rinse the shorter 3MM paper and place it on the top of the nylon membrane with no bubble in between. Stack at least 10-cm-thick 3MM blotting pads on the top of the short 3MM paper and cover the buffer surface by plastic wrap to prevent water vaporization. Put a 500-g weight on the top of the 3MM blotting pads to ensure a tight contact between the layers of materials used in the blotting system. Allow RNA transferring for at least 14 h.
3. Crosslinking: Remove all materials on the top of nylon membrane and carefully peel off the membrane from the gel. Soak the membrane in 6X SSC buffer for 5 min at room temperature; then, dry it on a paper towel for at least 15 min. Place the membrane with RNA face up in an ultraviolet irradiator (254 nm) and bake it for 20 s (1 J/cm²). Store the membrane in a –80°C freezer for up to 2 mo.

3.3. Preparation of Radioactive Probes

This protocol is modified from those of Feinberg and Vogelstein *(6)*. The method relies on the ability of random 9-mer oligonucleotides to anneal to multiple sites along the DNA template of interest. The Klenow fragment of DNA polymerase I then synthesizes new DNA by incorporating deoxynucleotide monophosphates at the free 3′–OH end provided by the 9-mers. The newly synthesized DNA is made radioactive by substituting a radiolabeled dATP for a nonradioactive one in the reaction mixture. The resulting labeled DNA has been proven to be useful for differential display *(7)*, Northern/Southern blotting *(8,9)*, and *in situ* hybridization techniques. The handling of radioactive materials should be performed under proper shielding protection (**Note 2**).

1. Primer annealing: Add 2 μL random 9-oligonucleotide primer mix into 31 μL DEPC-treated ddH$_2$O containing cDNA template of interest at 25 ng/μL. Heat the reaction at 94°C for 5 min and cool to the room temperature for 1 min.
2. Nick-translation labeling: Under proper shielding, sequentially add following materials into the reaction: 5 μL of 10X nick-translation buffer, 5 μL of [α-^{32}P] dATP, and 1 μL of Exo(−) Klenow enzyme; carefully mix well. Incubate the reaction at 37°C for 20 min and add 2 μL stop solution.
3. Probe purification: Unload the content solution from a purification spin column, spin for 2 min at 1000*g*, and discard the flowthrough (**Note 3**). Under proper shielding, load the labeled reaction into a purification spin column, spin for 4 min at 1000*g*, and collect the flowthrough. Store the purified probes in a −20°C freezer or perform the next step immediately.

3.4. Blot Hybridization with Probes

There are many commercial materials and solutions available to hybride radioactive probes to nucleic acids immobilized on a nylon membrane. To prevent the tedious preparation of traditional hybridization reagents, we use QuikHyb solution to achieve cleaner and more efficient results.

1. Blot prehybridization: Place the membrane blotted with RNA face up in a hybridization tube and add 10 mL QuikHyb solution to rinse the membrane. Heat 90 μL of fragmented salmon sperm DNA to 94°C for 7 min and place on ice immediately. Add the denatured salmon sperm DNA into the tube and mix well. Make sure that there is no leakage and place the tube into a hybridization chamber with rotation at 60°C for 20 min.
2. Probe denaturation: Under proper shielding, add 10 μL of the fragmented salmon sperm DNA into the labeled radioactive probes and mix well. Heat to 94°C for 7 min and place on ice immediately.
3. Hybridization: Under proper shielding, add the denatured probes into the tube without touching the membrane and gently mix well after closing the tube. Make sure that there is no leakage and place the tube into a hybridization chamber with rotation at 60°C for 4 h.
4. Washing: Under proper shielding, trash the QuikHyb solution into a certified container and add 200 mL of low-stringent washing buffer to the tube. Then, make sure that there is no leakage and place the tube into a hybridization chamber with rotation at 25°C for 20 min. Trash the low-stringent washing buffer into the certified container and add 200 mL of high-stringent washing buffer to the tube. Make sure that there is no leakage and place the tube into a hybridization chamber with rotation at 25°C for 20 min. Trash the high-stringent washing buffer into the certified container and add 200 mL of high-stringent washing buffer to the tube. Make sure that there is no leakage and place the tube into a hybridization chamber with rotation at 50°C for 20 min. Trash the high-stringent

washing buffer into the certified container and rinse the membrane shortly by DEPC ddH$_2$O. Dry it on a paper towel for 15 min.

3.5. Autoradiography

Perform this procedure in a dark room with a very limited red light source.

1. Film loading and exposure: Under proper shielding, place the semidry nylon membrane into a plastic bag and stick the margin of the bag onto the back surface inside a cassette. Place a film on top of the bag and close the cassette tightly to prevent any leakage of light. Place the cassette in a –80°C freezer and allow the film to be exposed by the isotope radiation on the membrane for about 24–48 h. Then, develop the film.

4. Notes

1. Formaldehyde is toxic and can be vaporized at room temperature. Solutions containing formaldehyde should be prepared and used in a chemical hood. The electrophoresis tank for running a formaldehyde gel should be covered whenever possible.
2. All the procedure related to radioactive materials should be performed under proper shielding devices. All of the waste of radioactive materials should be trashed into certified containers with a clear isotope label display.
3. Relative Centrifugal Force (RCF) (g) = $(1.12 \times 10^{-5}) \cdot (\text{rpm})^2 \cdot r$, where r is the radius in centimeters measured from the center of the rotor to the middle of the spin column and rpm is the speed of the rotor in revolutions per minute.

References

1. Lin, S. L., Chuong, C. M., Widelitz, R. B., and Ying, S. Y. (1999) In vivo analysis of cancerous gene expression by RNA-polymerase chain reaction. *Nucleic Acid Res.* **27,** 4585–4589.
2. Zhang, Z., Zheng, J., Zhao, Y., Li, G., Batres, Y., Luo, M.P., et al. (1997) Overexpression of activin A inhibits growth, induces apoptosis, and suppresses tumorigeecity in an androgen-sensitive human prostate cancer cell line LNCaP. *Intl. J. Oncol.* **11,** 727–736.
3. Lin, S. and Ying, S. Y. (1999) Differenbtitially expressed genes in activin-induced aoiotituc LNCaP cells. *Biochem. Biophys. Res. Commun.* **257,** 187–192.
4. Lehrach, H., Diamond, D., Wozney, J. M., and Boedtker, H. (1977) RNA molecular weight determinations by gel electrophoresis under denaturing conditions, a critical reexamination. *Biochemistry* **16,** 4743–4751.
5. Goldberg, D. A. (1980) Isolation and partial characterization of the *Drosophila* alcohol dehydrogenase gene. *Proc. Natl. Acad. Sci. USA* **77,** 5794–5798.
6. Feinberg, A. P. and Vogelstein, B. (1983) A technique for radiolabeling DNA restriction endonuclease fragments to high specific activity. *Anal. Biochem.* **132,** 6–13.

7. Liang, P. and Pardee, A. B. (1992) Differential display of eukaryotic messenger RNA by means of the polymerase chain reaction. *Science* **257,** 967–971.

8. Smith, G. E. and Summers, M. D. (1980) The bidirectional transfer of DNA and RNA to nitrocellulose or diazobenyloxymethyl paper. *Anal. Biochem.* **109,** 123–129.

9. Southern, E. M. (1975) Detection of specific sequences among DNA fragments separated by gel eleectrophoresis. *J. Mol. Biol.* **98,** 503–517.

17

Generation of cDNA Libraries for Profiling Gene Expression of Given Tissues or Cells

Xin Zhang, Qiu-Hua Huang, and Ze-Guang Han

1. Introduction

Complementary DNA (cDNA) library construction is a foundational technique for various organism genome projects and molecular biological approaches. Numerous methods for generating a cDNA library have been applied for different purposes. To profile the gene expression of a given tissue or cells, the researchers often used an oligo-(dT) primed and directional cloned strategy for generating cDNA libraries, and then DNA sequencing to characterize novel genes from those libraries *(1–6)*. The cDNA library derived from the strategy may contain more abundance gene species without significant bias because polymerase chain reaction (PCR) is not usually used for amplifying the double-strain cDNA. However, the strategy needs more sample and abundant message RNA. If there is only a little mRNA, the PCR-based method must been adopted in cDNA library construction. Recently, some researchers utilized cap structure at the 5′ end of eukaryotic mRNA to design novel methods for PCR-based cDNA library construction *(7–11)*. Moreover, the strategy has been applied to identifying gene expression and isolating full-length cDNA *(12–14)*. In our laboratory, the conventional and PCR-based strategies have been utilized for generating some libraries from massive sample (human liver cancer, hypothalamus, pituitary, adrenal gland) and a small amount of RNA (human CD34+ cells), respectively.

2. Materials

1. TRIzol reagent (Gibco-BRL; Gaithersburg, MD).
2. Oligotex mRNA kit (Qiagen, Hilden, Germany).

From: *Methods in Molecular Biology, vol. 221: Generation of cDNA Libraries: Methods and Protocols*
Edited by: S.-Y. Ying © Humana Press Inc., Totowa, NJ

3. *Escherichia coli* strains XL1-Blue, SOLR, and BM25.8.
4. Lambda Uni-ZAP XR vector digested with *Eco*RI and *Xho*I, CIAP-treated and plasmid pBluescript SK vector (Stratagene, La Jolla, CA).
5. StrataScript reverse transcriptase (Stratagene) or SuperScript II transcriptase (Gibco-BRL).
6. RNase inhibitor.
7. Methyl nucleotide mixture (10 mM dATP, dGTP, and dTTP, and 5 mM 5-mthyl dCTP).
8. DEPC (diethyl pyrocarbonate)-treated water.
9. RNase H and *Pfu* DNA polymerase.
10. *Eco*RI adapter (Stratagene).
11. T4 DNA ligase, DNA polymerase I, and T4 polynucleotide kinase.
12. dNTP (10 mM) and ATP.
13. *Eco*RI, *Xho*I, and *Sfi*I restriction enzymes.
14. Sepharose CL-2B (Pharmacia Piscataway, NJ).
15. STE buffer: 1 M NaCl, 200 mM Tris-HCl (pH 7.5), and 100 mM EDTA.
16. SM buffer: 5.8 g NaCl, 2.0 g MgSO$_4$·7H$_2$O, 50 mL of 1 M Tris-HCl (pH 7.5), and 5.0 mL of 2% (w/v) gelatin, and add deionized H$_2$O to a final volume of 1 L.
17. Gigapack III Gold packaging extract (Stratagene).
18. ExAssist interference-resistant helper phage (Stratagene).
19. LB broth and LB agar plate.
20. Ampicillin and Kanamycin.
21. IPTG (isopropyl-1-thio-β-D-galactopyranoside) and X-Gal (5-bromo-4-chloro-3-indolyl-β-D- galactopyranoside).
22. λTriplEx2 (*Sfi*I A and B-digested arms) and pTriplEx2 (Clontech, Palo Alto, CA).
23. Oligonucleotide primer for oligo-(dT) and oligo-capping.
24. Proteinase K.
25. CHROMA SPIN-400 columns (Clontech)

3. Methods

The following methods outline (1) extraction of total RNA and message RNA, (2) vectors for the cDNA library, (3) construction of oligo-(dT) primed and directional cloned cDNA library (conventional cDNA library), (4) PCR-based cDNA library construction, and (5) library amplification.

3.1. Extraction of Total RNA and Message RNA

The RNA used to generate the cDNA library for gene expression profile analysis must be high-quality, nondegraded, and no genomic DNA contamination. The procedures for isolating RNA include (1) the extraction of total RNA and (2) the isolation of poly(A)$^+$ RNA from total RNA (*see* **Note 1**).

3.1.1. Isolation of Total RNA from Tissues or Cells

Recently, the modified single-step RNA isolation method with TRIzol reagent, which was developed by Chomczynski and Sacchi, was often used for extracting total RNA from tissues or cells *(15,16)*.

1. Take out the sample tissue wrapped with foil from liquid nitrogen and quickly weight on a balance. Grind the tissue in a mortar adding liquid nitrogen frequently. As the tissue become powder-like, TRIzol reagent (50–100 mg tissue/mL TRIzol) may be added into the mortar and transfer the liquid to a glass–Teflon homogenizer.
2. Homogenize tissue completely and incubate the homogenized samples for 5 min at room temperature. Transfer the sample to a polypropylene tube. Add 0.2 mL of chloroform per milliliter of TRIzol reagent. If extracting from cells, such as CD34⁺ cells, adequate TRIzol reagent may be mixed directly with cells without grinding and homogenizing *(12)*.
3. Cap sample tubes securely and shake tubes vigorously by hand or shaker for 15 s. Incubate them at room temperature for 2–3 min. Centrifuge the sample at 12,000g for 15 min at 4°C.
4. Carefully transfer the colorless upper aqueous phase to a fresh tube. Add 0.5 mL isopropyl alcohol per milliliter of TRIzol reagent to precipitate the RNA from the aqueous phase. Mix and incubate the sample at room temperature for 10 min.
5. Centrifuge the sample at 12,000g for 10 min at 4°C and RNA precipitate on the side and bottom of the tube.
6. Carefully remove the supernatant. Wash the RNA pellet with 75% ethanol (diluted in DEPC-treated water) and add 1 mL of 75% ethanol per milliliter TRIzol reagent. Centrifuge the samples at 7500g for 5 min at 4°C.
7. Carefully remove the supernatant with a pipet and briefly air-dry the RNA pellet. Dissolve RNA in RNase-free water. Examine the quality of the total RNA on a denaturing formaldehyde–agarose gel. The total RNA should appear as two bright bands (18S and 28S ribosomal RNA) at approx 4.5 and 1.9 kb (*see* **Fig. 1**). The ratio of intensities of 28S and 18S rRNA should be 1.5–2.5:1. The total RNA with high quality can be stored at –80°C for up to 1 yr.

3.1.2. Isolation of Poly(A)⁺ RNA from Total RNA

Usually, poly(A)⁺ RNA, not total RNA, is used for generating a cDNA library. We preferred to utilize the Oligotex mRNA column (Qiagen) to isolate high-quality poly(A)⁺ RNA from total RNA. Total RNA preparations are incubated with Oligotex resin, and oligo-(dT):mRNA complexes, which are linked covalently to the surface of resin, are collected by a brief centrifugation step. After washing, the mRNA is eluted in a small volume of Tris buffer or water. Commonly, 1 μg poly(A)⁺ RNA could be isolated from 100 μg total RNA.

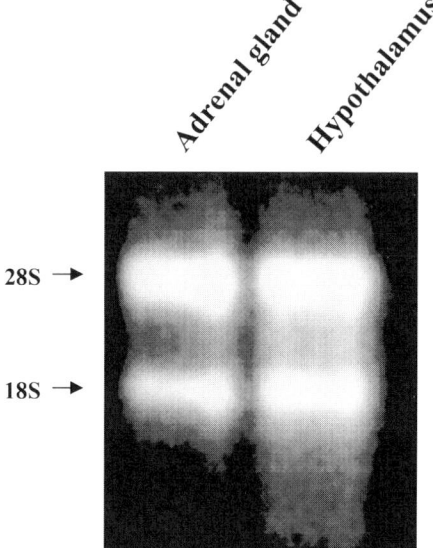

Fig. 1. Total RNA from the hypothalamus and adrenal gland on a denaturing formaldehyde–agarose gel. Total RNA was extracted by the modified RNA isolation method with TRIzol reagent *(15,16)*.

3.2. Vectors for the cDNA Library

The Uni-ZAP XR vector system (Stratagene) and TriplEx2 system (Clontech) were used in our studies. The two vector systems combine the high efficiency of phage library construction and the convenience of a plasmid system with blue-white color selection. The Uni-ZAP XR vector (*see* **Fig. 2**) may be double digested with *Eco*RI and *Xho*I restriction enzymes and accommodates DNA inserts from 0 to 10 kb in length. This vector allows in vivo excision of the matched pBluescript phagemid (*see* **Fig. 3**), allowing the inserts to be characterized in a plasmid system. The multiple-cloning-site (MCS) region of pBluescript phagemid has 21 unique cloning sites flanked by T3 and T7 promoters and many different primer binding sites. Like the Uni-ZAP XR vector system, the TriplEx2 vector also possess the following advantages for generating cDNA library in a phagemid vector: high titer libraries, blue/white screening for recombinants, regulated expression of cloned inserts, and ease of converting clones from phage to a plasmid vector via Cre-loxP-mediated subcloning (*see* **Fig. 4**). However, the two vector systems have distinct features (*see* **Note 2**).

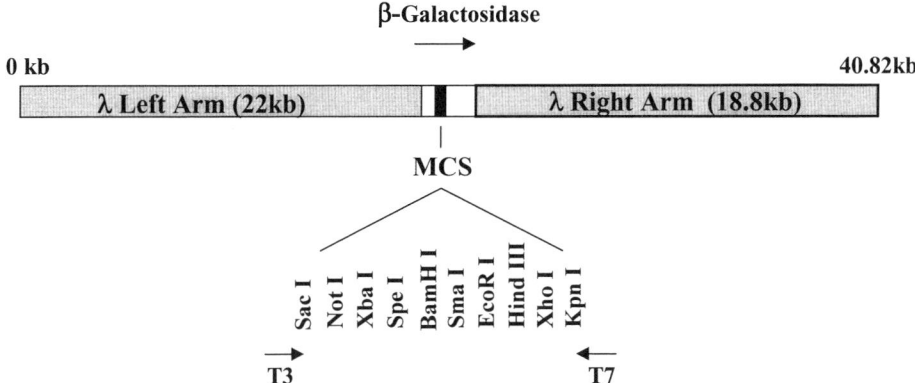

Fig. 2. Schematic drawing of the Uni-ZAP XR insertion vector from Stratagene (La Jolla, CA).

3.3. Construction of Oligo-(dT) Primed and Directional Cloned cDNA Library (Conventional cDNA Library)

The Uni-ZAP XR system was used for generating the oligo-(dT) primed and directional cloned cDNA library derived from massive samples such as liver, liver cancer, hypothalamus, pituitary, and adrenal gland in our laboratory *(4,5)*. The procedure includes (1) conversion of mRNA into double-stranded cDNA, (2) preparation of the cDNA insert, (3) size fractionating, (4) ligating cDNA into the Uni-ZAP XR vector, (5) packaging and titering, and (6) in vivo excision.

3.3.1. cDNA Synthesis

The most basic step in constructing a cDNA library is process of generating a double-strand cDNA copy derived from corresponding mRNA. For obtaining essentially the representative information of gene expression pattern and full-length cDNA copies from mRNA, the most important factors are the quality of the mRNA and efficiency of the reverse transcriptase that will effect the extension of the cDNA along mRNA. It is essential to start with the high-quality mRNA or total RNA and good reverse transcriptase.

1. To synthesize the first-strand cDNA, an oligo-(dT) linker–primer with an appropriate restriction enzyme recognition site should be used. For the lambda–ZAP cDNA library, the primer is a 50-base oligonucleotide with the following sequence: 5′-GAGAGAGAGAGAGAGAGAGAGAGA ACTAGTCTC GAGTTTTTTTTTTTTTTTTTTT-3′. This oligonucleotide contains a *Xho*I restric-

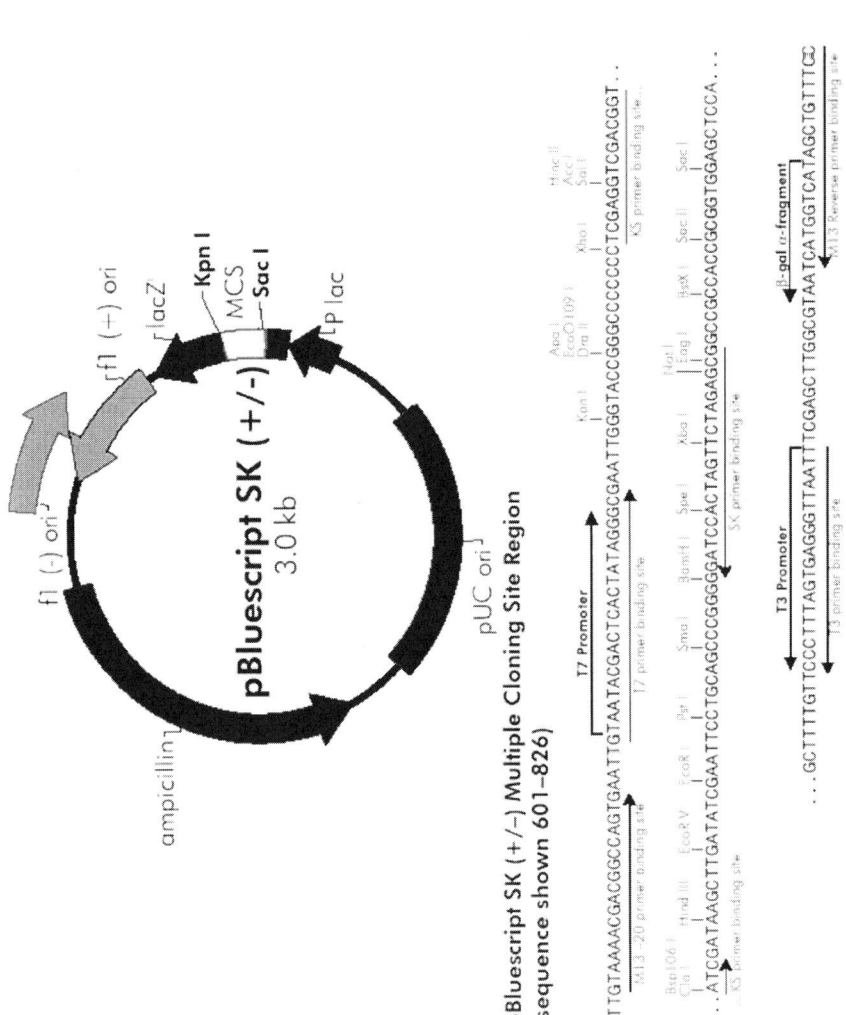

Fig. 3. Schematic drawing of the pBluescript SK(+/−) phagemid and its multiple-cloning-site region.

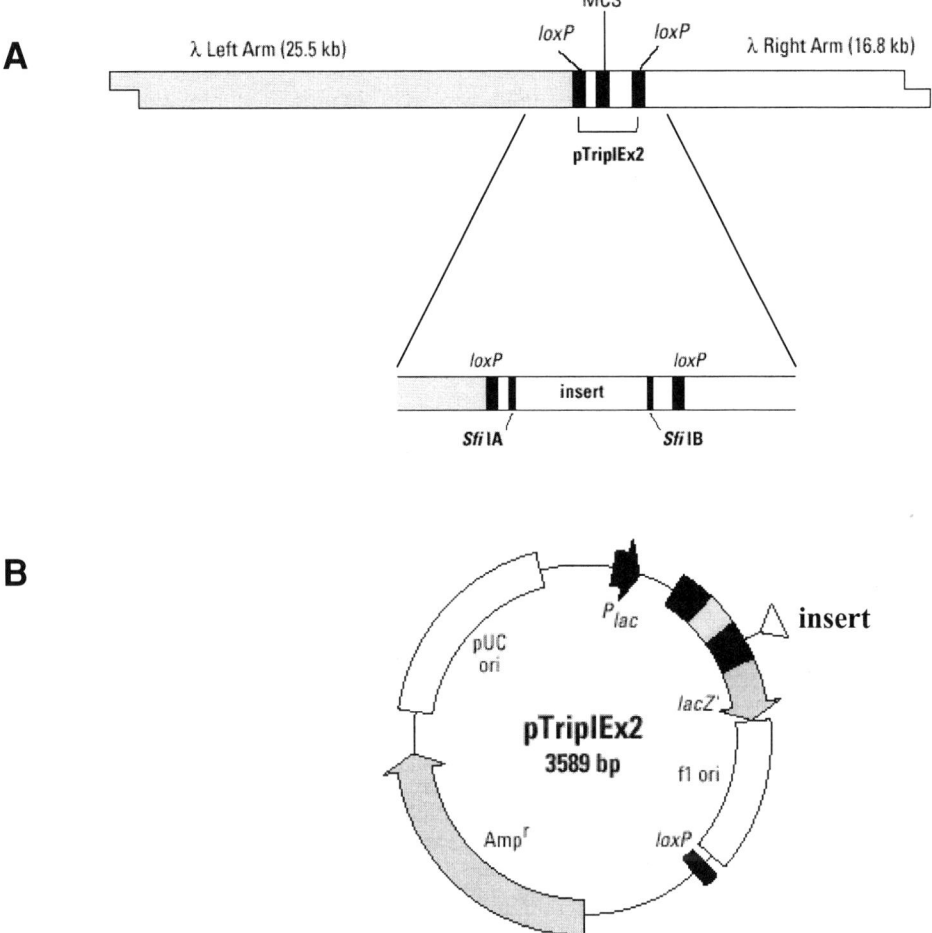

Fig. 4. Schematic drawing of the λTriplEx2 system from Clontech (Palo Alto, CA) (**A**) and the corresponding plasmid pTriplEx2 (**B**).

tion site and an 18-base poly(dT) sequence. Five to six micrograms of mRNA is used for first-strand cDNA synthesis by reverse transcriptase, such as StrataScript RT or SuperScript II, in a 50-µL reaction system at 37°C for 1 h. Three microliters of methyl nucleotide mixture, which contains normal dATP, dGTP, dTTP, and the analog 5-methyl dCTP, 2.8 µg linker primer and 40 U RNase block ribonuclease inhibitor from Stratagene, is added into the reaction. Both Stratascript RT and SuperScript II lack RNase H activity, but retain wild-type polymerase activity, so they can synthesize longer cDNA fragments than wild-type Moloney murine leukemia virus reverse transcriptase (MMLV RT). Methyl nucleotide mixture will prevent the cDNA from the following *Xho*I digestion. Usually, the start amount of poly(A)$^+$ RNA is about 5 μg (*see* **Notes 1** and **3**).

2. One hundred units of DNA polymerase I may be used for synthesizing the second-strand cDNA at 16°C for 2.5 h. Three units of RNase H are added to digest mRNA.

3. The uneven termini of the double-strand cDNA will be nibbled back and filled in with 5 U of cloned *Pfu* DNA polymerase at 72°C for 30 min. T4 DNA polymerase can be used at this step as an option. Then, add the equal volume of phenol–chloroform (1:1) to extract the cDNA from the reaction and precipitate DNA using 3 *M* sodium acetate and 100% ethanol and wash the DNA pellet with 75% ethanol.

3.3.2. cDNA Insert

To insert the cDNA fragment into the vectors directionally, adapters should first be ligated to the double-strand cDNA to provide restriction endonuclease sites.

1. *Eco*RI adapters, which are composed of 10- and 14-mer oligonucleotides and complementary with an *Eco*RI cohesive end and the 14-mer oligonucleotide is kept dephosphorylated, are ligated to the blunted cDNA ends by T4 ligase. The dephosphorylated end could eliminate self-ligation of the adapters during ligation to the cDNA. Then, the adapters are phosphorylated by T4 polynucleotide kinase.

2. To release the *Eco*RI adapter and residual linker–primer from the 3′ end of the cDNA, add 120 U of *Xho*I to digest the cDNA at 37°C for 1.5 h. Then, precipitate the cDNA with 3 *M* sodium acetate and 100% ethanol. Resuspended the cDNA pellet with 14 μL of 1X STE buffer. The sample is ready for size fraction after adding a 3.5-μL column loading dye.

3.3.3. Size Fractionation

Size fractionation could remove the unligated adapters and the low-molecular-weight materials from adapted cDNA. The different sized cDNAs for different purposes could also be separated using this procedure (*see* **Note 4**).

1. Connect a 1-mL sterile pipet and a 10-mL syringe with a small piece of connecting tube to assemble the drip column. Cut off the cotton plug of the pipet and leave approx 3–4 mm in the pipet before the connection. Thrust the cotton plug down into the tip of the pipet by pushing the plunger into the syringe.

2. Fill the drip column with 2 mL of 1X STE and immediately add a uniform suspension of Sepharose CL-2B slurry to the column with a Pasteur pipet by inserting the pipet as far into the column as possible. Continue adding the slurry as the resin settles. When the surface of the packed bed is 0.25 in. below the lip of the pipet, stop adding the resin. Avoid air bubbles forming in the column.

A

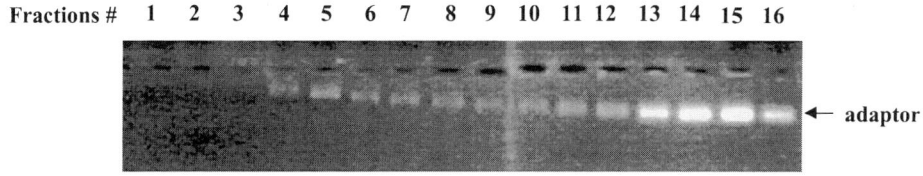

B

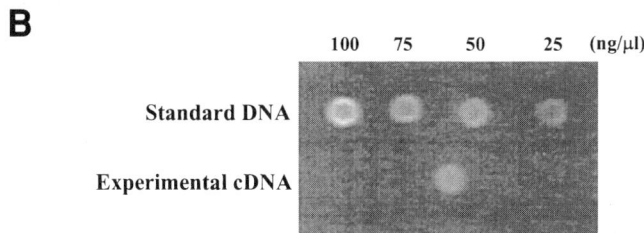

Fig. 5. The analysis of cDNA after size fractionation. Collect 2–12 fractions for Sepharose CL-2B column and electrophoresis on 1% agarose–EtBr gel to evaluate the cDNA size of each fraction (**A**). Measure the amount of the recovered cDNA by comparing with the concentration of standard DNA on an agarose–EtBr gel. The amount of the cDNA is about 40 ng/μL (**B**).

3. Wash the column with 10 mL of 1X STE. Immediately load the cDNA sample into the column when approx 50 μL STE buffer remains above the surface of the resin. Fill the connecting tube with STE buffer when the sample enters the gel.

4. Collect each fraction in 1.5-mL centrifuge tubes. Three drops per fraction are collected when the dye reaches the 0.4-mL graduation on the pipet, until the trailing edge of the dye reaches the 0.3-mL graduation. About 12–18 tubes of fraction could be collected, each tube containing approx 100 μL. Remove 8 μL of each collected fraction and analyze on a 1% agarose–EtBr gel for the size of each fraction (*see* **Fig. 5A**).

5. Pool the proper sizes of the fractions together and extracted the cDNA with phenol–chloroform (1:1) and precipitate using ethanol. Resuspend the cDNA pellet with 3.5 μL sterile water. To measure the amount of the recovered cDNA, a 0.5-μL cDNA solution is compared with 0.5 μL of the different concentrations of standard DNA by loading the DNA to an agarose gel and staining with ethidium bromide (*see* **Fig. 5B**).

3.3.4. Ligation

About 100 ng of resuspended cDNA is ligated into 1 µg of the Uni-ZAP XR vector with 2 U of T4 DNA ligase at 4°C for 2 d in a 5-µL volume.

3.3.5. Packaging

The ligated cDNA is packaged with the Gigapack III Gold packaging extract (Stratagene) at 22°C for 2 h. Add 500 µL SM buffer and 20 µL chloroform. Mix gently and spin down to sediment the debris. Store the supernatant containing the phage at 4°C for up to 1 mo. For long-term storage, the phage cDNA library should be stored as small aliquots in 7% (v/v) dimethyl sulfoxide (DMSO) at –80°C.

3.3.6. Titering

1. Incubate the XL-1 blue MRF′ strain in LB with 10 mM MgSO$_4$ and 0.2% (w/v) maltose at 37°C for 4–6 h, with shaking at 250 rpm. Spin the bacterial cells at 500g for 10 min, resuspend the cells to an optical density (OD$_{600}$) of 0.5 with sterile 10 mM MgSO$_4$.
2. Prepare two consecutive 10^{-2} dilutions of the packaging reaction with SM buffer. Add 10 µL and 1 µL, respectively, of the 10^{-4} dilution to 200 µL of the XL-1 blue cells from **step 1**. Incubate at 37°C for 15 min. Add 3 mL of LB top agar (0.7% agar in LB broth) that was melted and cooled to 48°C and quickly pour onto dry, prewarmed LB agar plates.
3. Incubate the plates for at least 12 h at 37°C. Count the plaques of the plates. Calculate the titer of the library using the following equation:

$$\frac{\text{No. of plaques} \times \text{Dilution factor} \times \text{Total packaging volume}}{\text{µL of diluted phage plated}}$$

   Usually, the titer of the libraries is 10^5–10^7 plaque-forming units (pfu/mL).
4. To verify the inserts, pick 25 plaques from the titering plates randomly. Inoculate the agar to 5 µL sterile water. Use the water containing phage as the template, PCR amplifying the inserted fragment with T3 and T7 primers. Calculate the average size of the inserted fragments after electrophoresis of the PCR product on an agarose–EtBr gel. In a high-quality cDNA library, the average of the inserted fragments should be longer than 1 kb (*see* **Fig. 6** and **Note 5**).

3.3.7. In Vivo Excision

The helper phage allows efficient excision of the pBluescript phagemid from the Uni-ZAP XR vector. The ExAssist helper phage contains an amber mutation that prevents replication of the phage genome in a nonsuppressing *E. coli* strain (e.g., SOLR cells). Only the excised phagemid can replicate in SOLR, removing the possibility of coinfection.

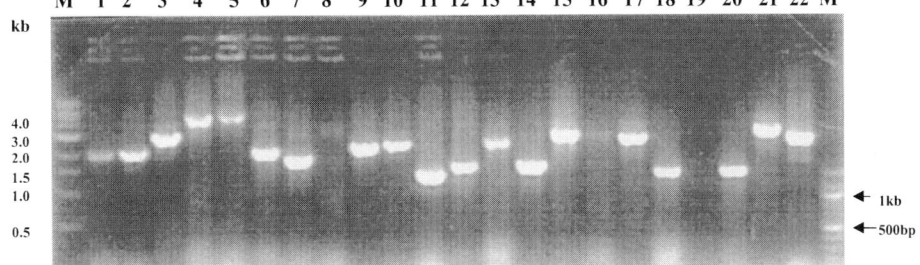

Fig. 6. Pick 22 plaques from the titering plates randomly and PCR amplifying the inserted fragment with T3 and T7 primers to evaluate the quality of our cDNA library. Among those plaques, PCR products 19–22 (86%) show significant bands on an agarose–EtBr gel. The average 2-kb size of the cDNA fragments in the library was estimated according to the size of all inserts. M: molecular marker.

1. Grow separate overnight cultures of XL-1 Blue MRF′ and SOLR cells in LB broth with 0.2% (w/v) maltose and 10 mM MgSO$_4$ at 30°C. Gently spin down the XL1-Blue and SOLR cells (100g) and resuspend the cells in 10 mM MgSO$_4$ to an OD$_{600}$ of 1.0 (8×10^8 cell/mL).

2. In a 15-mL tube, combine a portion of the λ phage library with 200 μL XL-1 Blue MRF′ cells at a 1:10 helper phage-to-cell ratio. Add ExAssist helper phage (Stratagene) at a 10:1 helper phage-to-cell ratio to ensure that every cell is coinfected with λ phage and helper phage.

3. Incubate the tube at 37°C for 15 min to allow absorption. Add 3 mL of LB broth and incubate the conical tube for 1 h at 37°C, shaking at 250 rpm. Do not incubate for more than 1 h because this will cause high redundancy.

4. Heat the tube at 65–70°C for 20 min to kill the bacterial cells. Spin down the debris at 1000g for 10 min and then decant the supernatant that contains the excised phagemid in a sterile conical tube. The phagemid can be stored at 4°C for 2 wk.

5. Add 10 μL and 100 μL, respectively, of the excised phagemid from **step 4** to 200 μL of SOLR cells from **step 1** in a 1.5-mL microcentrifuge tube. Incubate at 37°C for 15 min. Plate to LB agar plate containing IPTG and X-Gal. After incubation at 37°C overnight, the excised phagemid can be titered by counting the colony number. The white colonies contain the phagemid with cDNA insert and can be used for DNA sequencing.

3.4. PCR-Based cDNA Library Construction

The PCR-based strategy allows us to generate a cDNA library derived from a small amount of total RNA or poly(A)$^+$ RNA. In our laboratory, the strategy was used for constructing some cDNA libraries, such as a human CD34$^+$ cDNA library *(12–14)*. The method can produce high-quality, enriched full-length cDNA libraries from nanograms of total or message RNA. The procedures include (1) cDNA synthesis, (2) digestion with *Sfi*I, (3) size fractionation, (4) ligation of cDNA to λTripEx2 vector, (5) packaging and titering the library, and (6) converting the phage library to plasmid library.

3.4.1. cDNA Synthesis

1. Use 1 to 2 μg of total RNA or 50 ng to 1 μg mRNA (or even less) for first-strand cDNA synthesis by reverse transcriptase, such as Superscript II, at 37°C for 1 h with CDS III/3′ PCR primer and SMART III oligonucleotide (Clontech). CDS III/3′ PCR primer [5′-ATTCTAGAGGCCGAGGCGGCCGACATG-d(T)$_{30}$N–1N–3′] primes the first-strand synthesis reaction, and the SMART III oligonucleotide with oligo(G) at its 3′ end (5′-AAGCAGTGG TATCAACG CAGAGTGGCCATT ACGGCCGGG-3′) serves as a short, extended template at the 5′ end of the mRNA because reverse transcriptase's terminal transferase activity adds a few additional nucleotides, primarily deoxycytidine, to the 3′ ends of the cDNA. SMART III oligonucleotide and CDS III primer contains *Sfi*IA and *Sfi*IB restriction sites, respectively.
2. Use long-distance (LD) PCR to synthesize double-strand cDNA. Add high-quality DNA polymerase, such as Adervantage 2 polymerase (Clontech), an additional 2 μL CDS III primer to the reaction, 2 μL of first-strand cDNA from **step 1**, and 2 μL 5′ PCR primer (5′-AAGCAGTGGTATCAACGCAGT-3′). The optimal number of thermal cycles depends on the amount of RNA starting material. Usually, perform 18–25 thermal cycles for amplifying the cDNA at 95°C for 5 s and 68°C for 8 min using the 9600 Thermal reactor (Perkin-Elmer). If the starting material RNA for cDNA synthesis is more than 1 μg, 11 μL of first-strand cDNA and only a few cycles were used for synthesizing the ds cDNA. Analyze a 5-μL sample of the PCR product on a 1.1% agarose–EtBr gel. The double-strand (ds) cDNA should appear as a 0.1–4-kb smear on the gel, with some bright bands corresponding to the abundant mRNAs for that tissue or cell source (*see* **Fig. 7** and **Note 3**).

3.4.2. Digestion

The 2- to 3-μg amplified ds cDNA is digested with proteinase K at 45°C for 20 min to inactivate the DNA polymerase activity and then extracted with phenol–chloroform (1 : 1) and precipitated with 3 *M* sodium acetate and 100% ethanol. Resuspend the cDNA and digest with 10 μL *Sfi*I at 50°C for 2 h.

3.4.3. Size Fractionation

A CHROMA SPIN-400 column may be used for cDNA size fractionation according to the manufacturer's protocol. Check the fraction on a 1% agarose–EtBr gel. Pool the desired size fraction and precipitate the ds cDNA

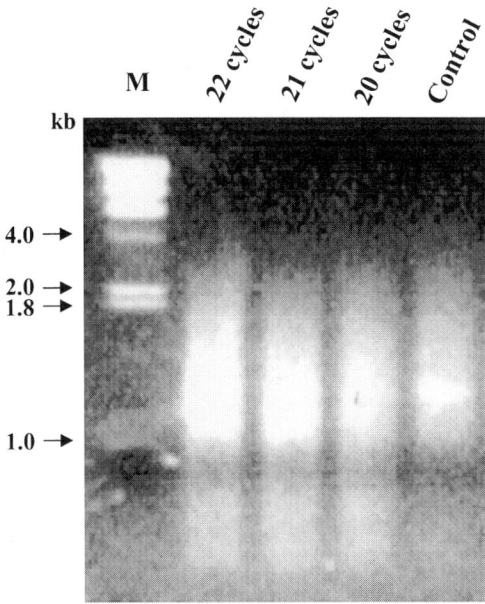

Fig. 7. Perform 20–22 thermal cycles for amplifying 1 µg total RNA derived from human CD34+ cells, in parallel with the 1-µg control human placenta poly(A)⁺ RNA provided by Clontech. Analyze a 5-µL sample of the PCR product on a 1.1% agarose–EtBr gel. The control and experimental double-strand cDNAs appear as a 0.2- to 4-kb smear on the gel, with similar bright bands and sizes. M: molecular marker.

with 3 *M* sodium acetate, 20 mg/mL glycogen, and 100% ethanol. Measure the amount of the cDNA (*see* **Subheading 3.3.3.**).

3.4.4. Ligation, Packaging, and Titering

Ligate cDNA to λTriplEx2 vector (or the Uni-ZAP XR vector) at an optimal ratio of cDNA to vector by T4 DNA ligase. Optimizing the ratio of cDNA to vector is a critical step in determining transformation efficiency and the titer of cDNA libraries. Packaging and titering methods are similar to that of the conventional cDNA library construction (*see* **Subheadings 3.3.5.–3.3.6.**).

3.4.5. Converting the Phage Library to Plasmid Library by In Vivo Excision

The conversion of a λTriplEx2 clone to a pTriplEx2 plasmid involves in vivo excision. pTriplEx2 is released as a result of Cre recombinase-mediated site-specific recombination at the loxP sites flanking the embedded plasmid (*see* **Fig. 4**). In this system, *E. coli* BM25.8 provides the necessary Cre recombinase activity. Therefore, release of the plasmid occurs automatically when the recombinant phage is transduced into BM25.8.

1. Incubate cultures of BM25.8 cells in LB/MgSO$_4$ broth at 31°C overnight, shaking until the OD$_{600}$ of 1.1–1.4 is attained. Add MgCl$_2$ (10 mM) to the cultures of BM25.8 host cells.
2. Combine a portion of the phage λTriplEx2 library in dilution buffer with BM25.8 host cell culture and then incubate at 31°C for 30 min. Continue to incubate with shaking after adding LB broth.
3. Spread the infected cell suspension on LB/carbenicillin plates with IPTG and X-Gal and grow at 31°C.

3.5. Library Amplification

1. Pick a colony from the primary working plate of the XL1-Blue strain and inoculate it with LB/10 mM MgSO$_4$ broth at 37°C overnight, shaking until the OD$_{600}$ of the culture reaches 2.0. Centrifuge the cells for 5 min, pour off the supernatant, and resuspend the pellet with 10 mM MgSO$_4$ at an OD$_{600}$ of 0.5 for each 150-mm plate.
2. Plan the number of LB/MgSO$_4$ agar plates you will need. Combine 500 µL of overnight bacterial culture from **step 1** and the packaged mixture or library suspension that can yield 1×10^5 plaques per 150-mm plate. Incubate at 37°C for 15 min.
3. Add 4.5 mL of melted LB/MgSO$_4$ soft top agar to each tube. Quickly mix and pour the mixture onto LB/MgSO$_4$ agar plates. Swirl the plate quickly and spread evenly in the plate. Cool the plates at room temperature for 10 min. Invert the plates and incubate at 37°C for 6–18 h or until the plaques become confluent.
4. Add 12 mL of SM buffer to each plate. Store plates at 4°C overnight. This allows the phage to diffuse into the SM buffer. On a platform shaker (approx 50 rpm), incubate the plates at room temperature for 1 h. This is the amplified-library lysate.
5. Mix the phage lysate well and pour it into a sterile 50-mL polypropylene tube. Add 10 mL of chloroform to the lysate. Vortex for 2 min. Centrifuge at 5000g for 10 min. Collect the supernatant into another sterile 50-mL tube. Determine the titer of the amplified library (*see* **Subheading 3.3.6.** and **Note 5**). The amplified library can be stored at 4°C for up to 6 mo. For long-term storage (up to 1 yr), make 1-mL aliquots, add DMSO to a final concentration of 7%, and place at –70°C. Avoid repeated freeze–thaw cycles.

4. Notes

1. The quality of total RNA and mRNA is crucial for construction of a high-quality cDNA library. In other words, the sequence complexity of the double-strand cDNA synthesized and the cDNA library constructed depends on the quality of the experimental RNA starting material. Avoiding RNA and cDNA degeneration is the most important for generating cDNA libraries. Wear gloves throughout the procedure to protect your RNA and cDNA samples from degradation by nucleases. Use only disposable plastic pipets and pipet tips with RNA. Checking the quality of total RNA and poly(A)$^+$ RNA, which should appear as a smear from 0.5 to 12 kb, is necessary for constructing cDNA libraries. If the ratio of

intensity of 28S RNA to 18S RNA (for total RNA) is less than 1:1 or if your experimental poly(A)$^+$ RNA appears significantly smaller than expected (e.g., no larger than 1–5 kb), we suggest that you prepare fresh RNA after checking your RNA purification reagents for RNase and other impurities. For generating a conventional cDNA library, the starting amount of mRNA could be from 1.5 μg to 6 μg. The larger starting amount of total RNA and mRNA will be valuable for profiling gene expression of a given tissue or cells. Less reverse transcriptase should be used for small amounts of mRNA. For PCR-based cDNA library construction, the strategy also needs the larger starting amoung of total RNA or mRNA and as few numbers of PCR thermal cycles as possible, in order not to lose the representativity of cDNA libraries. In general, the more RNA you start with, the fewer PCR cycles will be required for the second-strand synthesis. Fewer thermal cycles are less likely to generate nonspecific PCR products and, therefore, are best for optimal cDNA library construction. Usually, the minimum starting amount of the strategy is 50 ng of total RNA or 25 ng of poly(A)$^+$ RNA.

2. The cDNA fragments were inserted into the upstream of β-galactosidase under the *lacZ* promoter in the Uni-ZAP system. Probably, some of the genes in the frame are fused to the β-galactosidase. This will allow the antibody screening on the phage library or the excised phagemid library, in addition to DNA hybridizing screening. The phagemid from the Uni-ZAP system is pBluscript, which has the bacteriophage fl origin of replication, allowing rescue of single-stranded DNA and site-directed mutagenesis. Transcripts made from the T3 and T7 promoters could be useful in *in situ* hybridization as probes. The *lacZ* promoter may be used to drive expression of fusion protein for Western blotting or protein purification under inducing by IPTG. However, even if every cDNA molecule inserted into the MCS of λTriplEx2 are expressed in all three reading frames, the PCR-based cDNA libraries may not be suitable for immunoscreening because 5′ untranslated regions (UTRs) of the genes enriched in the cDNA libraries using the oligo-capping strategy are valuable for relatively large-scale, efficient, full-length cDNA cloning from small amount of RNA *(12,14)*. To generate cDNA plasmid library directly, the vectors, such as pBluescript, should be digested with restriction enzymes overnight and then treated with CIP (New England Biolab, Beverly, MA) to remove 5′-phosphate residues. Purify the large vector fragment using the Quiquick Gel Extraction kit (Qiagen). Follow the protocol for the construction of the conventional cDNA library (*see* **Subheading 3.3.**). Transform the plasmid into XL1-Blue stain instead of packaging and in vivo excision with electroporation.

3. The optimal temperature for the first-stand synthesis is 37°C; a higher reaction temperature may be used if the secondary structure makes reverse transcription difficult. If possible, we suggest that the beginners may use [α-^{32}P]-dATP as a tracer in the conventional cDNA library construction to control the quality of each procedure for synthesized cDNA, although a lower amount of cDNA may be observed using radioactive labels. In PCR-based cDNA library construction,

we strongly recommend that a positive control cDNA synthesis be performed in parallel with the experimental cDNA synthesis. This is especially important because it may not be possible to visualize your first-strand reaction product on a gel. Furthermore, the control reaction through the PCR step will allow us to evaluate the yield and size distribution of the ds cDNA synthesized from your RNA sample (*see* **Fig. 7**). If the size distribution of the experimental cDNA synthesis is smaller than that of the control, all of the components involved in the experiment might not be working properly.

4. It is very important to choose the fraction containing the larger-size cDNA because the smaller-size cDNA fraction may be unligated adapters, primer–dimer, and small-size cDNA inserts. The majority of clones in the cDNA library, if the small size fraction do not put away, will contain small-size cDNA inserts for the small-size cDNA fragments are easier to insert to the vectors. Therefore, the cDNA library generated will lead to bias for profiling gene expression of the given tissues and cells and to difficult in isolating full-length cDNA. The cDNA should be collected as large and small sized potions and then ligated to the vectors respectively. Somtimes, the fraction of the large-size cDNA could not be detected on the agarose gel because of the small amount. In this case, just collect the fractions, without the small-size DNA, and check the cDNA amount after recovering the cDNA. If the amount of DNA is enough for construction of the library, continue the procedure.

5. Both the conventional cDNA libraries and the PCR-based cDNA libraries may be amplified. The amplified cDNA libraries could be used for molecular biology. However, the amplified library could not be parallel with original cDNA library—in particular gene species could be less than original one. More importantly, the amplified library could lead to significant bias, in which some genes would be redundant while others would be lost. Therefore, the amplified library, which may be used for screening and isolating an individual gene, would not be suitable for gene expression profile analysis.

Acknowledgments

This work was supported by the Chinese National Key Program on Basic Research (973), the Chinese High Tech R&D Program (863), the National Natural Science Foundation of China, the National Foundation for Excellence Doctoral Project, and the Shanghai Commission for Science and Technology.

References

1. Adams, M. D., Kelley, J. M., Gocayne, J. D., Dubnick, M., Polymeropoulos, M. H., Xiao, H., et al. (1991) Complementary DNA sequencing: "expressed sequence tags" and the human genome project. *Science* **252,** 1651–1656.
2. Adams, M. D., Kerlavage, A. R., Fleischmann, R. D., Fuldner, R. A., Bult, C. J., Lee, N. H., et al. (1995) Initial assessment of human gene diversity and expression

patterns based upon 83 million nucleotides of cDNA sequence. *Nature* **377(6547 Suppl.)**, 3–174.

3. Xu, X. R., Huang, J., Xu, Z. G., Qian, B. Z., Zhu, Z. D., Yan, Q., et al. (2001) Insight into hepatocellular carcinogenesis at transcriptome level by comparing gene expression profiles of hepatocellular carcinoma with those of corresponding noncancerous liver. *Proc. Natl. Acad. Sci. USA* **98,** 15,089–15,094.

5. Hu, R. M., Han, Z. G., Song, H. D., Peng, Y. D., Huang, Q. H., Ren, S. X., et al. (2000) Gene expression profiling in the human hypothalamus–pituitary–adrenal axis and full-length cDNA cloning. *Proc. Natl. Acad. Sci. USA* **97,** 9543–9548.

6. Yu, Y., Zhang, C., Zhou, G., Wu, S., Qu, X., Wei, H., et al. (2001) Gene expression profiling in human fetal liver and identification of tissue- and developmental-stage-specific genes through compiled expression profiles and efficient cloning of full-length cDNAs. *Genome Res.* **11,** 1392–1403.

7. Kato, S., Sekinie, S., Oh, S., Kim, N. S., Umezawa, Y., Abe, N., et al. (1994) Construction of a human full-length cDNA bank. *Gene* **150,** 243–250.

8. Maruyama, K. and Sugano, S. (1994) Oligo-capping: a simple method to replace the cap structure of eucaryotic mRNAs with oligoribonucleotides. *Gene* **138,** 171–174.

9. Suzuki, Y., Yoshitomo, K., Maruyama, K., Suyama, A., and Sugano, S. (1997) Construction and characterization of a full length-enriched and a 5′-end-enriched cDNA library. *Gene* **200,** 149–156.

10. Chenchik, A., Diachenko, L., Moqadam, F., Tarabykin, V., Lukyanov, S., and Siebert, P. D. (1996) Full-length cDNA cloning and determination of mRNA 5′ and 3′ ends by amplification of adaptor-ligated cDNA. *BioTechniques* **21,** 526–534.

11. Zhu, Y. Y., Machleder, E. M., Chenchik, A., Li, R., and Siebert, P. D. (2001) Reverse transcriptase template switching: a SMART approach for full-length cDNA library construction. *BioTechniques* **30,** 892–897.

12. Mao, M., Fu, G., Wu, J. S., Zhang, Q. H., Zhou, J., Kan, L. X., et al. (1998) Identification of genes expressed in human CD34(+) hematopoietic stem/progenitor cells by expressed sequence tages and efficient full-length cDNA cloning. *Proc. Natl. Acad. Sci. USA* **95,** 8175–8180.

13. Gu, J., Zhang, Q. H., Huang, Q. H., Ren, S. X., Wu, X. Y., Ye, M., et al. (2000) Gene expression in CD34+ cells from normal bone marrow and leukemic origins. *Hematology J.* **1,** 206–217.

14. Zhang, Q. H., Ye, M., Wu, X. Y., Ren, S. X., Zhao, C. J., et al. (2000) Cloning and functional analysis of cDNAs with open reading frames for 300 previously undefined genes expressed in CD34+ hematopoietic stem/progenitor cells. *Genome Res.* **10,** 1546–1560.

15. Chomczynski, P. and Sacchi, N. (1987) Single-step method of RNA isolation by acid guanidinium thiocyanate–phenol–chloroform extraction. *Anal. Biochem.* **162,** 156–159.

16. Chomczynski, P. (1993) A reagent for the single-step simultaneous isolation of RNA, DNA and proteins from cell and tissue samples. *BioTechniques* **15,** 532–537.

18

Screening Poly [dA/dT(–)] cDNA for Gene Identification

San Ming Wang, Scott C. Fears, Lin Zhang, Jian-Jun Chen, and Janet D. Rowley

1. Introduction

The goal for developing the SPGI (screening poly [da/dT(–)] cDNA for gene identification technique was to generate an efficient tool for maximal identification of the expressed genes from eukaryotic genomes. The impetus for developing the SPGI method was triggered by the observation that a large number of human novel transcripts have not been identified in spite of intensive efforts in the past decades using the expressed sequence tag (EST) approach *(1–9)*. Based on our analysis, we believe that the use of complementary DNA (cDNA) libraries generated by regular oligo-(dT) primers contributes significantly to this problem *(10,11)*. The cDNAs generated by oligo-(dT) priming contain various lengths of poly(dA/dT) tail sequences at the 3′ end. Most of the cDNA libraries are processed through normalization/subtraction before being used for sequencing analysis *(7)*. During the process of normalization/ subtraction, poly(dA)–poly(dT) hybrids will form randomly between unrelated cDNA templates. The removal of these hybrids causes the loss of cDNA templates. This phenomenon affects especially the low-copy cDNA templates, which represent most of the genes. To overcome this problem, we developed the SPGI method. In this method, a set of anchored oligo-(dT) primers is used for mRNA priming to prevent the inclusion of the poly(dA/dT) sequences from the 3′ ends of cDNAs. Using the poly [dA/dT(–)] cDNA for subtraction/ normalization prevents the formation of poly(dA)–poly(dT) hybrids, therefore preventing the nonspecific loss of cDNAs because of the poly(dA)–poly(dT) hybrids. Screening the subtracted/normalized poly [dA/dT(–)] cDNA libraries should provide high rate of novel gene identification.

From: *Methods in Molecular Biology, vol. 221: Generation of cDNA Libraries: Methods and Protocols*
Edited by: S.-Y. Ying © Humana Press Inc., Totowa, NJ

2. Materials

1. TRIzol reagent (Invitrogen, Grand Island, NY).
2. Dynabeads Oligo-(dT)$_{25}$, Dynabeads M$_{280}$ streptavidin, and dynal MPC (Dynal, Lake Success, NY).
3. 2X Binding buffer: 20 mM Tris-HCl (pH 7.5), 1 M LiCl, and 2 mM EDTA.
4. Washing buffer: 10 mM Tris-HCl (pH 8.0), 150 mM LiCl, and 1 mM EDTA.
5. Elution buffer: 10 mM Tris-HCl.
6. Reverse transcription (RT) primers: 5′ biotin-ATCTAGAGCGGCCGC-T16-**R**, 5′ biotin-ATCTAG AGCGGCCGC-T16-C-**V** (**R**=A/G, **V**=A/G/C) (IDT, Integrated DNA Technologies, Coralville, IA).
7. cDNA synthesis kit (Invitrogen, cat. no. 18267-021).
8. DNase-free RNase A (Qiagen, Santa Clarita, CA, cat. no. 19101).
9. pGEM5Zf(+) vector (Promega, WI).
10. T4 ligase (Promega).
11. JM109 competent cell (Promega).
12. Qiagen Maxi plasmid preparation kit (Qiagen).
13. Gene II and buffer (Invitrogen).
14. Exon III (Invitrogen).
15. *Pvu*II (Promega).
16. Hydroxyapatite (Sigma, St. Louis, MO).
17. Sephadex G50 minicolumn (AmershaPharmacia).
18. Glycogen (Roche Diagnostics; Indianapolis, IN).
19. SP6 primer (IDT).
20. Sequenase (AmershaPharmacia).
21. DH5α (Invitrogen).
22. Qiagen REAL system for plasmid preparation (Qiagen).
23. GeneClean beads (Bio101, CA).

3. Methods

3.1. Isolate Total RNA

1. Isolate total RNA with TRIzol solution following the manufacturer's protocol and determine the yield of RNA through measuring at OD$_{260}$ (OD = optical density).

3.2. Isolate mRNA

1. Transfer 1 mL Dynal dT$_{25}$ beads (5 mg) to an Eppendorf tube and place on Magnetic Particle Concentrator (MPC). Remove supernatant.
2. Resuspend beads in 1000 µL of 2X binding buffer; place in MPC.
3. Remove supernatant; resuspend the beads in 200 µL of 2X binding buffer.
4. Add 200 µg/200 µL total RNA to the tube, mix well, and place the tube in MPC for 1 min.
5. Remove the supernatant and wash three times with 1000 µL washing buffer.
6. Resuspend beads with 20 µL elution buffer.

7. Keep the tube at 65°C for 2 min and recover the supernatant in MPC.
8. Resuspend the beads in 10 μL elution buffer and keep at 65°C for 2 min.
9. Recover the supernatant again.
10. Combine all mRNA elutes, use 1 μL for gel check and OD_{260} quantification, and use immediately for cDNA synthesis or store at –70°C for later use.

3.3. Synthesis of poly[dA/dT(–)] Double-Strand cDNA

3.3.1. Prepare the Primer Mixture (see **Note 1**).

1. Set the A/G anchored primer to 2 μg/μL.
2. Set the C-A/G/C anchored primer to 1 μg/μL.
3. Mix A/G and C-A/G/C anchored primers at a ratio of 1:1. The final concentration is 0.5 μg/μL for A and G anchored primers and 0.5 μg/μL (0.167 μg/μL for each) for CA/CG/CC anchored primers.

3.3.2. Synthesis of First-Strand cDNA (see **Note 2**)

1. Prepare the reaction mixture:

Items	1X	2X	3X	4X	5X
5X First-strand buffer	10	20	30	40	50
10 mM dNTP mix	2.5	5	7.5	10	12.5
RT oligo	2	4	6	8	10
RNasin	1	2	3	4	5
Dithiothreitol (DTT) (100 mM)	5	10	15	20	25
MMLV RT	2.5	5	7.5	10	12.5
mRNA(1 μg/μL)	5	10	15	20	25
H_2O^*	23	46	69	92	105

*Depending on the input of mRNA, the final volume can be adjusted with H_2O.

2. Keep the reaction at 37°C for 30 min, 55°C for 2 min, and back to 37°C again.
3. Add 2 μL RT, keep the tube at 37°C for 30 min.
4. Repeat **steps 2** and **3** twice.
5. Remove 5 μL from the reaction to a tube and add 1 μL DNase-free RNase A.
6. Aliquot the rest to several tubes, 50 μL/tube.

3.3.3. Synthesis of Second-Strand cDNA

1. Prepare the reaction mixture:

Items	1X	2X	3X	4X	5X
(DEPC) H_2O	290	580	870	1160	1450
dNTP mix (10 mM)	7.5	15	22.5	30	37.5
10X second-strand buffer	40	80	120	160	200
Escherichia coli DNA pol. I	10	20	30	40	50
E. coli DNA ligase	1.25	2.5	3.75	5	6.25

2. Mix 350 μL to each first-strand cDNA tube and mix well.
3. Keep the reaction at 16°C for 2 h.
4. Check 20 μL on an agarose gel together with the first strand cDNA.
5. Extract the reaction with 400 μL phenol–chloroform and recover the upper phase.
7. Aliquot the upper phase to new tubes, 300 μL/tube.
8. To each tube, add 3 μL glycogen, 150 μL of 7.5 *M* NH4oAC, and 600 μL cold ethanol; mix well
9. Spin for 15 min at maximal speed and at room temperature.
10. Combine all pellets into one tube and wash with 500 μL of 70% ethanol.
11. Dry and dissolve the pellets in 20–50 μL Tris-EDTA (TE).
12. Make double dilutions from 1:0 to 1:10 with 2 μL cDNA. Estimate the concentration by EBr-dot quantification with dilutions of DNA with known concentration.
13. Adjust the concentration to 200 ng/μL.

3.4. Purify 3′ cDNAs Using NlaIII *Digestion (see* Note 3)

3.4.1. Prepare the Digestion

Double-strand cDNA	75 (20 μg)
Buffer 4 (NEB)	10
10X bovine serum albumin (BSA) (NEB)	10
*Nla*III	5

Keep the digestion at 37°C for 1–2 h; check 1 μL on gel. The digested cDNA should be centered between 200 and 500 bps.

3.4.2. Recovery of 3′ cDNA (see **Note 4***)*

1. Aliquot 200 μL Dynal M280 beads into a tube for every 4 μg cDNA.
2. Wash beads with 500 μL binding/washing buffer; resuspend beads in 200 μL binding/washing (B/W) buffer.
3. Add 20 μL cDNA and 180 μL H_2O to the beads.
4. Rotate the tube at room temp for 20–60 min.
5. Isolate the beads in MRC; save the supernatant.
6. Wash the beads three times—twice with 500 μL of 1X B/W and once with 1000 μL TE.
7. Resuspend the beads in 50 μL TE.
8. Add 50 μL phenol and vortex for 5 min at maximal speed.
9. Keep the tube at 65°C for 30 min, vortex every 5 min.
10. Vortex at full speed for 10 min, spin, and recover the upper phase.
11. Add equal volume of phenol–chloroform, spin, and recover the upper phase.
12. Add 1/2 vol of 7.5 *M* NH4oAC, 2 vol of 100% ethanol, and 3 μL glycogen.
13. Spin for 15 min at room temperature, wash pellets with 70% ethanol, air-dry, and dissolve the pellets in 20 μL TE.

14. *Not*I digestion:

NEB buffer 3 +BSA	3
DNA	20
*Not*I (10 μ/μL)	3

    Keep at 37°C overnight

15. Extract with phenol–chloroform, precipitate, wash as before, and dissolve cDNA in 22 μL H$_2$O. Check 1 μL on a 1.5% agarose gel.

3.5. Prepare Vector for 3′ cDNA Cloning

1. Prepare the digestion:

pGEM5zf(+)	10 μg/10
NEB buffer 3 + BSA	4
*Not*I (10 μ/μL)	2
*Sph*I (10 μ/μL)	2
H$_2$O	22

    Digest at 37°C for more than 2 h.

2. Add 1 μL Calf Intestinal Alkaline Phosphatase (CIP) to the reaction and keep at 37°C for 1 h.

3. Purify the digested DNA with GeneClean beads following the manufacturer's instruction, resuspend the purified DNA to 200 ng/μL.

3.6. Clone 3′ cDNAs into Vector

1. Mix the following:

Linearized vector	1
3′ cDNA	5
Buffer	1
Ligase	1

    Keep the reaction at 16°C overnight.

2. Transformation: Add 5 μL ligation mixture to 50 μL JM109, keep on ice for 30 min, heat shock at 42°C for 2 min, cool on ice for 2 min, add 800 μL SOC, and shake at 250 rpm at 37°C for 30 min; alternatively, one can use electroporation for transformation. Plate 100 μL of transformants in a plate containing IPTG/Xgal/Amp and incubate overnight. Pull the remaining transformants into 400 mL LB (50 μg/mL Amp) and shake at 250 rpm at 37°C overnight.

3. Check the quality of clones: Prepare regular PCR mixtures with T7/SP6 primers, pick clones directly from plates to PCR mixture, perform regular PCR with 30 cycles, and check the PCR products on agarose gels. If necessary, purify the PCR products with a S-300 column (AmershamPharmacia) and sequence the PCR products using PE BigDye kit with SP6 primer.

4. Perform large-scale preparation of plasmids using Qiagen Maxi plasmid preparation kit following manufacture's protocol; adjust plasmid concentration to 1 μg/μL.

3.7. Prepare the Driver for Normalization

1. Release the inserts:
   Plasmid   30 µg/30
   *Not*I   3
   *Apa*II   3
   H₂O   18
   NEB buffer 3   6
   Keep the digestion overnight.
2. Run the digestion on a 1% agarose gel to separate inserts from vector.
3. Cut the inserts from the gel.
4. Purify the inserts using Gene Clean kit following the manufacturer's protocol.
5. Resuspend the purified inserts in TE.

3.8. Prepare the Tester for Normalization

3.8.1. Gene II Digestion

Plasmid   50
Gene II buffer   8
H₂O   24
Gene II   8
Keep the reaction at 30°C for 1 h, 65°C for 10 min, and then move back to 37°C.

3.8.2. Exon III Digestion

1. Add 8 µL Exon III to the sample, keep the reaction at 37°C for 60 min.
2. Extract the sample twice with phenol–chloroform.
3. Precipitate the DNA with 40 µL of 7.5 *M* NH4oAC and 200 ethanol; keep the tube on dry ice for 15 min.
4. Spin, wash with 70% ethanol, dry, and resuspend the pellets in 40 µL H₂O.

3.8.3. Pvu*II* Digestion to Linearize the Double-Strand DNA Escaping the Digestion

Single-strand DNA   40
NEB buffer 2   5
*Pvu*II   5
Keep the reaction at 37°C for 2 h.

3.8.4. Remove Double-Strand DNA

1. Aliquot 500 µL hydroxyapatite (HAP) (pH 6.8) in a tube and place at 65°C, add the *Pvu*II-digested DNA directly to the tube and vortex briefly, spin for 10 s, and recover 400 µL of supernatant containing the single-strand DNA.
2. Desalt the single-strand DNA. Add the supernatant to a Sephadex G50 minicolumn (20 µL/column), spin, and collect the elutes together (about 17 µL each, total about 370 µL).

3. Precipitate the purified single-strand DNA:

Single-strand DNA elutes   350
7.5 *M* NH4oAC   150
Ethanol   875
Glycogen   2

Keep on dry ice for 20 min, spin, wash pellet as before, resuspend DNA in 20 μL H₂O, use 1 μL for gel check and 2 μL for OD260 quantification, and adjust the final concentration to 100 ng/μL.

3.9. Perform Normalization

1. Prepare the following mixture:

Single-strand DNA (100 ng/μL)   1
Insert DNA (100 ng/μL)   4
4X hyb buffer (CloneTech)   2
H₂O   1

Heat the tube at 98°C for 3 min; then move and keep at 68°C overnight.

2. HAP absorption to remove the hybrids (*see* **Note 5**). Aliquot 50 μL HAP in an eppendorf tube and keep the tube at 60°C, resuspend the mixture to HAP, spin for 10 s, and recover the supernatant. Pass the supernatant through Sephadex G50 minicolumn twice to desalt, 20 μL/column. Pull all of the elutions together and precipitate with 7.5 *M* NH4oAC, as previously. Resuspend the purified DNA in 22 μL H₂O; use 1 μL for gel check.

3.10. Conversion of Single-Strand DNA to Double-Strand Plasmids (see Note 6)

1. Prepare the mixture:

Single-strand DNA   11
5X Sequenase buffer   4
SP6 primer (1 μg/μL)   1

Keep at 65°C for 5 min; move to 30°C for 5 min.

2. Add the following to the reaction:

DTT (100 m*M*)   1
dNTP (10 m*M*)   2
Sequenase   1

Keep the reaction at 30°C for 2 min and then switch to 37°C for 30 min.

3. Extract the sample with 100 μL phenol–chloroform and 50 μL H₂O. Vortex. spin, and collect 100 μL supernatant. Precipitate with glycogen as previously. Dissolve DNA in 10 μL H₂O.

4. Transformation of normalized plasmids (*see* **Note 7**).

   a. Mix 5 μL double-strand DNA with 100 μL DH5a.

   b. Transform cells with electroporation.

   c. Plate the transformants in IPTG/Xgal/Amp plate and incubate at 37°C overnight.

   d. Prepare clones with 96X Qiagen REAL system following the protocol.

3.11. Sequencing Plasmids

1. Sequence each clone with either T7 primer (5′ reading through) or SP6 primer (3′ reading through) using the Big-dye sequencing kit following the standard sequencing process.
2. Collect all the sequences in ABI377 or other auto sequencer.

3.12. Identify Genes

1. Match each sequence using BLAST to "Non Redundant" in GenBank. If matched, it is a known gene.
2. Match "human EST" for sequence without match in "Non Redundant." If matched, it is a known human EST.
3. For the sequences without match in both "Non Redundant" and EST, they are likely to be novel expressed sequences.
4. Match all of the sequences to Human Genome Sequences (http://genome.ucsc.edu/) for genomic confirmation.

4. Notes

1. The concentrations provide equal probability for each primer to match the expressed sequences theoretically. We analyzed the distribution of the last nucleotide in human reference sequences deposited in the RefSeq database and observed the similar rate of distribution for G, C, and T in the last position before the poly(A) sequence (C=2371, G=2212, T=2099). However, the number of A residues at the last position is 7353. This number can not be reliable because the uncertainty of the last A re-tailed in the original submitted sequences.
2. Repeating the heating/cooling process increases significantly the yield of cDNA by increasing the correct annealing of the anchored oligo-(dT) primers to the right position in mRNA templates. The efficiency of cDNA synthesis using this procedure is about $1:1.7$ (mRNA : double-strand cDNA) as determined by OD_{260}.
3. We used 3′ cDNA after the last CATG (*Nla*III restriction site) for the construction of the cDNA library, in order to combine the analysis with the SAGE system *(12,13)*. However, one can generate the cDNA library containing the full-length cDNAs following the procedures *(7)*.
4. This treatment directly effects the rate of 3′ cDNA recovery. The steps of repeated heating/vortexing are absolutely necessary to disrupt the biotin–streptoavidin complex to release 3′ cDNA.
5. The HAP absorption is very efficient. After mixing the sample with HAP, it needs to be spun quickly to recover the supernatant. Delay of spinning will result in a decrease of the recovery of single-strand DNA because of nonspecific binding.
6. In addition to the use of Sequenase, there are several methods for conversion of single-strand DNA into double-strand DNA, such as PCR-based methods. A recent method developed by Strategies is worth trying. In this method, the single-strand DNAs are directly transformed into competent cell (XL10-Gold

ultracompetent cells, QuickChange site-directed mutagenesis kits). This will simplify substantially the procedures if it works.

7. It is better not to amplify the normalized library in order to maintain the normalized clonal distribution. If the numbers of clones are not enough for large-scale sequencing, the amplification should be kept minimal.

Acknowledgments

This work was supported by NCI grants CA42557 and CA78862-01, the American Cancer Society (IRG-41-40), and the G. Harold and Lelia Y. Mathers Foundation.

References

1. Adams, M. D., Dubnick, M., Kerlavage, A. R., Moreno, R., Kelley, J. M., Utterback, T. R., et al. (1992) Sequence identification of 2,375 human brain genes. *Nature* **355,** 632–634.
2. Boguski, M. S. (1995) The turning point in genome research. *Trends Biochem. Sci.* **20,** 295–296.
3. Fields, C., Adams, M. D., White, O., and Venter, J. C. (1994) How many genes in the human genome? *Nat. Genet.* **7,** 345–346.
4. Gerhold, D. and Caskey, C. T. (1996) It's the genes! EST access to human genome content. *BioEssay* **18,** 973–981.
5. Mao, M., Fu, G., Wu, J. S., et al. (1998) Identification of genes expressed in human CD34(+) hematopoietic stem/progenitor cells by expressed sequence tags and efficient full-length cDNA cloning. *Proc. Natl. Acad. Sci. USA* **95,** 8175–8180.
6. Strausberg, R. L., Dahl, C. A., and Klausner, R. D. (1997) New opportunities for uncovering the molecular basis of cancer. *Nat. Genet.* **15,** 415–416.
7. Bonaldo, M. F., Lennon, G., and Soares, M. B. (1996) Normalization and subtraction: two approaches to facilitate gene discovery. *Genome Res.* **6,** 791–806.
8. Wang, S. M. and Rowley, J. D. (1998) A strategy for genome-wide gene analysis: integrated procedure for gene identification. *Proc. Natl. Acad. Sci. USA* **95,** 11,909–11,914.
9. Velculescu, V. E., Madden, S. L., Zhang, L., et al. (1999) Analysis of human transcriptomes. *Nat. Genet.* **23,** 387–388.
10. Wang, S. M., Fears, S. C., Zhang, L., Chen, J.-J., and Rowley, J. D. (2000) Screening poly(dA/dT)– cDNAs for gene identification. *Proc. Natl. Acad. Sci. USA* **97,** 4162–4167.
11. Martin, K. J. and Pardee, A. B. (2000) Identifying expressed genes. *Proc. Natl. Acad. Sci. USA* **97,** 3789–3791.
12. Velculescu, V. E., Zhang, L., Vogelstein, B., and Kinzler, K. W. (1995) Serial analysis of gene expression. *Science* **270,** 484–487.
13. Zhou, G., Chen, J., Lee, S., Clark, T., Rowley, J. D., and Wang, S. M. (2001) The pattern of gene expression in human CD34+ hematopoietic stem/progenitor cells. *Proc. Natl. Acad. Sci. USA* **98,** 13,966–13,971.

19

Generation of Longer cDNA Fragments from SAGE Tags for Gene Identification

Jian-Jun Chen, Sanggyu Lee, Guolin Zhou, Janet D. Rowley, and San Ming Wang

1. Introduction

Serial analysis of gene expression (SAGE) is a powerful technique for genomewide analysis of gene expression *(1–14)*. However, almost two-thirds of SAGE tags cannot be used directly for gene identification for two reasons. First, many of SAGE tags match to multiple known expressed sequences because of the short length of SAGE tag sequences *(12–14)*. Second, many SAGE tags do not match any known expressed sequences because the sequences corresponding to these SAGE tags have not been identified yet *(2–14)*. These problems substantially diminish the power of the SAGE technique. The GLGI (Generation of Longer complementary DNA [cDNA] fragments from SAGE tags for Gene Identification) technique was designed to solve these two problems *(15)*. The basic principle of GLGI is to use the SAGE tag as the sense primer, and anchored oligo-(dT) primers as the antisense primer to amplify the original 3′ cDNA from which the SAGE tag was derived. The size of 3′ cDNA will be hundreds of bases longer, which is long enough for solving the two problems. In a typical SAGE project, hundreds or thousands of SAGE tags need to be further analyzed. To facilitate such large-scale performance, we developed the GLGI method into a high-throughput procedure *(16)*. In this high-throughput GLGI procedure, 3′ cDNAs starting from the last CATG are used as the templates for GLGI amplification, a SAGE tag sequence is used as the sense primer, and a universal sequence located at the 3′ end of all the cDNA templates generated from anchored oligo(dT) primers is used as antisense primer to amplify the original 3′ cDNA template from which the SAGE tag was

From: *Methods in Molecular Biology, vol. 221: Generation of cDNA Libraries: Methods and Protocols*
Edited by: S.-Y. Ying © Humana Press Inc., Totowa, NJ

derived. The whole process is simple, rapid, and low cost, with high efficiency *(12–14,16)*. In addition to its use in identifying the correct genes for SAGE tags with multiple matches, GLGI can be used for large-scale identification of novel genes by converting novel SAGE tags into novel 3′ cDNAs. This GLGI method provides a powerful tool for gene identification in the human and other eukaryotic genomes.

2. Materials

1. TRIzol reagent (Invitrogen, Grand Island, NY).
2. Dynabeads oligo-$(dT)_{25}$, Dynabeads M-280 streptavidin, and Dynal MPC (Dynal, Lake Success, NY).
3. 2X Binding buffer: 20 mM Tris-HCl (pH 7.5), 1M LiCl, and 2 mM EDTA.
4. Washing buffer: 10 mM Tris-HCl (pH 8.0), 150 mM LiCl, and 1 mM EDTA.
5. DNA oligonucleotide primers (IDT, Integrated DNA Technologies, Coralville, IA): Anchored oligo-(dT) reverse transcription (RT) primers polyacrylamide gel electrophoresis ([PAGE] purified):
   5′ Biotin–ATCTAGAGCGGCCGC-T16-**R** (**R**=A/G);
   5′ Biotin–ATCTAGAGCGGCCGC-T16-C-**V** (**V**=A/G/C);
   Linker A (PAGE gel purified):
    A1: 5′-TTTGGATTTGCTGGTGCAGTACAACTAGGCTTAATAGGGACA TG-3′;
    A2: 5′-pTCCCTATTAAGCCTAGTTGTACTGCACCAGCAAATCC[amino-modifiedC7]-3′;
   Linker B (PAGE gel purified):
    B1: 5′-TTTCTGCTCGAATTCAAGCTTCTAACGATGTACGGGGACAT G-3′;
    B2: 5′-pTCCCCGTACATCGTTAGAAGCTTGAATTCGAGCAG [amino-modified C7]-3′;
   SAGE primer-P1: 5′-GGATTTGCTGGTGCAGTACA-3′;
   SAGE primer-P2: 5′-CTGCTCGAATTCAAGCTTCT-3′;
   Antisense primer: 5′-ACTATCTAGAGCGGCCGCTT-3′;
   GLGI sense primers: 5′-GGATCCCATG XXXXXXXXXX-3′ (XXXXXXXXXX is the 10-base SAGE tag sequence);
   A positive control GLGI sense primer for human libraries: 5′-GGATTCCATGCT GTTGGTGA-3′;
   M13 Forward (–20): 5′-GTTGTAAAACGACGGCCAGT-3′;
   M13 Reverse: 5′-ACAGGAAACAGCTATGACCA-3′.
6. CDNA synthesis kit (Invitrogen; cat. no. 18267-021).
7. Recombinant ribonuclease inhibitor (rRNasin) (40 U/μL; Promega, Madison, WI; cat. no. N2511).
8. RNase ONE™ Ribonuclease (5–10 U/μL; Promega, cat. no. M4261) or DNase-free RNase A (100 mg/mL; 7 U/μL; Qiagen, Santa Clarita, CA; cat. no. 19101).
9. Glycogen (20 mg/mL) (Roche Diagnostics, Indianapolis, IN).

10. LoTE: 3 mM Tris-HCl (pH 7.5) and 0.2 mM EDTA (pH 7.5), in H$_2$O.
11. *Nla*III (New England Biolabs, Beverly, MA).
12. 2X Binding/washing buffer (2X B/W): 10.0 mM Tris-HCl (pH 7.5), 1.0 mM EDTA, and 2.0 M NaCl.
13. Platinum *Taq* DNA polymerase (Invitrogen).
14. MicroAmp caps, MicroAmp optical 96-well reaction plate, PE GeneAmp PCR Systems 9600 or 9700 (Applied Biosystems; Foster City, CA).
15. Square-well block for plasmid preparation (Qiagen, cat. no. 26173) and tape pads (Qiagen, cat. no. 19570).
16. Centrifuge for 96-well plate or block (SORVALL RC5C plus; rotor: SH3000; Kendro Laboratory Products LP, Newtown, CT).
17. pCR4-TOPO vector, TOPO10 chemically competent cells, and SOC medium (TOPO TA Cloning Kit for Sequencing; Invitrogen, cat. no. K4575-40).
18. LB medium, kanamycin, agar powder (Sigma; St. Louis, MO).
19. TaKaRa *Taq* DNA polymerase (TaKaRa Shuzo, Tokyo, Japan).
20. Agarose and electrophoresis equipment.
21. Big-Dye premixture, sequencing loading dye, ABI377 sequencer, and sequencing analysis V3.3 program (Applied Biosystems).

3. Methods

Figure 1 shows the schematic of the GLGI method. **Figure 2** shows the high-throughput procedure. Several important changes have been made in this GLGI procedure compared with the original GLGI protocol *(15,16)*. Our data showed that these changes significantly increase the efficiency and specificity for GLGI amplification *(12–16)*. The changes include the following:

1. Use of 3′ cDNAs digested by *Nla*III rather than full-length cDNAs as the templates for GLGI amplification. This will decrease the complexity of the templates and thus reduce nonspecific amplification.
2. Amplifying the 3′ cDNAs by polymerase chain reaction (PCR) to provide sufficient templates for GLGI amplification. We tested the 3′ cDNAs amplified twice from the original 3′ cDNAs and the results showed that these templates maintain high specificity for GLGI analysis.
3. Use of a single antisense primer (5′-ACTATCTAGAGCGGCCGCTT-3′) instead of the combination of three anchored oligo-(dT) primers (dT11A, G, C) for all GLGI reactions (*see* **Note 1**).
4. Use of platinum DNA polymerase rather than *Pfu* DNA polymerase for GLGI amplification. This DNA polymerase generates a higher yield of PCR products while keeping a high specificity of amplification. This feature further increases the efficiency of amplification for low-copy templates.
5. The GLGI amplified products are directly precipitated and cloned into vectors without gel purification to prevent loss of amplified products. This is especially important for small-size or low-copy GLGI products.

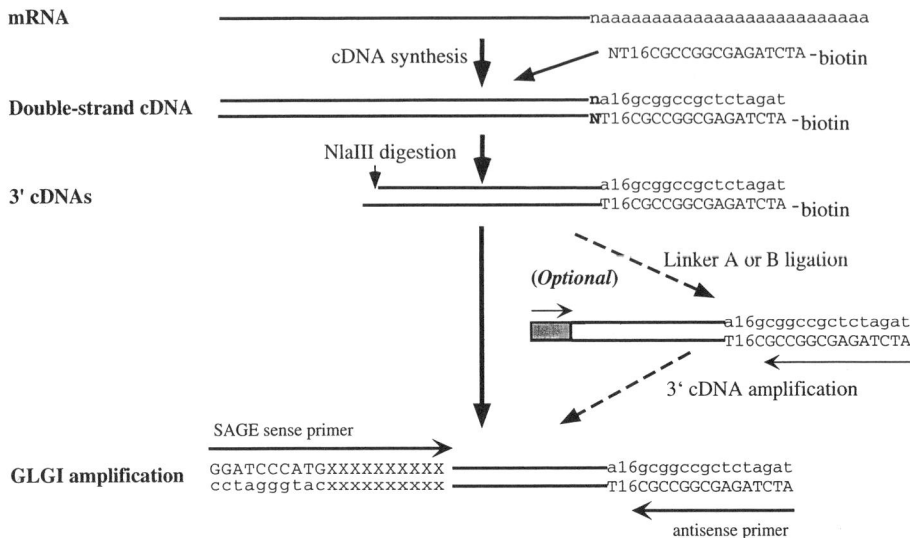

Fig. 1. Schematic of the GLGI procedure. Biotin-labeled double-strand poly(dA/dT)– cDNAs are synthesized and digested with *Nla*III. SAGE linker A or B is ligated to the 3′ cDNAs bound to the streptavidin beads, and the 3′ cDNAs can be amplified by PCR (optional). A sense primer based on each unique SAGE tag sequence (5′-GGATCCCATGXXXXXXXXXX-3′) and a universal antisense primer (5′-ACTATCTAGAGCGGCCGCTT-3′) are used for GLGI amplification.

3.1. cDNA Synthesis

The same RNA samples used for SAGE analyses should be used for GLGI analysis. The steps for cDNA synthesis and 3′ cDNA generation in the GLGI procedures are similar to the steps in our SAGE process *(17)*. The poly [dA/dT(–)] 3′ cDNAs generated in SAGE *(16–19)* (*see* **Note 2**) can be amplified to produce more 3′ cDNA templates (starting from **step 3** or **4** of **Subheading 3.2.**) or used directly for GLGI amplification (starting from **step 3** of **Subheading 3.2.**). The following are detailed descriptions.

3.1.1. Isolation of mRNA

1. Prepare total RNA from tissues or cells with TRIzol reagent (Life Technology) following the manufacturer's protocol and dissolve every 1–10 μg of the total RNA sample in 100 μL diethyl pyrocarbonate (DEPC)-treated H_2O. Check RNA quality on a mini RNase-free 1% agarose gel.
2. Transfer 200 μL of Dynabeads oligo-(dT)$_{25}$ (6.6×10^7 beads) to a RNase-free 1.5-mL Eppendorf tube and place on a magnet particle concentrator (MPC); then, remove the supernatant and resuspend beads in 200 μL of 2X binding buffer.

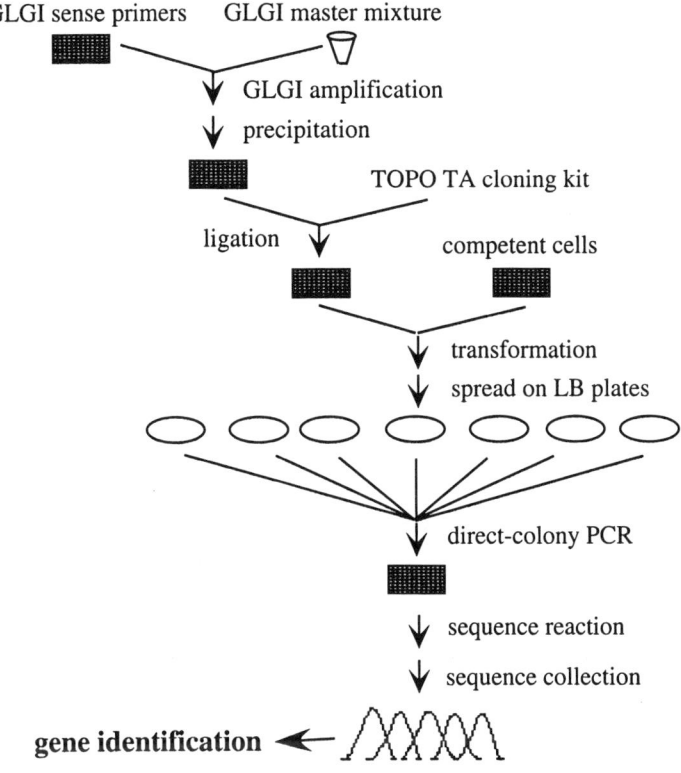

Fig. 2. Schematic of the high-throughput GLGI procedure. Ninety-six sense primers based on each SAGE tag are synthesized in a 96-well plate. The GLGI master mixture contain the universal antisense primer and 3′ cDNAs are prepared. The specific 3′ cDNA templates are amplified in a new 96-well plate. GLGI-amplified products are precipitated directly and cloned into the pCR4-TOPO vector. After transformation, clones from each GLGI product are screened by direct-colony PCR in a 96-well and sequencing in 96-lane format. The qualified sequences are matched to the databases for gene identification.

 Place in the MPC and remove the supernatant again; then, resuspend the beads in 100 µL of 2X binding buffer.
3. Heat 100 µL total RNA (1–10 µg) at 65°C for 2 min and move to ice to disrupt secondary structure. Then, mix the total RNA thoroughly with the prewashed 100-µL beads and rotate the tube on a roller for 5 min at room temperature.
4. Place the tube in the MPC for 1 min and remove the supernatant. Wash the beads two times with 300 µL washing buffer.
5. Resuspend beads with 20 µL of DEPC-treated H$_2$O and heat at 65°C for 2 min. Recover the supernatant in the MPC and transfer it to a new RNase-free 1.5-mL Eppendorf tube.

6. Resuspend beads with 10 µL of DEPC-treated H_2O again and keep at 65°C for 2 min; recover the supernatant in MPC.
7. Combine all mRNA elutes (total about 30 µL) and check 1–2 µL on a mini RNase-free agarose gel.

3.1.2. Synthesis of Poly[dA/dT(–)] Double-Strand cDNA

Two sets of anchored RT primers *(19)* (*see* **Note 2**) are synthesized by IDT for the generation of poly[dA/dT(–)] cDNAs: RTP-R (5′ biotin-ATCTAGAGC GGCCGC-T16-**R**) (**R**=A/G) and RTP-CV (5′ biotin-ATCTAGAGCGGCCGC-T16-C-**V**) (**V**=A/G/C) (PAGE gel purified). Primer RTP-R is prepared to 2 µg/µL and primer RTP-CV is prepared to 1 µg/µL with TE buffer (pH 8.0). These two sets of anchored primers are mixed at a volume ratio of 1:1. The final concentration for A/G anchored primers is 0.5 µg/µL, and final concentration for CA/CG/CC anchored primers is 0.5 µg/µL (0.167 µg/µL for each). The first- and second-strand cDNAs are synthesized with cDNA synthesis kit (Invitrogen).

3.1.2.1. SYNTHESIS OF THE FIRST-STRAND CDNAS

1. Add the following items to a 1.5-mL Eppendorf tube cooled on ice:

Items	Volume (µL)
5X First-strand buffer	10.0
dNTP mix (10 m*M*)	2.5
Anchored RT primers mixture	2.0
DTT (100 m*M*)	5.0
mRNA in DEPC-treated H_2O	27.0

2. Incubate at 65°C for 3–5 min and place on ice for 1–2 min.
3. Add 1 µL of rRNasin (40 U/µL) and 2.5 µL of Molony murine leukemia virus (MMLV) RT (200 U/µL) to above components. Mix well the reaction gently and incubate at 37°C for 40 min.
4. Spin the tube briefly and heat the tube at 65°C for 2 min. Then, set to 37°C again, add 2 µL RT, mix well, and keep for 30 min.
5. Repeat **step 4** two more times (65°C 2 min → add 2 µL RT → 37°C 30 min) (*see* **Note 3**).
6. Remove 3 µL from the reaction, add 0.2 µL RNase ONE™ Ribonuclease (5–10 U/µL) or DNase-free RNase A (100 mg/mL) and keep at 37°C for 30 min. Use the rest of the reaction (53 µL) for synthesizing second-strand cDNAs.

3.1.2.2. SYNTHESIS OF THE SECOND-STRAND CDNAS

1. Prepare the following mixture and add to the remainder of the first-strand reaction (48 µL):

Items	1X
DEPC-treated H_2O	288 µL
dNTP mix (10 mM)	7.5 µL
10X Second-strand buffer	40.0 µL
Escherichia coli DNA pol. I	10.0 µL
E. coli DNA ligase	1.5 µL

2. Mix the sample gently and incubate the reaction at 16°C for 2 h. (Do not let the temperature rise above 16°C). Add 1–2 µL RNase ONE™ Ribonuclease (5–10 U/µL) or DNase-free RNase A (100 mg/mL) and keep at 37°C for 30 min.
3. Check 5 µL RNase-treated double-strand cDNAs on an 1% agarose gel, together with the RNase-treated first-strand cDNAs to confirm the success syntheses of the first- and second-strand cDNAs (*see* **Note 4**).
4. Treat the sample with 400 µL phenol–chloroform–isoamyl alcohol (25:24:1; pH 8.0). Recover the upper phase and divide cDNA into two Eppendorf tubes (200 µL/tube).
5. Precipitate cDNAs by adding 3 µL glycogen (20 mg/mL), 100 µL 10 M NH$_4$OAc, and 700 µL cold 100% ethanol to each tube. Vortex the tube; spin for 15 min at maximal speed at room temperature.
6. Combine pellets into one tube and wash the pellets twice with 800 µL of 70% cold ethanol.
7. Air-dry the pellets. Dissolve the pellets in 21 µL LoTE. Check 1 µL double-strand cDNAs on a 1% agarose gel.

3.2. Generation of 3′ cDNAs

3.2.1. NlaIII Digestion of Double-Strand cDNAs

1. Digest the 20 µL double-strand cDNAs in 100 µL volume at 37°C for 1–2 h, with 5 µL of *Nla*III (10 U/µL), 10 µL of 10X buffer 4, 1 µL of 100X BSA (bovine serum albumin), and 64 µL of H_2O.
2. Check 4 µL of digestion on a 2% agarose gel. The cDNA should be centered at about 200–500 bp.

3.2.2. Recovery of 3′ cDNAs

1. Transfer 100 µL of Dynal M280 beads into a 1.5-mL Eppendorf tube.
2. Wash beads with 300 µL of 2X binding/washing buffer (2X B/W) and resuspend beads in 100 µL 2X B/W.
3. Add 4 µL H_2O to the 96 µL of *Nla*III-digested cDNAs and then mix with the 100 µL of beads. Rotate or shake at room temperature for 60 min.
4. Place the tube in the MPC and remove the supernatant. Rinse beads with 300 µL of 1X B/W three times and 300 µL TE one time. Resuspend the beads in 50 µL LoTE.

3.2.3. Linker Ligation to 3′ cDNAs (Optional; see **Note 5**)

3.2.3.1. PREPARATION OF LINKERS

1. Either linker A or linker B used in SAGE procedures (*3,17*) can be used in this step. Each oligo is dissolved in LoTE at 100 pmol/μL. Mix 50 μL A1 and 50 μL A2 (or B1 and B2) and add 5 μL buffer 2 (NEB) (the concentration of the resulting linker is about 50 pmol/μL [1.5 μg/μL]).
2. Heat the tube at 95°C for 2 min and cool down the sample at room temperature for 2 h.
3. Remove 20 μL of annealed linkers (1.5 μg/μL) to a new tube, add 80 μL LoTE to make 10X diluted annealed linkers (150 ng/μL).

3.2.3.2. LIGATING LINKER TO 3′ CDNA

1. Resuspend thoroughly the 3′ cDNA-bound beads and transfer 30 μL into a new 1.5-mL Eppendorf tube (total is 50 μL; save 20 μL). Place the tube in the MPC and remove the supernatant.
2. Wash the beads with 200 μL of 1X T_4 ligase buffer (Invitrogen) two times.
3. Resuspend the beads in 14 μL H_2O, add 2 μL of 10X T_4 ligase buffer (Invitrogen) and 2 μL of 10X diluted annealed linkers (150 ng/μL) to the tube, and mix well.
4. Heat the tube at 50°C for 2 min and keep at room temperature for 5–10 min. Then, move to ice. Add 2 μL T_4 ligase (4 U/μL; Invitrogen), mix well, and incubate at 16°C for 2 h.
5. Place the ligation tube in the MPC and remove the supernatant. Rinse three times with 300 μL of 1X B/W.
6. Rinse the beads one times with 200 μL LoTE. Then, resuspend beads with 50 μL LoTE.

3.2.4. PCR Amplification of the 3′-cDNAs (Optional)

The 3′ cDNAs linked with linker A or B can be amplified by SAGE sense primer (SAGE primer-P1 for linker A, or SAGE primer-P2 for linker B) (*3*) and the antisense primer (5′-ACTATCTAGAGCGGCCGCTT-3′). The amount of input beads and numbers of PCR cycles need to be pretested for large-scale PCR amplification.

3.2.4.1. PRETESTING OF 3′ CDNA AMPLIFICATION

1. Premix the following items and aliquot 48 μL per PCR tube:

Items	1X	4X
10X PCR buffer	5.0 μL	20.0 μL
MgCl2 (50 m*M*)	1.5 μL	6.0 μL
dNTPs (10 m*M*)	1.0 μL	4.0 μL
Sense primer (SAGE-P1, or P2, 100 ng/μL)	1.0 μL	4.0 μL
Antisense primer (100 ng/μL)	1.0 μL	4.0 μL
Platinum Taq (5 U/μL)	0.5 μL	2.0 μL
H_2O	38 μL	152 μL

2. Dilute 2 µL of the beads binding linker-ligated 3′ cDNAs by 2, 4, 8, and 16 times. Add 2 µL of each of the diluted beads to one PCR tube.
3. Perform the PCR reactions in the PE GeneAmp PCR Systems 9600 or 9700. Heat at 94°C for 2 min and then perform PCR at 94°C for 30 s, 55°C for 30 s, and 72°C for 30 s. Pick 5 µL of the PCR reaction at 15/20/25 cycles and check on 2% agarose gel to determine the optimal PCR conditions.

3.2.4.2. LARGE-SCALE 3′ CDNA AMPLIFICATION

Based on the pretesting results, perform large-scale PCR amplification. Normally, 600 µL (12X 50 µL) of PCR products are enough to provide templates for large-scale GLGI amplification. Upon finishing the reaction, maintain the tubes at 72°C for 10 min for extension. The amplified templates are purified by phenol–chloroform extraction, ethanol precipitation (the process is similar to the **step 2** of **Subheading 3.1.2.**), and then resuspended in 150 µL TE (pH 8.0).

3.3. GLGI Amplification

Target genes can be specially amplified from the above 3′ cDNA pool by GLGI, with a special sense primer (5′-GGATCCCATGXXXXXXXXXX-3′) and the universal antisense primer (5′-ACTATCTAGAGCGGCCGCTT-3′) (*see* **Note 1**).

3.3.1. Pretesting of GLGI Amplification

Pretest using PCR to determine the optimal amount of 3′ cDNAs and the number of PCR cycles for large-scale GLGI amplification. A sense primer (GGATTCCATGCTGTTGGTGA) can be used as a positive control in most of cases. This primer is generated from SAGE tag CTGTTGGTGA (UniGene ID. Hs.3463). Using this sense primer for GLGI, a special band of 150 bp has been amplified from the cDNAs generated from 24 different human tissues *(15)*.

1. Set PCR conditions as follows (*see* **Note 6**):

Items	1X	XX	
10X PCR buffer (Invitrogen)	3.0 µL	XX	3.0 µL
MgCl2 (50 m*M*) (Invitrogen)	1.2 µL	XX	1.2 µL
dNTPs (10 m*M*)	0.6 µL	XX	0.6 µL
Sense primer (50 ng/µL)	1.4 µL	—	
Antisense primer (50 ng/µL)	1.4 µL	XX	1.4 µL
3′ cDNAs (diluted 2, 4, 8 times, respectively)	1.5 µL	—	
Platinum *Taq* (5 U/µL; Invitrogen)	0.3 µL	XX	0.3 µL
H₂O	20.6 µL	XX	20.6 µL

2. Dilute 1.5 µL of the 3′ cDNAs by 2, 4, and 8 times with H₂O. Add 1.5 µL of each of the diluted 3′ cDNAs to one PCR tube.

3. Perform the PCR reactions in a PE GeneAmp PCR System 9600 or 9700. Heat at 94°C for 2 min, followed by five cycles at 94°C for 30 s, 55°C for 30 s, and 72°C for 30 s. The condition is then changed to 94°C for 30 s, 60°C for 30 s, and 72°C for 30 s for 15/20/25/28 more cycles (20/25/30/33 cycles in total, respectively) (*see* **Note 7**). At each point of the cycle, pick 3 μL from the reaction and check on a 2% agarose gel to determine the optimal conditions.

3.3.2. Small-Scale GLGI Amplification (see **Note 8**)

1. If there are only a few GLGI reactions, perform the GLGI reactions in small numbers with the optimal PCR conditions.
2. After PCR, transfer PCR reactions to 1.5-mL Eppendorf tubes and add LoTE to 200 μL.
3. Extract with an equal volume with phenol–chloroform–isoamyl alcohol (25:24:1; pH 8.0) and recover the upper phase into a new Eppendorf tubes.
4. Precipitate the PCR products by adding 1.5 μL glycogen (20 mg/mL), 100 μL of 10 M NH$_4$OAc, and 800 μL 100% ethanol. Mix well by vortex and spin for 15 min at maximal speed at room temperature.
5. Wash pellet twice with 0.5 mL of 70% cold ethanol. Air-dry and dissolve the pellets with 6 μL H$_2$O for each sample.

3.3.3. Large-Scale GLGI Amplification

1. Ninety-six sense primers are synthesized and deposited in a 96-well plate and dissolved in TE at 1 μg/μL. Transfer 10 μL of each 1-μg/μL primer into a new 96-well plate and adjust the concentration of primers to 50 ng/μL by adding 90 μL of LoTE.
2. Prepare GLGI master mixtures with 28.6 μL per reaction, including 3 μL of 10X PCR buffer, 1.2 μL of 50 mM MgCl$_2$, 0.6 μL of 10 mM dNTPs, 1.4 μL of 50 ng/μL antisense primer, 0.5–5 ng of 3′ cDNAs, 0.3 μL of 5 U/μL of Platinum *Taq* DNA polymerase (Invitrogen), and H$_2$O to 28.6 μL. Prepare 100 aliquots of the 28.6-μL GLGI master mixtures together for each 96 reactions.
3. Aliquot the master mixtures into a 96-well PCR plate (Applied Biosystems) by 28.6 μL/well.
4. Transfer 1.4 μL of each sense primer (50 ng/μL) into each well containing 28.6 μL GLGI master mixture by using a multiple-channel pipet. Spin the GLGI plate briefly (centrifuge: SORVALL RC5C plus; rotor: SH3000).
5. Perform GLGI reactions in the PE GeneAmp PCR System 9600 or 9700. The PCR condition is 94°C for 2 min, followed by five cycles at 94°C for 30 s, 55°C for 30 s, and 72°C for 30 s. The conditions are then changed to 20–25 cycles at 94°C for 30 s, 60°C for 30 s, and 72°C for 30 s. Maintain the reactions at 72°C for 8 min after the last cycle. Check 3 μL of each PCR reaction on a 2% agarose gel if necessary.
6. Directly precipitate the GLGI amplified products in the PCR plate as the follows (*see* **Note 8**):

a. Add 100 µL of precipitation mixtures to each well, containing 1 µL of glycogen (20 mg/mL; Roche), 15 µL of 7.5 *M* NH$_4$oAc, and 84 µL of 100% ethanol, by using a multiple-channel pipet.
b. Seal the plate with a tape pad (Qiagen), vortex, and keep at room temperature for 15 min.
c. Spin at 4000 rpm for 35 min at 4°C (centrifuge: SORVALL RC5C plus; rotor: SH3000), pour out the supernatants and dry with paper towels.
d. Add 150 µL of 70% cold ethanol to each well by using a multiple-channel pipet, seal the plate with a new tape pad, and invert the plate several times.
e. Spin at 4000 rpm for 15 min, pour out the supernatants, and dry with paper towels.
f. Air-dry the pellets and add 6 µL H$_2$O to each well by using a multiple-channel pipet. Then, vortex thoroughly and spin the plate briefly.
g. Check 1.0 µL of purified PCR samples on a 2% agarose gel (optional).

3.4. Cloning GLGI Products

The GLGI-amplified cDNAs are directly cloned into pCR4-TOPO vector (Invitrogen) as the follows (*see* **Note 9**):

1. Prepare master ligation mixtures with 2 µL per reaction, which contains 0.5 µL of pCR4-TOPO vector (10 ng/µL), 0.7 µL of salt solution (1.2 *M* NaCl, 0.06 *M* MgCl$_2$), and 0.8 µL of sterile water. Prepare 100 aliquots of the 2-µL master ligation mixtures together for each 96 reactions.
2. Aliquot the ligation mixtures into a 96-well plate at 2 µL/well by using a repeating pipet, then add 2 µL of each purified GLGI product to each well and mix gently by using a multiple-channel pipet. Spin the plate briefly.
3. Keep the ligation reaction at room temperature for 30 min; then, keep on ice.
4. Aliquot TOPO10 chemically competent cells (Invitrogen) into a 96-well PCR plate (Applied Biosystems) or a Square-Well Block (Qiagen) at 25 µL/well by using a repeating pipet; then, transfer 1.2 µL of each ligation mixture into each well containing competent cells and mix gently by using a multiple-channel pipet.
5. Keep the plate on ice for 25 min, heat at 42°C for 50 s, and then keep on ice for 2 min.
6. Add 125 µL of SOC medium to each well by using a repeating pipet. Seal the plate and shake at 37°C for 60 min at 225 rpm. Spread the transformants on LB plates containing 50 µg/mL of kanamycin and incubate at 37°C overnight.

3.5. Direct-Colony PCR Amplification

Direct colony-PCR is performed for amplification of the GLGI clones. Four colonies are screened for each SAGE tag with 50 or more copies, and 6 colonies are screened for each SAGE tag with less than 50 copies (*see* **Note 10**):

1. Prepare PCR master mixtures with 25 μL per reaction, containing 2.5 μL of 10X PCR buffer (TaKaRa), 1.0 μL of 2.5 mM dNTPs (TaKaRa), 0.1 μL of *Taq* polymerase (5 U/μL; TaKaRa), 60 ng of M13 reverse primer, and 60 ng of M13 forward [–20] primer. Prepare 100 aliquots of the 25-μL PCR master mixtures together for each 96 reactions.

2. Aliquot the reaction mixtures into 96-well PCR plates at 25 μL/well by using a repeating pipet. Colonies are picked directly from the LB plates into each well with sterile pipet tips (*see* **Note 11**).

3. Perform PCR in the PE GeneAmp PCR System 9600 or 9700; the PCR condition is 94°C for 2 min, followed by 25 cycles at 94°C for 30 s, 55°C for 30 s, and 72°C for 60 s. Maintain the reactions at 72°C for 7 min after the last cycle.

4. Precipitate the PCR products by adding of 75 μL of the precipitation mixture per well by using a multiple-channel pipet, containing 22 μL of H_2O, 15 μL of 2 M $NaClO_4$, and 38 μL of 2-propanol. Seal the plates, vortex, and keep at room temperature for 5 min.

5. Spin at 4000 rpm for 35 min at 4°C, then pour out the supernatants and dry with paper towels.

6. Add 150 μL of 70% cold ethanol to each well by using a multiple-channel pipet, seal the plate with a new tape pad, and invert the plate two to three times.

7. Spin at 4000 rpm for 25 min, pour out the supernatants, and dry with paper towels.

8. Air-dry the pellets and add 10 μL H_2O to each well by using a multiple-channel pipet. Then, vortex thoroughly and spin the plates briefly.

9. Check 1.5 μL of several purified PCR samples on a 2% agarose gel (optional).

3.6. Sequencing

The sequencing reaction is performed in a total volume of 7 μL for each reaction (*see* **Note 12**):

1. Prepare master sequencing mixtures with 4 μL for each reaction, containing 0.7 μL of Big-Dye premixture (Applied Biosystems), 1.5 μL of dilution buffer (400 mM Tris•HCl [pH 9.0] and 10 mM $MgCl_2$), 0.3 μL of 100 ng/μL sequencing primer (M13 forward [–20] primer or M13 reverse primer), and 1.5 μL of H_2O. Prepare 100 aliquots of the 4-μL master sequencing mixtures together for each 96 reactions.

2. Aliquot the reaction mixtures into 96-well PCR plates at 4 μL/well by using a repeating pipet. Add 3 μL of DNA template to each well and mix gently by using a multiple-channel pipet.

3. Perform sequencing reactions at 96°C for 10 s, 50°C for 5 s, and 60°C for 4 min for 99 cycles.

4. Precipitate the sequencing products by adding 75 μL of precipitation mixture per well by using a multiple-channel pipet, containing 64 μL of 100% ethanol/3 M NaOAc mixture (25:1), 1 μL glycogen (20 mg/mL; Roche), and 10 μL of H_2O. Seal the plates, vortex, and keep at room temperature for 5 min.

5. Spin at 4000 rpm for 35 min at 4°C; then, pour out the supernatants and dry with paper towels.

6. Add 150 µL of 70% cold ethanol to each well by using a multiple-channel pipet, seal the plate with a new tape pad, and invert the plate several times.

7. Spin at 4000 rpm for 15 min, pour out the supernatants, and dry with paper towels.

8. Air-dry the pellets and add 3 µL loading dye to each well by using a repeating pipet. Then, vortex thoroughly and spin the plates briefly.

9. Heat sequencing sample at 96°C for 2 min and load 0.7–1.2 µL sequencing products on a 5–6% sequencing gel with 96 lanes; then, collect sequences in a ABI377 sequencer (Applied Biosystems).

3.7. Sequence Analysis

Sequences are extracted by the Sequencing Analysis V3.3 program (Applied Biosystems). All sequences without a 14-base pair SAGE tag sequence (CAT GXXXXXXXXXX) should be excluded from further analysis. The qualified sequences are matched to the GenBank Database (NR and ESTs, http:// www.ncbi.nlm.nih.gov/BLAST/) through BLAST. For the sequences generated from SAGE tags with a known match, any mismatch between the SAGE tag sequence used for GLGI amplification and the SAGE tag sequence of the matched sequence in the database is classified as artifact, and these sequences are eliminated from further analysis. The matched sequence ID is used to search the UniGene database to obtain the UniGene cluster ID. A sequence generated from a novel SAGE tag sequence is defined as a real novel sequence if it did not match to any known expressed sequences with higher than 80% homology in the same orientation. A sequence generated from a novel SAGE tag sequence is reclassified as a known sequence if it is matched to a known sequence with over 80% homology in the same orientation, including the same 14-base pair SAGE tag sequence.

4. Notes

1. In this revised protocol (*16*), we use a single antisense primer (5′-ACTATC TAGAGCGGCCGCTT-3′) for GLGI reaction. This sequence is located at the 3′ end of all the cDNA templates generated from anchored oligo-(dT) primers. Comparing with the three anchored oligo-(dT) primers (dT11A, G, C) used in the original protocol (*15*), this change increases the efficiency of GLGI amplification significantly, particularly for low-copy templates.

2. Regular unanchored oligo-(dT) RT primers have been widely used for cDNA synthesis. Using regular oligo-(dT)-generated cDNAs for GLGI will generate several problems. (1) Regular oligo-(dT)-generated cDNAs include various lengths of poly(dA/dT) sequences at the 3′ end of the cDNA, even from the same mRNA template. These undefined poly(dA/dT) sequences will affect GLGI

amplification, resulting in multiple fragments with different sizes or smears. (2) The presence of longer poly(dA/dT) sequences will largely affect the quality of sequencing signals.

3. This treatment can significantly increase the cDNA yield because repeating dissociation/association of anchored RT primers with mRNA templates increases the rate of correct annealing and, therefore, increase the yield of cDNAs.

4. Compared with that of the first-strand cDNAs, the distribution of double-strand cDNAs should shift up on the 1% agarose gel.

5. If the *Nla*III-digested 3′ cDNA sample is enough for GLGI application, **steps 3** and **4** of **Subheading 3.2.** can be skipped.

6. Although satisfactory results can be obtained for most GLGI reactions by PCR with of 2 mM [Mg]$^{2+}$, optimal concentration of [Mg]$^{2+}$ can be tested to obtain better result for some specific target sequences. Platinum *Taq* DNA polymerase is better than *Pfu* DNA polymerase for GLGI amplification, generating a higher yield of PCR products and keeping a high specificity of amplification. Other DNA *Taq* polymerase also can be tested for GLGI amplification.

7. The number of PCR cycles is important to maintain the specificity of the amplification. For high-abundant templates, 20–25 cycles (in total) are sufficient; for low-abundant templates, 25–33 cycles (in total) are enough. Overamplification can generate false products.

8. The GLGI-amplified products are directly precipitated and cloned into vectors without gel purification in order to prevent the loss of amplified products. This is especially important for small-size or low-copy GLGI products. Moreover, it largely simplifies the cloning process.

9. Compared with the manufacturer's protocol, only half of the amount of ligation reagents, competent cells, and SOC medium is used in this protocol.

10. The efficiency of GLGI amplification for the targeted template parallels with the abundance of the SAGE tags. Amplification of SAGE tags with high copy numbers (≥50 copies) usually generates a single intense band, whereas amplification of SAGE tags with lower-copy numbers (<50 copies) yields fewer products and may contain additional bands. Compared with standard PCR, only the sense primer provides specificity for the GLGI amplifications. When the level of targeted templates is very low, partial annealing of the sense primer with other templates can result in nonspecific amplification. This limitation can be overcome by screening several clones for each GLGI product. Direct-colony PCR amplification is used to collect templates for sequencing reactions. Screening four to six colonies usually generated more than 85% qualified 3′ cDNAs for SAGE tags with high copies and between 65% and 80% qualified 3′ cDNAs for SAGE tags with low copies. If necessary, a second round of screening four to six colonies usually generates an additional 10–15% more qualified 3′ cDNAs from SAGE tags with low copies.

11. In regular colony PCR protocols, colonies should be heated and broken before they were added into PCR reactions, but our experiment results showed that this step is not necessary.

12. In our protocol, only 0.7 µL of Big-Dye premixture (Applied Biosystems) is used in each sequencing reaction, and sequencing cycles increase to 99 cycles. Such a setting decreases the sequencing cost for large-scale studies.

Acknowledgments

This work was supported by the G. Harold and Lelia Y. Mathers Charitable Foundation (SMW), a Fellowship from the University of Chicago Committee on Cancer Biology (JC), and CA84405 (JDR).

References

1. Velculescu, V. E., Zhang, L., Vogelstein, B., and Kinzler, K. W. (1995) Serial analysis of gene expression. *Science* **270,** 484–487.
2. Velculescu, V. E., Madden, S. L., Zhang, L., Lash, A. E., Yu, J., Rago, C., et al. (1999) Analysis of human transcriptomes. *Nat. Genet.* **23,** 387–388.
3. Zhang, L., Zhou, W., Velculescu, V. E., Kern, S. E., Hruban, R. H., Hamilton, S. R., et al. (1997) Gene expression profiles in normal and cancer cells. *Science* **276,** 1268–1272.
4. Chen, H., Centola, M., Altschu, S. F., and Metzger, H. (1998) Characterization of gene expression in resting and activated mast cells. *J. Exp. Med.* **188,** 1657–1668.
5. Virlon, B., Cheval, L., Buhler, J. M., Billon, E., Doucet, A., and Elalouf, J. M. (1999) Serial microanalysis of renal transcriptomes. *Proc. Natl. Acad. Sci. USA* **96,** 15,286–15,291.
6. Welle, S., Bhatt, K., and Thornton, C. A. (1999) Inventory of high-abundance mRNAs in skeletal muscle of normal men. *Genome Res.* **9,** 506–513.
7. Angelastro, J. M., Klimaschewski, L., Tang, S., Vitolo, O. V., Weissman, T. A., Donlin, L. T., et al. (2000) Identification of diverse nerve growth factor-regulated genes by serial analysis of gene expression (SAGE) profiling. *Proc. Natl. Acad. Sci. USA* **97,** 10,424–10,429.
8. Charpentier, A. H., Bednarek, A. K., Daniel, R. L., Hawkins, K. A., Laflin, K. J., Gaddis, S., et al. (2000) Effects of estrogen on global gene expression: identification of novel targets of estrogen action. *Cancer Res.* **60,** 5977–5983.
9. El-Meanawy, M. A., Schelling, J. R., Pozuelo, F., Churpek, M. M., Ficker, E. K., Iyengar, S., et al. (2000) Use of serial analysis of gene expression to generate kidney expression libraries. *Am. J. Physiol. Renal Physiol.* **279,** F383–F392.
10. Hough, C. D., Sherman-Baust, C. A., Pizer, E. S., Montz, F. J., Im, D. D., Rosenshein, N. B., et al. (2000) Large-scale serial analysis of gene expression reveals genes differentially expressed in ovarian cancer. *Cancer Res.* **60,**6281–6287.
11. Welle, S., Bhatt, K., and Thornton, C. A. (2000) High-abundance mRNAs in human muscle: comparison between young and old. *J. Appl. Physiol.* **89,** 297–304.
12. Lee, S., Zhou, G., Clark, T., Chen, J., Rowley, J. D., and Wang, S. M. (2001) The pattern of gene expression in human CD15+ myeloid progenitor cells. *Proc. Natl. Acad. Sci. USA* **98,** 3340–3345.

13. Chen, J., Rowley, D. A., Clark, T., Lee, S., Zhou, G., Beck, C., et al. (2001) The pattern of gene expression in mouse Gr-1(+) myeloid progenitor cells. *Genomics* **77,** 149–162.
14. Zhou, G., Chen, J., Lee, S., Clark, T., Rowley, J. D., and Wang, S. M. (2001) The pattern of gene expression in human CD34(+) stem/progenitor cells. *Proc. Natl. Acad. Sci. USA* **98,** 13,966–13,971.
15. Chen, J., Rowley, J. D., and Wang, S. M. (2000) Generation of longer cDNA fragments from serial analysis of gene expression tags for gene identification. *Proc. Natl. Acad. Sci. USA* **97,** 349–353.
16. Chen, J., Lee, S., Zhou, G., and Wang, S. M. (2002) High-throughput GLGI procedure for converting a large number of serial analysis of gene expression tag sequences into 3 complementary DNAs. *Genes Chromosomes Cancer* **33,** 252–261.
17. Lee, S., Chen, J., Zhou, G., and Wang, S. M. (2001) Generation of high quality and quantity of tag/ditag for SAGE analysis. *BioTechniques* **31,** 348–354.
18. Wang, S. M. and Rowley, J. D. (1998) A strategy for genome-wide gene analysis: integrated procedure for gene identification. *Proc. Natl. Acad. Sci. USA* **95,** 11,909–11,914.
19. Wang, S. M., Fears, S. C., Zhang, L., Chen, J., and Rowley, J. D. (2000) Screening poly(dA/dT)– cDNAs for gene identification. *Proc. Natl. Acad. Sci. USA* **97,** 4162–4167.

20

Generation of Full-Length cDNA Libraries Enriched for Differentially Expressed Genes

Bakhyt Zhumabayeva, Cynthia Chang, Joseph McKinley, Luda Diatchenko, and Paul D. Siebert

1. Introduction

With sequence analysis of the human genome well underway, the scientific focus is shifting toward understanding the fundamentals of gene function. Sequence information alone is insufficient for a full understanding of gene function, expression, and regulation. Modern approaches such as subtractive hybridization *(1–5)*, differential display *(6,7)*, serial analysis of gene expression (SAGE) *(8)*, or complementary DNA (cDNA) arrays *(9,10)* allow researchers to monitor the differential expression of many genes in parallel. However, cDNAs identified by these techniques are often only short fragments. Further analyses, such as functional assays, require full-length cDNA clones. Two dominant approaches for cloning full-length cDNAs are cDNA library screening and PCR-based rapid amplification of cDNA ends (RACE) technology. Both techniques require knowledge of the cDNA sequence and are only useful for cloning a few genes at a time. Cloning full-length cDNAs for multiple genes and generation of full-length cDNA libraries enriched for differentially expressed genes in an efficient high-throughput manner still remain challenging.

A potential method for overcoming this limitation is to use a RecA-based cloning technology to convert either defined cDNA fragment mixtures or whole subtracted cDNA populations into expression-ready full-length cDNA sublibraries. The application of the RecA-technology for cloning single full-length cDNAs from cDNA libraries has been described previously *(11)*.

From: *Methods in Molecular Biology, vol. 221: Generation of cDNA Libraries: Methods and Protocols*
Edited by: S.-Y. Ying © Humana Press Inc., Totowa, NJ

RecA is a protein required for DNA repair and genetic recombination in *Escherichia coli* *(12–14)*. In the presence of ATP, RecA coats single-stranded DNA that can then form triple-stranded hybrid structures with homologous sequences in double-stranded DNA (**Fig. 1**) *(12–14)*. In RecA cloning technology, a DNA fragment is biotinylated, coated with RecA and used as an "enrichment probe" to isolate target cDNA clones from a plasmid-based cDNA library using streptavidin–biotin affinity interactions to isolate the resulting D-looped plasmids *(11,15,16)*. The enriched plasmid population is then transformed into *E. coli* and screened by colony hybridization with a gene-specific probe. Although the RecA-based conversion of subtracted cDNA into full-length clones has been proposed *(17,18)*, thorough optimization of reaction conditions and evaluation of reaction efficiencies have not been conducted.

Here, we describe an optimized protocol for obtaining full-length low-complexity cDNA libraries enriched for multiple-target genes using the RecA-method. We validated this method by sequentially increasing the number of probes used in individual RecA reaction up to library-scale experiments and showed that target cDNAs were efficiently enriched in each case.

2. Materials

1. cDNA libraries:
   a. ClonCapture™ mammalian cDNA expression libraries from the human liver, brain, heart, testis, and placenta.
   b. Human retroviral cDNA libraries from the mammary gland, leukocyte, brain, skeletal muscle, and prostate (all from BD Biosciences Clontech, Palo Alto, CA) (*see* **Note 1**).
2. Enrichment probes:
   a. HLH, NF45, RPA, SKY, TNFR1, Dad1, TRADD, CIAP2, ETS, and caspase.
   b. Independent clones from several subtracted cDNA libraries corresponding to differentially expressed genes.
   c. Subtracted cDNA populations enriched for upregulated or downregulated genes in lung carcinoma, adenocarcinoma of prostate, liver carcinoma, and postcolorectal adenocarcinoma.
3. RecA protein (New England Biolabs, Beverly, MA).
4. 10X RecA buffer: 250 mM Tris-acetate (pH 7.5), 80 mM MgOAc, 20 mM CoCl$_2$, and 10 µg/µL bovine serum albumin (BSA) (CoCl$_2$ should be prepared as a 0.1-M stock solution and added separately just before use).
5. ATP mix: 2 mM ATPγS, and 1 mM ATP (ATP mix should be kept at –80°C; Sigma, St. Louis, MO) (*see* **Note 2**).
6. Proteinase K (0.2 µg/µL, Sigma).
7. 10% Sodium dodecyl sulfate (SDS).
8. 0.1 M Phenylmethylsulfonyl fluoride (PMSF), Sigma.
9. Streptavidin magnetic beads (Roche Applied Science, Indianapolis, IN).

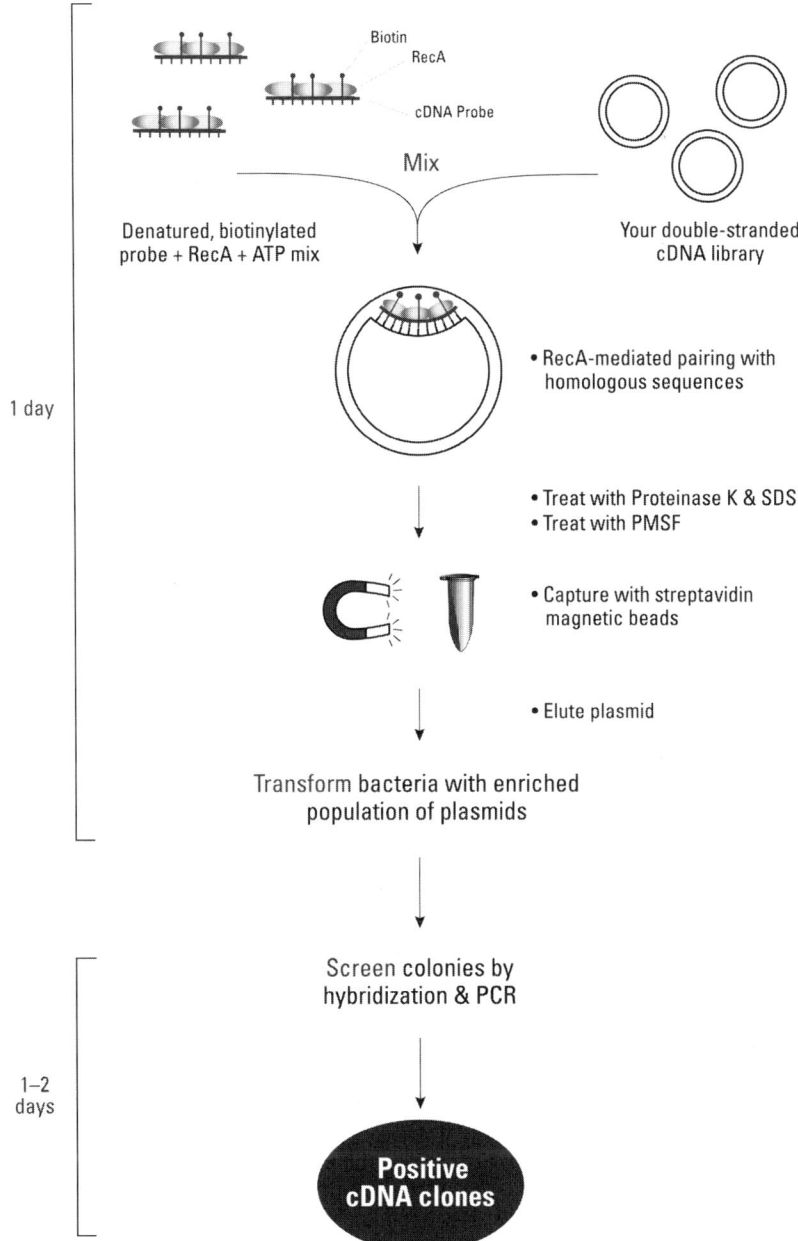

Fig. 1. Schematic diagram of recA protein-mediated affinity capture technology. The entire procedure, from synthesis of biotinylated probe to screening of the enriched cDNA library could be performed within 2–3 d.

10. Biotin-21-dUTP (BD Biosciences Clontech).
11. Binding buffer: 10 mM Tris-HCl (pH 7.5), 1 mM EDTA, and 1 M NaCl.
12. Washing buffer: 10 mM Tris-HCl (pH 7.5), 1 mM EDTA, and 2 M NaCl.
13. Elution buffer: 0.1 N NaOH and 1 mM EDTA.
14. Precipitation mix: 12 µL of 3 M NaOAc (pH 5.0), 0.5 µL of glycogen (20 mg/mL), and 10 µL of 1 M acetic acid.
15. [α–^{32}P]-dATP (Amersham Pharmacia Biotech, Piscataway, NJ).
16. Advantage™ cDNA PCR kit (BD Biosciences Clontech).
17. 10X Advantage buffer: 40 mM Tricine-KOH (pH 9.2 at 25°C), 3.5 mM Mg(OAc)$_2$, 15 mM KOAc, and 3.75 µg/mL BSA.
18. 5′ and 3′ gene-specific primers (each 0.2 µM).
19. NucleoSpin® Extraction kit (BD Biosciences Clontech).
20. DH-5α electrocompetent cells (BD Biosciences Clontech).
21. Ethanol (70% and 100%).
22. Luria Bertani (LB)-agar plates containing 50 µg/mL ampicillin.
23. DECAprime™ II DNA labeling kit (Ambion, Austin, TX).
24. ExpressHyb™ hybridization solution (BD Biosciences Clontech).
25. PCR-Select™ cDNA subtraction kit (BD Biosciences Clontech).
26. NYTRAN® SuperCharge nylon membranes (Schleicher & Schuell, Keene, MA).

3. Methods

RecA-mediated enrichment of cDNA libraries consists of three distinct steps. The schematic overview of RecA-mediated affinity capture is outlined in **Fig. 1**. First, a single-stranded, biotinylated enrichment probe is incubated with RecA to allow polymerization on the single-stranded DNA (ssDNA). In the second step, the RecA-coated enrichment probe is incubated with plasmid DNA in the presence of γ[S] ATP to allow formation of stable triple-stranded complexes between the enrichment probe and homologous portions of the duplex DNA. In the final stage, RecA is proteolytically removed, and the triple-helical hybrid molecules are recovered by capture on streptavidin-coated magnetic beads.

3.1. Preparation of Biotinylated Enrichment Probes

Biotinylated enrichment probes were prepared by polymerase chain reaction (PCR) incorporation of biotin-21-dUTP. For efficient enrichment, we generated PCR fragments of 200–600 bp. Gene-specific PCR fragments generated from poly(A$^+$) or total RNA (using reverse transcription [RT]-PCR) may be contaminated with fragments of other genes. For best results, we amplified enrichment probes from plasmid clones containing a particular gene of interest. Amplification of such clone yields a single PCR product. Enrichment probes should not contain repetitive sequences, poly-(A) sequences, or sequences sharing homology with other genes. Therefore, we subdivided the probe preparation stage into two steps: (1) preparation of a "pure" template for biotinylation and (2) biotin incorporation by PCR using a "pure" template from the previous step. In order to prepare a "pure" template, we performed an analytical PCR and visualized the PCR products on 2% TAE agarose gel. Then, a small piece (approx 1 mm^3) of the band was picked using a sterile Pasteur pipet. The excised gel fragment was placed in a 1.5-mL microcentrifuge tube with 200 μL of deionized water and incubated at 70°C for 30 min with occasional vortexing. This extract was then used as a template for further probe generation. Subtracted cDNAs were generated according to the PCR-Select Subtraction kit using NP2 nested primers *(3)*. Biotinylated probes were generated as follows:

Template (Gel extract)	2 μL
dNTP (10 m*M*)	2 μL
10X PCR buffer	10 μL
5′ Primer (10 m*M*)	2 μL
3′ Primer (10 m*M*)	2 μL
Biotin-21-dUTP (0.5 m*M*)	10 μL
H$_2$O	70 μL
Advantage PCR enzyme mix	2 μL

Polymerase chain reaction parameters were as follows: 94°C for 1 min, followed by 25 cycles of 94°C for 30 s and 68°C for 2 min, followed by a final incubation at 68°C for 5 min. Overcycling of PCR products should be avoided. The purity of the biotinylated PCR product is essential for efficient enrichment. The biotinylated PCR fragments were purified from unincorporated nucleotides and primers using a NucleoSpin® Extraction kit. Purified PCR fragments should be visualized on a 2% TAE agarose gel and checked for the concentration on a spectrophotometer.

3.1.1. Checking the Biotinylation Efficiency of the Probe Using a Scintillation Counter

In order to check the biotinylation efficiency of the enrichment probe, a probe should be generated by PCR incorporation of 1 μL of [α-^{32}P]-dCTP along with biotin-21-dUTP.

1. The magnetic beads are mixed thoroughly and 15 μL is aliquoted into a 1.5-mL tube. The storage buffer is removed using the magnetic stand.
2. The beads are then washed three times with 100 μL binding buffer and resuspended in 30 μL of binding buffer.
3. Then, 50 ng of the biotinylated probe (radioactively labeled) is added to the prewashed beads and mixed well.

4. At this time, the reaction tube containing beads and the biotinylated probe is placed in a scintillation vial and the signal is determined in a scintillation counter. This signal represents total cpm or the *preincubation signal*.

5. Then, the beads with the probe are incubated for 30 min at room temperature with shaking to allow beads to bind the biotinylated probe.

6. After incubation, we separate the bound probe from unbound supernatant using the magnetic stand.

7. Then, we measure each tube separately in a scintillation vial to determine (the *postincubation signal*) what percent of radioactively labeled probe is biotinylated or not biotinylated. (*see* **Note 3**.)

3.2. RecA-Mediated Enrichment of cDNA Libraries

The use of RecA-mediated affinity capture technology for cloning full-length cDNAs has been described previously *(11)*. The purpose of the present study was to explore the use of RecA-based cloning for functional genomics: (1) high-throughput cloning of full-length cDNAs when only partial cDNA fragments are available and (2) obtaining full-length, subtracted cDNA libraries enriched for differentially expressed genes. These applications are extremely important for cloning novel genes and functionally screening genes of interest.

3.2.1. The Enrichment of cDNA Libraries with Multiple Probes

First, we demonstrated the feasibility of cloning multiple full-length cDNAs in a single experiment by enriching cDNA libraries for several different genes using a mixture of 10 distinct probes. We prepared biotinylated probes for the following genes: HLH, NF45, RPA, SKY, TNFR1, Dad1, TRADD, CIAP2, ETS, and Caspase; we then pooled them to enrich a human liver ClonCapture cDNA library and a mixture of five ClonCapture libraries (*see* **Fig. 2**). Before the enrichment, we determined the relative abundance of these genes in the human liver cDNA library using PCR with β-actin and glyceraldehyde 3-phosphate dehydrogenase (GAPDH), which were assumed to be 0.1–0.2% of the total population *(11)*. Three out of 10 genes (SKY, TRADD, and CIAP2) were very low abundance (35 PCR cycles); the remaining 7 were medium to low abundance (*see* **Table 1**).

3.2.1.1. The Enrichment Procedure

For each multiple probe enrichment experiment, we used 10–15 ng of each biotinylated probe (total 100–150 ng) and 5 µg of the cDNA library. For experiments using the cDNA library mixtures, 1 µg of each library was used.

1. The necessary volumes of 1X RecA buffer, ATP mix, enrichment probe, cDNA library, RecA protein, and sterile water are determined to bring up a total reaction volume to 30 µL.

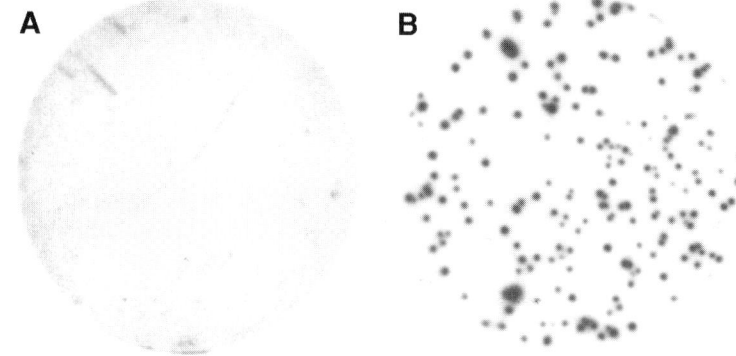

Fig. 2. Colony hybridization screening of LB-agar plates of an enriched human liver cDNA library with a mixture of 10 probes: (**A**) unenriched cDNA library after colony screening; (**B**) enriched cDNA library after colony screening. Approximately 1200 colonies were plated onto 90-mm LB-ampicillin plates. The filters were hybridized with a ^{32}P-labeled probe prepared from the mixture of 10 probes, as described in the Methods section and exposed to X-ray film.

2. Then, 100–150 ng of purified, biotinylated PCR fragments are mixed with water and denatured by boiling for 5 min, immediately chilled on ice water, and centrifuged for 20 s at 20,000g in a microcentrifuge.
3. During the denaturation, a reaction mix containing 3 µL of 1X RecA buffer (including 0.6 µL of 0.1 M CoCL$_2$), 4 µL of ATP mix, and 2 µg RecA per each 50 ng of probe is prepared and added to the denatured probe (*see* **Note 3**).
4. After incubating this mixture at 37°C for 15 min, 5 µg of purified cDNA library is added and mixed by pipetting to give a final volume of 30 µL.
5. RecA is then removed enzymatically by adding 0.6 µL of 0.2 µg/µL Proteinase K and 0.4 µL of 0.2% SDS and incubating at 37°C for 10 min. We recommend mixing them first and then adding them to the reaction mixture.
6. Proteinase K is inactivated by adding 1 µL of 0.1 M PMSF. This step prevents proteolytic digestion of the streptavidin on the magnetic beads.
7. Then, the prepared streptavidin magnetic beads are transferred to this reaction mixture and incubated at room temperature for 30 min, with gentle shaking to capture the enriched cDNA clones (*see* the following details).

3.2.1.2. Capture of Enriched cDNA Clones on Streptavidin-Coated Magnetic Beads

Using a biotinylated ssDNA as an "enrichment probe" allows the isolation of enriched clones from a cDNA library using streptavidin–biotin affinity interactions.

1. The magnetic beads are mixed well before pipeting the 30 mL of beads into 1.5-mL tubes. The storage buffer is removed using the magnetic stand.
2. Streptavidin beads (10 mg/mL) are then washed with 100 μL binding buffer by gentle vortexing. The supernatant is removed using the magnetic stand. This washing step is repeated three more times and then beads are resuspended in 30 μL binding buffer.
3. Prewashed streptavidin beads are then added to the mixture containing the enriched ssDNA–dsDNA complexes generated earlier, and incubated for 30 min at room temperature on a shaker to keep beads in suspension.
4. After incubation, the supernatant is removed using a magnetic stand and the beads are washed three times in 1 mL washing buffer with gentle vortexing for 1–2 min.
5. Beads are washed once more with 1 mL of deionized water.
6. After the last wash, the captured plasmids are eluted in 100 μL elution buffer by incubating the tube for 5 min at room temperature with vortexing.
7. The eluate containing the enriched plasmids is transferred to a fresh tube containing 22.5 μL precipitation mix.
8. The eluate is precipitated with 2.5 vol of 100% ethanol, mixed well, and incubated for 45 min at –70°C.
9. After incubation, the tube is centrifuged for 20 min at 20,000g.
10. The pellet is washed with 70% ethanol, air-dried at room temperature, and resuspended in 10 μL deionized water.
11. Then, 2–5 μL of the resuspended pellet is electroporated into electrocompetent DH5-α cells.
12. Then, 50 and 100 μL of transformed cells are plated on LB-agar plates containing 50 μg/mL ampicillin.

3.2.1.3. Screening the Enriched cDNA Libraries

After completing the RecA procedure, we verified enrichment for all 10 probes by hybridizing colony lift replica filters with a radiolabeled probe mixture (*see* **Fig. 2**). After analyzing the enrichment for all 10 probes as a group, we determined the enrichment for each of the 10 probes individually. For this purpose, we plated 1000–1200 colonies on each of 10 LB-ampicillin agar plates. Replica filters were prepared from each plate and hybridized with the individual labeled probes. NYTRAN® SuperCharge nylon membranes were used for colony lifting according to the manufacturer's protocol. The hybridization probes were prepared by random priming using [α-^{32}P]-dATP and the DECAprime™ II DNA labeling kit. Prehybridization and hybridization were carried out at 68°C with ExpressHyb™ hybridization solution containing 100 ng/mL sheared salmon sperm DNA. Membranes were washed three times in 2X standard saline citrate (SSC) and 0.5% SDS for 30 min each and once in 0.2X SSC and 0.5% SDS for 30 min. Finally, membranes were briefly washed in 2X SSC and exposed to X-ray film.

Table 1
Enrichment of a Human Liver cDNA Library and Mixture
of Five ClonCapture Libraries Using Probes for 10 Different Genes

Enrichment probe	Representation in unenriched library	Human liver ClonCapture cDNA library		Mixture[a] of five ClonCapture cDNA libraries	
		Positive clones after hybridization (%)	Enrichment fold	Positive clones after hybridization (%)	Enrichment fold
Mixture of 10 probes					
HLH	0.007	1.3	186	2.2	314
NF45	0.006	2.2	367	3.2	533
RPA	0.007	0.4	57	0.5	71
SKY	0.0002	0.8	4000	0.7	3500
TNFR1	0.001	4.1	4100	1.4	1400
DAD1	0.005	4.3	860	2.2	440
TRADD	0.0005	0.7	1400	0.3	600
CIAP2	0.001	0.5	500	0.4	400
ETS	0.007	0.5	71	0.4	57
Caspase	0.003	0.3	100	0.2	67
Total positive clones		**15.1**		**11.5**	
A single probe:					
TNFR1				10.0	10,000

[a]Mixture of human brain, heart, liver, testis, and placenta libraries. Enrichment fold was calculated by dividing the percentage of positive clones in the enriched library by that in the original library.

3.2.1.4. RESULTS

Table 1 shows the enrichment efficiencies of the human liver and the mixture of 5 cDNA libraries for the combined set of 10 probes as well as for a parallel experiment using the TNFR1 probe individually. Of the clones in the enriched human liver cDNA library, 15.1% were positive after hybridization with the combined set of probes. For the mixture of libraries, about 11.5% of clones were positive. In both experiments, the most highly enriched genes were TNFR1, SKY1, TRADD, and Dad1. The least enriched genes, with less than 100-fold enrichment, were RPA, ETS, and Caspase. The highest percentages of positive clones were identified for TNFR1, DAD1, NF45, and HLH. Enrichment of the library mixture with the individual TNFR1 probe yielded about 10% positive clones with 10,000-fold enrichment efficiency. Comparison of enrichment efficiencies for TNFR1 when probed in a mixture or individually suggests that the relative abundance of each cDNA in the enriched library decreases as the number of individual cDNAs in probe is increased (*see* **Table 1**).

These results clearly demonstrate that low-abundance genes with representation frequencies as low as 0.0002% can be enriched along with medium-abundance genes when probes are used in a mixture. Nevertheless, it should be mentioned that enrichment efficiency for any given cDNA might vary significantly regardless of its abundance. For example, CIAP2 (low abundance) as well as RPA and ETS (medium abundance) were not highly enriched. However, NF45, which has about the same abundance as RPA and ETS was enriched fivefold to sixfold further. This variance could be the result of sequence specificity or quality of biotinylated probes.

In spite of this variability, the overall enrichment picture for all genes was sufficient to pick several colonies for further analyses. Next, we analyzed the insert sizes of several positive clones for TNFR1, CIAP2, and TRADD to see if full-length cDNAs could be identified. The longest clones were for 2.3 kb, 2.8 kb, and 1.8 kb for TNFR1, CIAP2, and TRADD, respectively (data not shown). The sequence identities of all three clones were confirmed by PCR as well as by secondary hybridization. Additionally, TNFR1 clones were verified by sequencing. All three analyzed clones contained complete open reading frames.

3.2.2. Generation of cDNA Libraries Enriched for the Differentially Expressed Genes

As a further extension of the analysis, we used the RecA method to convert entire subtracted cDNA populations—representing genes differentially expressed in tumor samples—into full-length retroviral cDNA libraries. These differentially expressed cDNA populations included genes that are upregulated or downregulated in lung carcinoma, adenocarcinoma of prostate, liver carcinoma, and colorectal adenocarcinoma (*1*). For this study, we tested specific mixes of subtracted cDNA fragments as well as entire subtracted populations (*see* **Table 2**). These subtracted cDNAs were obtained using suppression subtractive hybridization as described previously (*1*).

For enrichment, we used retroviral cDNA libraries constructed from human normal prostate, mammary gland, liver, and a mixture of five different libraries. Ideally, a plasmid cDNA library based on the tissue or sample from which the subtracted clones were identified should be used for the conversion of subtracted cDNA into full-length clones. However, cDNA libraries constructed from particular diseased tissues are not always available. Therefore, for RecA-conversion experiments, we used cDNA libraries from corresponding normal tissues. To increase the probability of obtaining target cDNAs, we used a mixture of several different cDNA libraries.

The enrichment procedure was the same as described in **Subheading 3.2.1.** However, we used 400 ng of biotinylated subtracted cDNA with 16 µg of

Table 2
Enrichment of Human Prostate, Mammary Gland, Liver, and Mixture
of Five Retroviral cDNA Libraries with Multiple Subtracted cDNA Probes

| Experiment | No. of clones from subtracted library | Origin of subtracted cDNAs | Positive clones after enrichment | | | |
			Prostate (%)	Mammary gland (%)	Mixture of five libraries	Liver (%)
1	5[b]	Prostate tumor	10	10	—	—
2	9[b]	Prostate tumor	20	10	—	—
3	19[c]	Prostate tumor	10	—	12	—
4	20[c]	Prostate tumor	13	—	—	—
5	30[c]	Prostate tumor	14	—	—	—
6	10[b]	Lung tumor	—	—	18	—
7	Subtracted cDNA	Prostate tumor	29	—	—	—
8	Subtracted cDNA	Liver tumor	—	—	—	25

[a]Mixture of human liver, prostate, leukocyte, brain, and skeletal muscle cDNA libraries.
[b]Subset of high-abundance cDNAs.
[c]Subset of low-abundance cDNAs.
Note: — indicates experiment has not been performed.

RecA. Because of the amount of used probes, we increased the reaction volume to 50 µL and other reaction components, respectively.

3.2.2.1. RESULTS

Colony hybridization of a RecA-enriched human prostate library using a subtracted probe containing genes upregulated in a prostate adenocarcinoma produced about 29% positive clones (*see* **Table 2**, experiment 7). In contrast, hybridization to the unenriched human prostate retroviral library produced only 2% positives (data not shown). A comparable result (25%) was obtained in the analysis of a human liver retroviral library after enrichment and subsequent probing for a population of genes downregulated in a liver-metastasized tumor (*see* **Table 2**, experiment 8). To evaluate whether this increased library complexity can affect the efficiency of RecA-mediated enrichment, we compared two sublibraries enriched with a pool of 19 different probes: from the prostate retroviral library (10%) and a mixture of five human retroviral libraries (12%; *see* **Table 2**, experiment 3). We also investigated the effect of increasing the number of individual probes in a mixture on the enrichment efficiency of cDNA libraries. The results show a correlation between the number of species in the probe mix and the percentage of positive clones after enrichment when pools of five or nine high-abundance probes were used to enrich the human prostate cDNA library (*see* **Table 2**, compare lanes 1 and 2). However, the correlation was less pronounced in the low-abundance cDNA mixture (*see* **Table 2**, experiments 3, 4, and 5). It is true for cDNA clones abundant in prostate cDNA library, but not in mammary gland (*see* **Note 5**).

Next, we evaluated how the abundance of target cDNAs affects the percentage of positive clones in the enriched sublibraries. In these experiments, we grouped the subtracted cDNA probes according to their abundance. The abundance level of differentially expressed cDNAs was evaluated using a differential screening procedure, as described previously (*1*). Clones were subdivided according to their abundance into two populations and used separately to enrich for full-length clones (*see* **Table 2**). We observed a good correlation between the abundance of target cDNAs and the percentage of positive clones in the enriched sublibraries. The results for the enrichment of human prostate and human mammary gland retroviral cDNA libraries with a pool of nine high-abundance cDNA probes derived from a subtracted prostate cancer cDNA library are shown in **Table 2**. Twenty percent of clones were identified as positive for enriched human prostate and about 10% were positive for enriched human mammary gland cDNA libraries. As was expected, the population of high-abundance cDNAs clones yielded a higher percentage of positive clones, than did the low-abundance cDNAs (compare experiments 1–3) (*see* **Note 6**).

4. Notes

1. The important criteria of successful enrichment is related to quality and integrity of dsDNA (cDNA library), which is to be used as a target molecule. In order to minimize the level of nonspecific binding and to maximize the stability of the triple-helix complex after removal of RecA protein, the plasmid library should be purified through the CsCl gradient.When compared to alkaline-based commercially available plasmid isolation kits, enrichment efficiency was more than 10-fold higher using CsCl. However, some commercial plasmid isolation kits such as Nucleobond (Macherey-Nagel GmbH, Duren, Germany) with minor modifications in the protocol, allow isolation of higher-quality superhelical DNA. Using the Nucleobond plasmid isolation kit, we have achieved comparable results to those obtained by CsCl purification.

2. One of the important factors of RecA-based enrichment technology is the requirement for the nucleoside triphosphate cofactor ATP. RecA protein is a DNA-dependent ATPase and ATP is hydrolyzed by RecA throughout each nucleoprotein filament. RecA-promoted dissociation of ATP directly affects stability of the deproteinized triple-helix complexes in vitro. ATP hydrolysis reduces the affinity of RecA protein to DNA; therefore, an ATP-regenerating system is required in vitro. A nonhydrolyzable analog of ATP, $\gamma[S]ATP$, blocks disassembly of triple-helix complex. We have tested different ratios of $\gamma[S]ATP$ and ATP as well as $\gamma[S]ATP$ alone on the stability and recovery of enriched

clones. γ[S]ATP used alone for the enrichment gave a much higher background and subsequently reduced percentage of specific DNA. However, a 1:2 ratio of ATP:γ[S]ATP achieved high recovery of homologous clones with lower levels of nonspecific background.

3. Another important factor is the quality of biotinylated PCR probe. We have found that 50–75% incorporation of biotin-21-dUTP into the PCR product allows successful enrichment of desired cDNA clones. Among different sources of biotin-dUTP, biotin-21-dUTP was found to give better enrichment by RecA protein as a result of its 21-mer-long spacer arm bio-dUTP than with analogs with shorter arms. The optimal size of a biotinylated probe was found between 200 and 600 bp. In addition, DNA probes as small as 100 bp could also be used with comparable efficiency. However, the biotinylated oligonucleotide probes with 28- and 48-mer length were not efficient and generate high level of nonspecific background.

4. An accurate estimation of the ratio of biotinylated probe and RecA protein in the reaction has also been found to be an important factor. We have calculated the exact amount of RecA protein required based on the amount of biotinylated probe according to the following formula: Amount of RecA (mol) = Amount of probe/990. According to this formula, we have determined that 51 pmol of RecA protein could be used for every 50 ng of biotinylated probe, which is approximately equal to 2 μg of RecA protein. This formula is based on the general knowledge that one monomer of RecA protein binds to three nucleotide bases on the single-stranded probe. This formula requires only the correct estimation of concentration of biotinylated probe, which could be done by a spectrophotometer or fluorometer.

5. Our experiments show that background is still a problem of this technology. Background is caused by many different factors such as sequence specificity of a biotinylated probe, quality of cDNA library, and nonspecific binding of streptavidin beads. Background may occur if an enrichment probe has some sequence similarity to some repetitive elements or poly(A) sequences. Our experiments with gel-shifting assays showed that up to 30% of the triple-helix complex could form the concatamers. The concatamers could be one source of nonspecific background that cannot be eliminated by washing. Our experiments on the model systems showed that a higher nonspecific background is directly correlated to greater complexity of the cDNA library.

6. In summary, the RecA enrichment technique has several benefits over existing approaches. First, it dramatically reduces the time and effort for screening compared to conventional library screening approaches. The RecA method is also superior to RACE techniques because it is not subject to PCR mutagenesis and it can be used to isolate many cDNAs at the same time. Furthermore, the RecA method does not require any sequence information for the target cDNA(s)—a characteristic that enables its use with complex cDNA mixtures such as uncloned, subtracted cDNA. We see a great deal of potential for this approach in converting defined mixtures of differentially expressed cDNA fragments into full-length

sublibraries. The entire subtracted cDNA mixture can also be used for obtaining cDNA libraries enriched for differentially expressed genes. Sublibraries enriched for populations of full-length cDNAs can be used for subsequent functional analysis as well as for high-throughput cloning of full-length cDNAs.

Acknowledgments

We thank Eric Machleder and Claire Granger for critical editing of the manuscript and Anna Sayre for the preparation of figures and tables.

References

1. Desai, S., Hill, J. E., Trelogan, S., Diatchenko, L., and Siebert, P. D. (2000) Identification of differentially expressed genes by suppression subtractive hybridization (SSH), in *Functional Genomics: A Practical Approach* (Hunt S. P. and Livesey F. J., eds.), Oxford University Press, Oxford, pp. 81–112.
2. Diatchenko, L., Chenchik, A., and Siebert, P. D. (1998) Suppression subtractive hybridization: a method for generating subtracted cDNA libraries starting from poly(A+) or total RNA, in *Gene Cloning and Analysis by RT-PCR* (Siebert, P. D. and Larrick, J. W., eds.), Eaton, Natick, MA, pp. 213–239.
3. Diatchenko, L., Lau, Y.-F. C., Campbell, A., Chenchik, A., Moqadam, F., Huang, B., et al. (1996) Suppression subtractive hybridization: a method for generating differentially regulated or tissue-specific cDNA probes and libraries. *Proc. Natl. Acad. Sci. USA* **93,** 6025–6030.
4. Gurskaya, N. G., Diatchenko, L., Chenchik, A., Siebert, P. D., Khaspekov, G. L., Lukyanov, K. A., et al. (1996) Equalizing cDNA subtraction based on selective suppression of polymerase chain reaction: cloning of jurkat cell transcripts induced by phytohemaglutinin and phorbol 12-myristate 13-acetate. *Anal. Biochem.* **240,** 90–97.
5. Schoen, T. J., Mazuruk, K., Chader, G. J., and Rodrigues, I. R. (1995) Isolation of candidate genes for macular degeneration using an improved solid-phase subtractive cloning technique. *Biochem. Biophys. Research Commun.* **213,** 181–188.
6. Ikomonov, O. C. and Jacob, M. H. (1996) Differential display protocol with selected primers that preferentially isolates mRNAs of moderate- to low-abundance in a microscopic system. *BioTechniques* **20,** 1030–1042.
7. Liang, P. and Pardee, A. B. (1992) Differential display of eukaryotic messenger RNA by means of the polymerase chain reaction. *Science* **257,** 967–971.
8. Velculescu, V. E., Zhang, L., Vogelstein, B., and Kinzler, K. W. (1995) Serial analysis of gene expression. *Science* **270,** 484–487.
9. DeRisi, J., Penland, L., Brown, P. O., Bittner, M. L., Meltzer, P. S., Ray, M., et al. (1996) Use of a cDNA microarray to analyze gene expression patterns in human cancer. *Nat. Genet.* **14,** 457–460.
10. Lockhart, D. J., Dong, H., Byrne, M. C., Follettie, M. T., Gallo, M. V., Chee M. S., et al. (1996) Expression monitoring by hybridization to high-density oligonucleotide arrays. *Nat. Biotechnol.* **14,** 1675–1680.

11. Zhumabayeva, B., Chenchik, A., and Siebert, P. D. (1999) RecA-mediated affinity capture: a method for full-length cDNA cloning. *BioTechniques* **27,** 834–845.
12. Hsieh, P., Camerini-Otero, C. S., and Camerini-Otero, R. D. (1992) The synapsis event in the homologous pairing of DNAs: RecA recognizes and pairs less than one helical repeat of DNA. *Proc. Natl. Acad. Sci. USA* **89,** 6492–6496.
13. Kowalczykowski, S. C. and Eggleston, A. K. (1994) Homologous pairing and DNA strand-exchange proteins. *Annu. Rev. Biochem.* **63,** 991–1043.
14. Rao, B. J., Dutreix, M., and Radding, C. M. (1991) Stable three-stranded DNA made by RecA protein. *Proc. Natl. Acad. Sci. USA* **88,** 2984–2988.
15. Honigberg, S. M., Rao, B. J., and Radding, C. M. (1986) Ability of RecA protein to promote a search for rare sequences in duplex DNA. *Proc. Natl. Acad. Sci. USA* **83,** 9586–9590.
16. Rigas, B., Welcher, A. A., Ward, D. C., and Weissman, S. M. (1986) Rapid plasmid library screening using RecA-coated biotinylated probes. *Proc. Natl. Acad. Sci. USA* **83,** 9591–9595.
17. Hakvoort, T. B. M., Spijkers, J. A. A., Vermeulen, J. L. M., and Lamers, W. H. (1996) Preparation of a differentially expressed, full-length cDNA expression library by RecA-mediated triple-strand formation with subtractively enriched cDNA fragments. *Nucleic Acid Res.* **24,** 3478–3480.
18. Hakvoort, T. B. M., Vermeulen, J. L. M., and Lamers, W. H. (1998) Enriched full-length cDNA expression library by RecA-mediated affinity capture, in *Gene Cloning and Analysis by RT-PCR* (Siebert P. D. and Larrick J. W., eds.), BioTechniques Books, Natick, MA, pp. 259–269.

21

Subtractive Hybridization for the Identification of Differentially Expressed Genes Using Uracil–DNA Glycosylase and Mung-Bean Nuclease

Tsen-Yin Lin and Shao-Yao Ying

1. Introduction

Subtractive hybridization is a technique for detecting differences of gene expression between two populations of RNAs or complementary DNAs (cDNAs). It is based on base pair complementary that nucleic acid sequences in common with the two populations can form hybrids. After hybridization, the hybrids are removed and the sequences in only one population are identified. This technique can be used for detecting differences between the mRNA in different cells, tissues, and organisms, or cells under two different conditions, tissues of various stages of development and growth, and treatments of hormone or other modulators. One of the common uses of subtractive hybridization is for the detection of DNA differences between different genomes or between cell types to identify certain types of genomic rearrangement as well as differentially expressed genes. For example, a recently developed method of subtractive hybridization has been used to identify cDNAs associated with activin-mediated inhibition of cell growth in human prostate cancer cells *(1)*. In addition to the detection of genes differentially expressed in the same cell type under two different conditions, this technique can be used to ascertain the quantity of mRNAs and cDNAs generated *(2)*.

1.1. Principles

There are two different populations of nucleic acids, mRNA and cDNA, in a subtractive hybridization. The cDNA generated from the control group is called the driver, whereas the experimental group is called the tester. The tester

From: *Methods in Molecular Biology, vol. 221: Generation of cDNA Libraries: Methods and Protocols*
Edited by: S.-Y. Ying © Humana Press Inc., Totowa, NJ

is supposed to contain certain sequences the driver lacks. By hybridization between the tester and an excessive amount of the driver (approx 10-fold), the vast majority of the cDNA molecules in the driver, which are identical or homologous to those in the tester, would form driver–tester hybrids rather than double-stranded tester DNAs, therefore leaving only those present in the tester that can be enriched as the specific cDNA species. Basically, this technique includes the following steps: (1) production of the tester and driver, (2) hybridization, (3) subtraction; removing of driver–tester hybrids and excess driver, and (4) isolation of the complete sequence of the remaining target DNA.

1.2. Production of the Tester and Driver

Generally, the tester is produced with the isolation of RNA, which is reverse-transcribed into double-stranded cDNA. To facilitate the elimination of unwanted homologous sequences between two cDNA libraries, modifications of the driver for specific enzymatic digestion *(3)* or by chemical carboxylation of the pyrimidines *(4)* was proposed. To assure efficient subtraction and amplification, cDNA from tester and driver sources is ligated to different primer "adapters"; double-stranded cDNAs are digested with different restriction enzymes to produce a few hundred base pair nucleotides that are amplified by PCR to form the representative amplicons *(5)*.

1.3. Hybridization

After the initiation hybridization, the tester–driver hybrids are removed and excess fresh driver is added to allow second hybridization. The remaining cDNA (the cDNA targeted to be isolated is called target cDNA), is either cloned or used to make a probe. For limited or complex starting material, multiple rounds of the hybridization–subtraction cycling procedure are used. During subtractive hybridization, the hybridization step is driven by the excess driver amplicons, so tester amplicons that have complementary sequences in the driver population rapidly form driver–tester hybrids, whereas sequences unique to the tester population remain single-stranded or form tester–tester pairs more slowly. Rare sequences from either population take longer to pair up than abundant sequences. The ratio of driver to tester, the overall concentration of driver, the temperature, and the length of hybridization should be chosen based on the complexity of the driver and tester populations, the abundance of the target nucleic acids, and the length of the driver and tester sequences used.

1.4. Subtraction and Removing of Driver–Tester Hybrids and Excess Driver

The purpose of the subtraction step is to remove driver–tester hybrids formed during the hybridization step, leaving behind a tester enriched for the target sequences. Many different methods are used for subtraction, depending on the nature of the driver and the tester. In one method, deoxyuridylate is incorporated into the driver cDNA; uracil–DNA glycosylase is used to remove uracil from driver–driver and driver–tester hybrid duplexes, leaving behind a partially single-stranded sequence that is cleaved by single-strand-specific digestion, leave nondigested sequences for enrichment by PCR.

Immobilizing medium such as hydroxyapatite chromatography is used to bind double-stranded driver and driver–tester hybrids, leaving single-stranded nucleic acids behind, particularly if the driver can be removed chemically or enzymatically, leaving only single-stranded cDNA tester after the subtraction.

Another procedure is to use biotin–streptavidin binding to separate nucleic acids. Streptavidin binds to biotinylated driver sequences, and phenol extraction is used to remove the streptavidin protein and the bound driver and driver–tester hybrids. Streptavidin can also be attached to beads or to a column and used to remove excess driver and driver–tester hybrids.

1.5. Isolation of the Complete Sequence Target DNA

After one or more hybridization and subtraction steps, the resulting tester nucleic acids should be greatly enriched for target sequences. However, it is still possible that rare sequences common to both the driver and the tester remain, and in many cases, the sequences isolated are only partial gene sequences. The remaining tester sequences are isolated and analyzed in a variety of ways. It is necessary to analyze the tester sequences by Northern blotting, *in situ* hybridization, or competitive polymerase chain reaction (PCR) methods to determine whether the sequences are truly tester-specific.

1.6. Subtractive Hybridization by Uracil–DNA Glycosylase and Mung-Bean Nuclease

A process to eliminate unwanted homologous sequences between compared cDNA libraries were developed using nonmodified cDNAs as tester and uracil-incorporated cDNAs (U-DNAs) as the driver has been described previously *(3)*. In this method, the elimination of nondifferentially expressed homologs is achieved by hybridization of the tester to the compared driver and subsequent digestion of driver-bound hybrids with uracil-dependent endonucleases. In this way, only differentially expressed genes in the tester can be preserved after the enzymatic digestion and final PCR amplification. To prevent crossover digestion among gene transcripts with a small homologous domain, the compared cDNA libraries are preferentially restricted by a four-cutting enzyme (such as Hpa2) and ligated to a specific adaptor to yield sequences ranging from 150 to 1000 bp with low complexity. This step also allows for a greater completeness during subtractive hybridization and differential amplification (*see* **Fig. 1**) *(4)*.

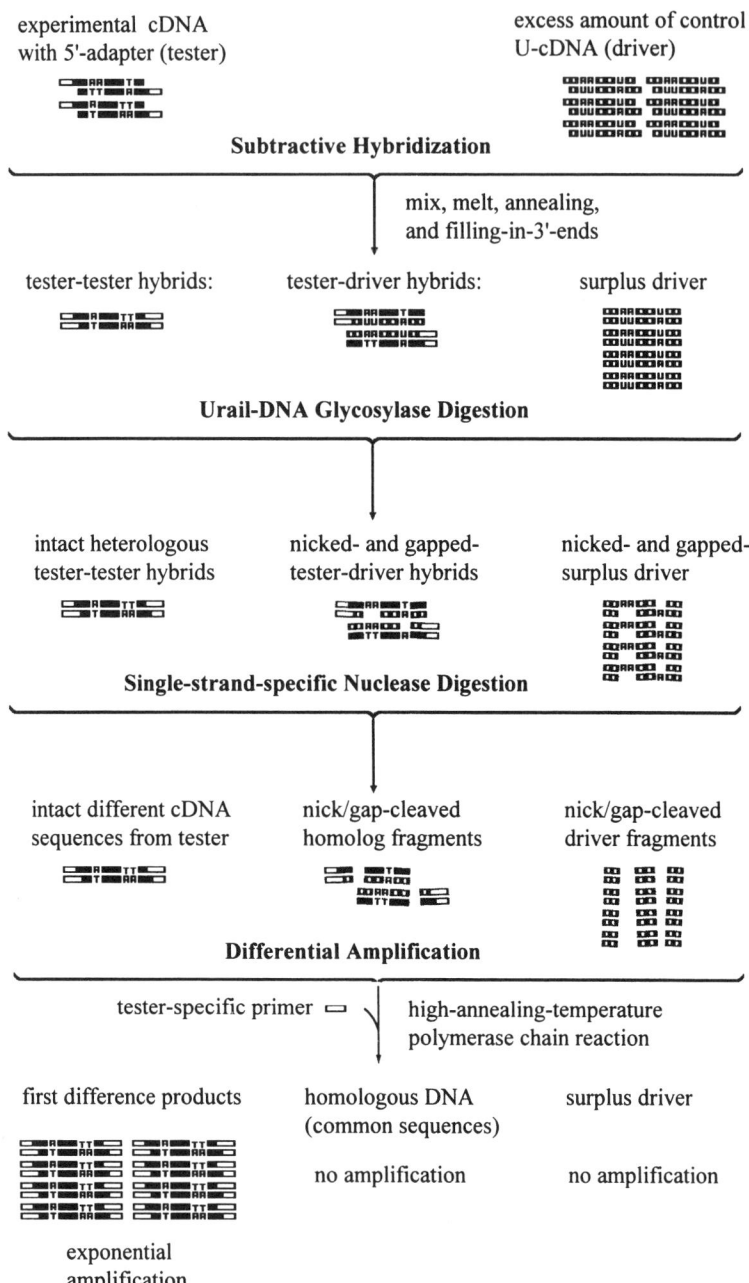

Fig. 1. Schematic protocol for uracil–DNA subtraction assay (USA), illustrating the sequential enzymatic digestion and differential amplification steps after subtractive hybridization. The enzymatic digestion contains two substeps: uracil–DNA *(continued)*

In this method, an activin-treated LNCaP cDNA library was used as tester and an untreated one as driver, and vice versa, in which the driver always contained uridine. After subtractive hybridization of tester to driver and then enzymatic digestion, the differentially expressed genes were amplified and displayed on an electrophoresis gel, from which the results were extracted and confirmed by Northern blot analysis (*see* **Fig. 2**). Based on the sequence data of the results from a Genbank search, 12 highly differentially expressed genes between the activin-treated and untreated LNCaP cells were identified (*see* **Fig. 3**).

1.7. Other Subtractive Hybridization Methods

1.7.1. Differential Display

Differential display is a technique capable of detecting very small changes in the expression of RNA species. However, it is a labor-intensive technique, potentially time-consuming with some limitations.

1.7.2. Representational Difference Analysis

Representative difference analysis (RDA) is a positive selection technique employing polymerase chain reaction (PCR). RDA was originally used to identify differences between complex genomes, such as those caused by chromosomal rearrangements or losses resulting from cancer, infections with pathogens, and polymorphisms between individuals *(6)*, and it was later adapted to analyze differences in gene expression *(7)*. In both cases, the tester and driver are ligated to adapters and amplified by PCR, the original adapters are removed, and new adapters (T2) are ligated only to the tester. After hybridization, only tester–tester DNA is amplified by using primers specific for the T2 adapter. The amplified tester is used again in further rounds of hybridization.

1.7.3. Suppression Subtractive Hybridization

In this positive selection technique, both driver and tester are digested with a frequent-cutting restriction enzyme to give blunt ends. The tester is divided into two samples, which are ligated to different adapters, P1 and P2, and then hybridized to excess driver *(8)*. Then, the two tester populations are mixed,

---

Fig. 1. *(continued)* glycosylase digestion, which nicks all driverlike homologs, and single-strand-specific nuclease digestion, which cleaves the homologs into unamplifiable fragments. The process is outlined through the final products of the first round of the USA method. To reiterate another round of subtraction, the first difference products are used as the tester following the same scheme to generate the second difference products and so on. (From **ref. *1*** with permission.)

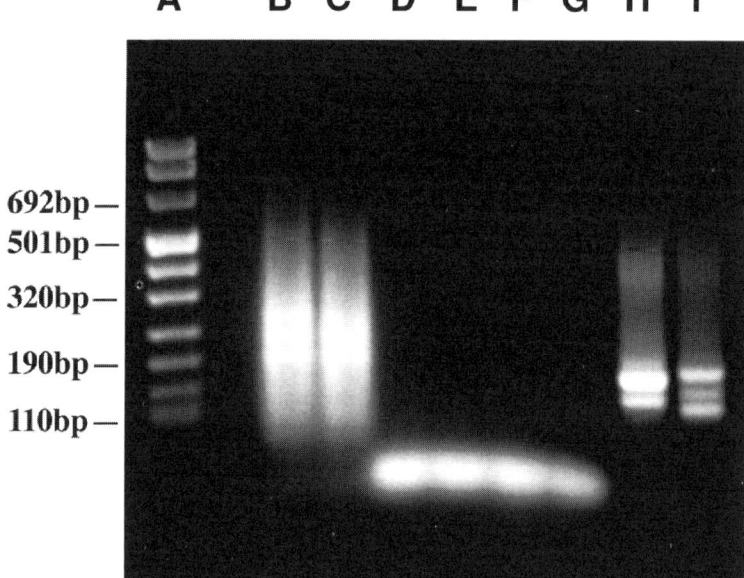

Fig. 2. Agarose gel electrophoresis of PCR-subtracted products. From left to right: lane A, DNA markers; lane B, driver amplicon from untreated LNCaP cells amplified by driver-24 primer; lane C, tester amplicon from activin-treated cells amplified by tester-24 primer; lane D, driver amplicon amplified by tester primer; lane E, tester amplicon amplified by driver primer; lane F, driver amplicon subtracted with driver and then amplified by driver primer; lane G, tester amplicon subtracted with tester and then amplified by tester primer; lane H, driver amplicon subtracted with tester and then amplified by driver primer; lane I, tester amplicon subtracted with driver and then amplified by tester primer. The self-subtraction of driver to driver (lane F) and tester to tester (lane G) shows complete elimination of all sequences, whereas the mutual subtraction between tester and driver (lanes H and I) presents difference products on the gel, indicating differential gene expressed in the tester and driver, respectively. The misuse of PCR primer (lanes D and E) will result in no amplification because of the specific affinity of the primer to its own adaptor. (From **ref. 1** with permission.)

and additional driver is added. Hybrids formed between members of the two subtracted tester populations are selectively amplified by PCR using primers specific to P1 and P2. Molecules that have either P1 or P2 adapters at both ends form "panhandles" as the adapters hybridize to each other. and these molecules are not amplified by PCR (this results in the "suppression").

Suppression subtractive hybridization is better suited to examine potential difference in rare transcripts but only when their expression is changed

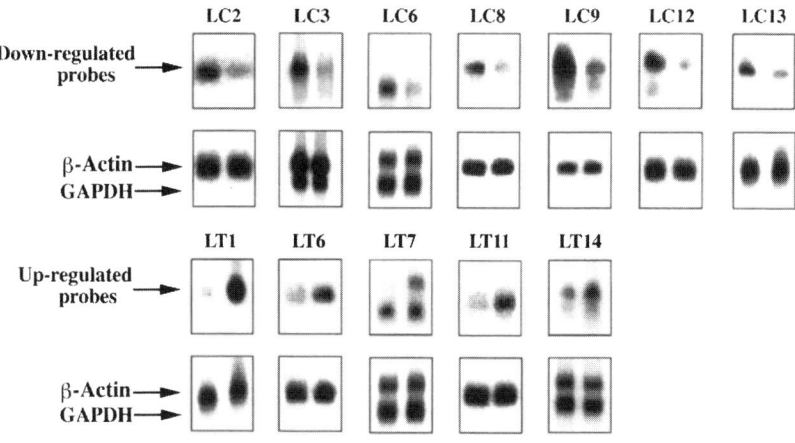

USA Probes:	Gene (Size):	Homology (%):	Change % (σ):	Function:
Down-regulated:				
probe 1 (LC2)	myosin-like (1.1kb)	(99%)	-47.6 (1.46)*	cytoskeleton
probe 2 (LC3)	RHAMM (2.8kb)	+2~+63 (95%)	-55.9 (1.54)*	cytoskeleton
probe 3 (LC6)	novel (1.0kb)		-60.9 (3.05)*	?
probe 4 (LC8)	novel (2.0kb)		-64.9 (3.09)**	?
probe 5 (LC9)	helicase motif-like (1.3kb)	+993~+1162 (95%)	-60.5 (4.66)*	replication
probe 6 (LC12)	Pax2 (3.7kb)	(97%)	-77.0 (2.37)**	proliferation
probe 7 (LC13)	eIF-4A1 (1.7kb)	+487~+557 (100%)	-53.5 (0.00)*	translation
Up-regulated:				
probe 8 (LT1)	novel (0.8kb)		+728 (1.53)**	?
probe 9 (LT6)	CCPK-like (1.8kb)	+432~+470 (100%)	+265 (4.38)**	spindle lesion apoptosis
probe 10 (LT7)	p16 (0.7kb)	+434~+573 (97%)	+354 (0.67)*	G1 arrest
probe 11 (LT11)	p53 (1.7kb)	+112~+262 (100%)	+213 (5.35)**	G1 arrest
probe 12 (LT14)	Siva (1.0kb)	+281~+333 (100%)	+191 (3.19)*	apoptosis

*   n=3, p<0.01
** n=4, p<0.01

Fig. 3. Autoradiogram of positive Northern blots hybridized to the final difference products of USA. Upper panel (LC2 to LC13) indicates seven downregulated genes mainly present in untreated LNCaP cells but not in the activin-treated cells, and the lower panel (LT1 to LT14) shows five upregulated genes significantly increased after activin treatment. p16 (LT7) and p53 (LT11) have been known to be upregulated in the activin-treated LNCaP cells. The downregulated known genes (LC2, 3, 9, 12, 13) are related to physiological functions of cells, and the upregulated known genes (LT6, 7, 11, 14) are involved in either cell cycle regulation, apoptosis or both. Genes listed are transcriptionally altered above twofold. The size of each identified gene transcript is deduced from individual Northern blots, and the homology shown here indicates the sequence similarity between the identified fragment and its deduced gene, rather than the full identified sequence. (From **ref. 1** with permission.)

substantially (>fivefold). Total RNA isolated from two sources of tissues (such as diseased vs normal) is reverse-transcribed to generated cDNA and subtracted, the suppression PCR is combined with subtraction, and the cDNA is digested with a restriction enzyme to generate small fragments (approx 500 bp). Specific DNA adapters are ligated onto one of the cDNAs (usually the tester), which followed by two rounds of hybridization and PCR.

1.7.4. Serial Analysis of Gene Expression

Serial analysis of gene expression (SAGE) is based on the fact that 9 bp of sequence located at its 3′ end is all the sequence information needed to identify a gene unambiguously *(9)*. The first step in SAGE involves generating a 9-bp cDNA tag for each of the mRNAs in a population. Then, many unrelated tags are concatenated, the concatenated tags are cloned, and random clones are sequenced. These sequences give a spectrum of the genes expressed in the tissue and indicate their relative abundance. Many genes can be analyzed at once, because only 9 bp of each gene are sequenced and many tags are sequenced in a single reaction. To be useful for most purposes, however, full-length genes corresponding to the tags must be subsequently identified and isolated.

1.7.5. Microarrays

Complementary DNAs representing either known or unknown genes can be spotted onto glass and probed with different sources of fluorescent-labeled mRNA *(10)*. Alternatively, 20-mer oligonucleotides (oligos) can be synthesized *in situ* in high-density arrays *(11)*. Oligo sequences are derived from known gene sequences and a number of oligos are prepared for each gene, so that there are many internal controls. Arrays can be synthesized that contain hundreds of thousands of oligos and can thus simultaneously monitor differences in the expression of tens of thousands of different genes corresponding to the cDNAs. A number of different probes can be analyzed at once because differently colored fluorescent labels are available. As the full repertoire of expressed sequences becomes available for different organisms, this technique will be extremely useful in monitoring genomewide changes in gene expression in different tissues, developmental states, and mutant backgrounds.

2. Materials

2.1. Oligonucleotides

1. Tester-24-mer (5′-GCCACCAGAAGAGCGTGTACGTCC-3′).
2. Tester-11-mer (5′-CGGGACGTACA-3' with a dephosphorylated 5′-end).
3. Driver-24-mer (5′-CGGTAGTGACTCGGTTAAGATCGC-3′).
4. Driver-11-mer (5′-CGGCGATCTTA-3' with a dephosphorylated 5′-end).

5. The 24-mer oligonucleotides were used as 5′ adaptors and primers for differential PCR, whereas the 11-mer oligonucleotides functioned as linkers for ligation of 5′-end adaptors.

2.2. Generation of Double-Stranded cDNA Libraries and Representative Amplicons

1. Diethyl pyrocarbonate (DEPC) H_2O: Stir double distilled water with 0.1% DEPC for more than 12 h and then autoclave at 120°C under about 1.2 kgf/cm^2 for 20 min, twice.
2. Washing buffer: 50 m*M* sodium citrate (pH 7.6), 0.5 *M* NaCl, 1 m*M* EDTA, 0.1% sodium laurylsarcosinate, and 0.1% DEPC.
3. cDNA double-stranding buffer: 500 m*M* Tris-HCl (pH 9.2) at 25°C, 160 m*M* $(NH_4)_2SO_4$, and 20 m*M* $MgCl_2$; prepare fresh.
4. Incubation chamber: 70°C.
5. Incubation mixer: 54°C, 100*g* vortex for 30 s betweem every 3-min interval.
6. Poly(dT)24 primer: dephosphrylated 5′-dTTTTTTTTTTT TTTTTTTTTT TTTT-3′ (100 pmol/mL).
7. AMV reverse transcriptase (50 U/mL) and 10X reverse transcription buffer (500 m*M* Tris-HCl [pH 8.5] at 25°C, 80 m*M* $MgCl_2$, 300 m*M* KCl, and 10 m*M* dithiothreitol [DTT]).
8. Reverse transcriptase mix: 7 μL DEPC-treated ddH_2O, 2 μL 10X reverse transcription buffer, 2 μL 10 m*M* dNTP mix (10 m*M* each of dATP, dGTP, dCTP, and dTTP), 1 μL Rnasom (25 U/mL), and 2 μL AMV reverse transcriptase; prepare just before use.
9. Incubation chamber: 65°C, 42°C, and 50°C.
10. Enzyme cocktail: DNA polymerase I, RNase H, and T4 ligase mixture.

2.3. Generation of Uracil–cDNA Driver

1. Enzymatic digestion: Hpa2 (3 h, 37°C) for double-stranded cDNA digestion.
2. 100-bp-cutoff microconcentrator columns (Amicon, Beverly, MA).
3. Ligase buffer.
4. dNTP mixture 1: 0.2 m*M* dATP, 0.2 m*M* dGTP, 0.2 m*M* dCTP, 0.05 m*M* dTTP, and 0.5 m*M* dUTP.
5. *Taq* DNA polymerase (5 U) in 1X PCR buffer.
6. Microconcentrator column.
7. Subtractive hybridization buffer: 3 m*M* *N*-(2-hydroxyethyl)piperazine-*N*-3-propansal fonic acid (EPPS) (pH 8.0), 0.5 *M* NaCl, and 3 m*M* EDTA.
8. 2% Agarose gel.

2.4. Subtractive Hybridization, Enzymatic Digestion, and Differential Amplification

1. Subtractive hybridization buffer: 3 m*M* EPPS (pH 8.0), 0.5 *M* NaCl, and 3 m*M* EDTA.

2. Incubation mixer: 37°C, 100*g* vortex for 30 s between every 30-min interval.
3. Phenol.
4. NaSCN.
5. Uracil–DNA glycosylase.
6. Mung-bean nuclease (MBN) and S1 nuclease.

3. Methods

In this section, we describe a detailed protocol for subtraction hybridization, including the isolation of mRNAs, preparation of cDNA libraries, and formation of representative amplicons,

3.1. Generation of Double-Stranded cDNA Libraries and Representative Amplicons

All routine techniques and DNA manipulations, including gel electrophoresis, plasmid preparations, and transformations, were performed according to standard procedures. All enzymes and buffer treatments were applied following the manufacture's recommendations (Boehringer-Mannheim, Indianapolis, IN). For Northern blots, mRNAs were fractionated on 1% formaldehyde–agarose gels and transferred onto nylon membranes (Schleicher & Schuell, Keene, NH). Probes were labeled with the Prime-It II kit (Stratagene, La Jolla, CA) by random primer extension in the presence of [^{32}P]-dATP (>3000 Ci/m*M*; Amersham International, Arlington Heights, IL) and purified with Micro Bio-Spin chromatography columns (Bio-Rad, Hercules, CA). Hybridization was carried out in the mixture of 50% freshly deionized formamide (pH 7.0), 5X Denhardt's solution, 0.5% sodium dodecyl sulfate (SDS), 4X saline sodium phosphate EDTA buffer (SSPE) (Sigma), and 250 mg/mL denatured salmon sperm DNAs (18 h, 42°C). Membranes were sequentially washed twice in 2X SSC, 0.1% SDS (15 min, 25°C), and once each in 0.2X SSC, 0.1% SDS (15 min, 25°C); and 0.2X SSC, 0.1% SDS (30 min, 65°C) before autoradiography.

1. The first strand of cDNAs is prepared by reverse transcription of the mRNAs with oligo-(dT) primers following the protocol of a cDNA Cycle kit (Invitrogen, Carlsbad, CA).
2. The quality is assessed on 2% agarose gel.
3. The second strand of cDNAs is synthesized with an enzyme cocktail containing DNA polymerase I, RNase H, and T4 ligase, as reported by Gubler and Hoffman *(12)*.
4. To generate adequate lengths of cDNA amplicons for efficient subtraction and amplification, double-stranded cDNAs (1.5 mg) are digested with Hpa2 (3 h, 37°C).
5. The products are recovered by 100-bp-cutoff microconcentrator columns (Amicon, Beverly, MA), then ligated to either Tester-24/11-mer or Driver-24/

11-mer adaptors in a mixture containing the 24/11-mer oligo (0.75 nmol each) in 30 mL of 1X ligase buffer.

6. For precise ligation between restricted cDNAs and the adaptors, the mixture is held by gradually cooling from 50°C to 10°C over a period of 1 h and then T_4 ligase is added to anneal the 24-mer oligonucleotides onto the 5′ ends of the restricted cDNAs, at 16°C for 14 h.

7. The products are the representative amplicons for both the tester and driver, depending on the distinctive adaptor used.

3.2. Generation of Uracil–cDNA Driver

1. For incorporation of deoxyuridylate into the driver cDNAs, multiple PCR reactions are set up as follows: Each 50 mL reaction contains driver amplicon (10 ng), 1 mM Driver-24-mer oligo, dNTP mixture 1 (0.2 mM dATP, 0.2 mM dGTP, 0.2 mM dCTP, 0.05 mM dTTP, and 0.5 mM dUTP), and *Taq* DNA polymerase (5 U) in 1X PCR buffer.

2. The Driver-11-mer linker is melted away (3 min, 72°C), and the recessed 3′ ends are filled in with *Taq* DNA polymerase (3 min, 72°C).

3. Thirty cycles of amplification are performed (1 min, 95°C; 3 min, 72°C), and the PCR products are combined and recovered by a microconcentrator column in Tris buffer (10 mM, pH 7.0).

4. The driver–adaptor is removed by Hpa2 cleavage and the digest was recovered by a microconcentrator column in EPPS and EDTA (EE) × 3 buffer (30 mM EPPS (pH 8.0) at 20°C, 3 mM EDTA) at 1 mg/mL to form the driver.

5. The quality of the driver (2 mg) is assessed on 2% agarose gel, ranging from 100 bp to about 1 kb.

3.3. Subtractive Hybridization, Enzymatic Digestion, and Differential Amplification

1. For the first subtractive hybridization, 0.2 mg of tester amplicon (tester) is mixed with 10 mL restricted driver, overlaid with mineral oil, denatured (5 min, 98°C), and immediately cooled on ice.

2. Hybridization is performed with the phenol emulsion reassociation technique in a 400 mL solution containing 1.5 M NaSCN and 8% phenol at 25°C for 48 h *(9,10)*.

3. An emulsion of the phenol and aqueous phases is maintained throughout the hybridization by continuous agitation on a vortex mixer.

4. The hybridized DNAs are recovered by a microconcentrator column in 20 mL of Tris buffer and treated with DNA polymerase I–T_4 DNA polymerase 3:1 mixture (3 min, 37°C without dNTPs; 25 min, 37°C with dNTP mixture 2 in 0.2 mM each for dATP, dGTP, dCTP, and dTTP) to fill in the 3′ end of the tester.

5. In order to eliminate driver homologs in the hybridized DNAs, uracil–DNA glycosylase (UNG) is added (30 min, 25°C) to remove uracil from driver–driver and driver–tester hybrid duplexes, This will result in a partially single-stranded conformation with abasic nicks and gaps.

6. The abasic homolog duplexes are then subjected to digestion by a single-strand-specific endonuclease, such as MBN and S1 nuclease (2 U at 25°C for 20 min), and finally cleaved into unamplifiable fragments (<50 bp).

7. To ensure the completeness of homolog elimination, the hybridization and enzymatic digestion described in **steps 1–6** are repeated at least once.

8. The final digest is recovered by a microconcentrator column in 20 mL of Tris buffer and prepared for differential PCR in a 50-mL reaction, containing 2 mL of the digest, 1 m*M* Tester-24-mer oligo, dNTP mixture 2, and *Taq* DNA polymerase.

9. Final PCR products are phenol-extracted, ethanol-precipitated, resuspended in 20 mL of Tris buffer, and assessed on 3% agarose gel.

10. The DNA bands shown on the electrophoresis gel are excised, recovered by a gel extraction kit (Qiagen), and further purified by 4% nondenaturing polyacrylamide gel.

11. The processes of hybridization, enzymatic digestion, and amplification are repeated until a clear banding pattern was observed.

3.4. Cloning and Sequencing of Differential Products

1. Final products of USA are ligated to the pCR2.1 plasmid and transformed into INVaF' cells using a TA cloning kit (Invitrogen).

2. Double-stranded plasmid DNAs are purified by miniprep spin columns (Qiagen) and sequenced by a Sequenase v.2 DNA sequencing kit (Amersham) with dideoxy-mediated chain termination.

3. Resulting sequences are searched and compared to the Genbank database using the BLAST program from the National Institutes of Health (NIH).

References

1. Lin, S. L. and Ying, S. Y. (1999) Differentially expressed genes in activin-induced apoptotic LNCaP cells. *Biochem. Biophys. Res. Commun.* **257,** 187–192.

2. Ying, S. Y. and Lin, S. L. (1999) High-performance subtractive hybridization of cDNAs by covalent bonding etween specific complementary nucleotides. *BioTechniques* **26,** 966–979.

3. Liang, P. and Pardee, A. B. (1992) Differential display of eukaryotic messenger RNA by means of the polymerase chain reaction. *Science* **257,** 967–971.

4. Lisitsyn, N., Lisitsyn, N., and Wigler, M. (1993) Cloning the differences between two complex genomes. *Science* **259,** 946–951.

5. Hubank, M. and Schatz, D. G. (1994) Identifying differences in mRNA expression by representational difference analysis of cDNA. *Nucleic Acids Res.* **22,** 5640–5648.

6. Zhang, Z., Zheng, J., Zhao, Y., Li, G., Batres, Y., Luo, M. P., et al. (1997) Overexpression of activin A inhibits growth, induces apoptosis, and suppression tumorigenecity in an androgen-sensitive human prostate cancer cell line LNCaP. *Int. J. Oncol.* **11,** 727–736.

7. Wang, Q. F., Tilly, J. L., Preffer, F., Schneyer, A. L., Crowley, W. F., Jr., and Sluss, P. M. (1996) Activin inhibits basal and androgen-stimulated proliferation and

induces apoptosis in the human prostatic cancer cell line, LNCaP. *Endocrinology* **137,** 5476–5483.

8. Sambrook, J., Fritsch, E. F., and Maniatis, T. (1989) *Molecular Cloning, A Laboratory Manual.* Cold Spring Harbor Laboratory, Cold Spring Harbor, NY.

9. Kohne, D. E., Levison, S. A., and Byers, M. J. (1977) Room temperature method for increasing the rate of DNA reassociation by many thousandfold: the phenol emulsion reassociation technique. *Biochemistry* **16,** 5329–5341.

10. Travis, G. H. and Sutcliffe, J. G. (1988) Phenol emulsion-enhanced DNA-driven subtractive cDNA cloning: isolation of low-abundance monkey cortex-specific mRNAs. *Proc. Natl. Acad. Sci. USA* **85,** 1696–1700.

11. Medori, R., Tritschler, H.J., and Gambetti, P. (1992) Production of single-stranded DNA for sequencing: an alternative approach. *BioTechniques* **12,** 132–135.

12. Gubler, U. and Hoffman, B. J. (1983) A simple and very efficient method for generating cDNA libraries. *Gene* **25,** 263–269.

22

Subtractive Cloning of Differential Genes Using RNA-PCR

Shao-Yao Ying and Shi-Lung Lin

1. Introduction

The ability to compare two different messenger RNA (mRNA) or complementary DNA (cDNA) libraries has permitted studies into the role of differentially expressed genes involved in the mechanisms of neoplastic transformation, developmental regulation, therapeutic effects, pathological disorders, and cell-physiological phenomena. Understanding the alterations of gene expression and chromosomal rearrangement between normal and disordered cells is important for gene therapy, eugenical improvement, pharmaceutical design, and etiological investigation.

Several methods have been designed to detect and isolate different cDNA sequences that are present in one cDNA *(1)* or mRNA *(2,3)* library but absent in another. The most commonly used one is subtractive hybridization cloning, which involves the elimination of homologous sequences from the mixture of two mutually compared nucleotide libraries. This method relies upon the formation of homologous hybrid duplexes between a control cell library (subtracter) and another cell library of experimental treatments, disorders, or morphological/functional changes (tester). After the homolog is removed, the remaining cDNA sequences are the desired differentially expressed genes only present in the tester, which is highly related to the treatments, disorders, or changes of interest.

The feasibility of subtractive cloning, however, is usually limited by the generation of abundance of mRNA, limited number of cells, and heterogenous tissue sources. To reach sufficient subtraction force, a minimal amount of 20 µg mRNA or 200 µg total RNA is needed for the library of subtracter. The use

From: *Methods in Molecular Biology, vol. 221: Generation of cDNA Libraries: Methods and Protocols*
Edited by: S.-Y. Ying © Humana Press Inc., Totowa, NJ

of RNA-PCR (polymerase chain reaction)-derived poly(A⁺) RNA library can overcome such problem and provides reproducible mRNA/cDNA sources for a more consistent library comparison. Moreover, the RNA-PCR is able to amplify mRNA/cDNA libraries from a few homologous tissue cells, which facilitate in vivo gene analysis at the single-cell level. Furthermore, the resulting cDNA is amplified by RNA-PCR, which has been proven to be of high fidelity, lineage, and reproducibility *(4)*. Use a few homogenous cells microdissected, subtractive hybridization, and the RNAPCR technique to clonc differentially expressed genes is called subtractive cloning. For the subtractive cloning of unknown differential genes, the present method provides a convenient and reliable approach.

2. Materials

2.1. Immobilization of a Poly(A⁺) RNA Library by Oligo-(dT)-Linked Beads

1. Diethyl pyrocarbonate (DEPC) H_2O: Stir double-distilled water (ddH_2O) with 0.1% DEPC for more than 12 h and then autoclave at 120°C under about 1.2 kgf/cm^2 for 20 min, twice.
2. Olig-o(dT)-linked beads (such as Oligotex resin, Qiagen, Valencia, CA).
3. Washing buffer: 50 mM sodium citrate (pH 7.6), 0.5 M NaCl, 1 mM EDTA, 0.1% sodium lauryl sarcosinate, and 0.1% DEPC.
4. Crosslinking buffer: 50 mM sodium citrate (pH 7.0), 0.5 M NaCl, 1 mM EDTA, 20% formamide, 5% dimethyl sulfoxide (DMSO), and 4 mM ascorbic acid; prepare just before use.
5. Crosslinking agent: 5 mM 2,5-diaziridinyl-1,4-benzoquinone, prepare just before use.
6. Subtractive hybridization buffer: 3 mM EPPS (N-(2-hydroxyethyl)piperazine-N-3-propansulfonic acid) (pH 8.0), 0.5 M NaCl, and 3 mM EDTA.
7. Incubation chamber: 70°C.
8. Incubation mixer: 54°C, 100g vortex for 30 s between every 3-min interval.

2.2. Preparation of a cDNA Library

1. Poly(dT)$_{24}$ primer: dephosphorylated 5′-dTTTTTTTTTTT TTTTTTTTTT TTTT-3′ (100 pmol/μL).
2. AMV reverse transcriptase (50 U/μL) and 10X reverse transcription buffer (500 mM Tris-HCl [pH 8.5] at 25°C, 80 mM MgCl$_2$, 300 mM KCl, and 10 mM dithiothreitol [DTT]).
3. Reverse transcriptase mix: 7 μL DEPC-treated ddH_2O, 2 μL 10X reverse transcription buffer, 2 μL of 10 mM dNTP mix (10 mM each of dATP, dGTP, dCTP, and dTTP), 1 μL RNasin (25 U/μL), and 2 μL AMV reverse transcriptase; prepare just before use.
4. Incubation chamber: 65°C, 42°C, and 50°C.

5. Purification spin column: 100 bp cutoff filter, such as a Microcon-50 centrifugal filter (Amicon, Beverly, MA).

2.3. Hybridization of the cDNA Library to the Immobilized Poly(A⁺) RNAs

1. Subtractive hybridization buffer: 3 mM EPPS (pH 8.0), 0.5 M NaCl, and 3 mM EDTA.
2. Incubation mixer: 37°C, 100g vortex for 30 s between every 30 min interval.

2.4. Amplification of Nonhybridized cDNAs Using RNA-PCR

1. Terminal transferase reaction mix: 4 µL DEPC-treated ddH$_2$O, 3 µL 10X terminal transferase tailing buffer (500 mM Tris-HCl [pH 8.0] at 25°C, 400 mM KCl, 80 mM MgCl$_2$, 100 mM DTT; prepare fresh), 1 µL 10 mM dCTP, and 2 µL terminal deoxynucleotidyl transferase (25 U/µL); prepare just before use.
2. Purification spin column: 100 bp cutoff filter.
3. Oligo-(dG)$_{10}$N-T7 RNA promoter primer mix: dephosphorylated 5′-GGCAGT GAAT TGTAATACGA CTCACTCACT ATAGGGAAGG CGGGGGGGGN-3′ (N = A, T, or C; total 100 pmol/µL).
4. 10X RT&T buffer: 600 mM Tris-HCl (pH 8.3) at 25°C, 300 mM KCl, 80 mM MgCl$_2$, 100 mM DTT, and 5 M betaine.
5. Double-stranding reaction mix: 6 µL DEPC-treated ddH$_2$O, 1 µL 10X RT&T buffer, 1 µL 10 mM dNTP mix (10 mM each for dATP, dGTP, dCTP, and dTTP), 1 µL *Taq* and Pwo DNA polymerase mixture (5 U/µL); prepare just before use.
6. RNA-PCR reaction mix: 4 µL 10X RT&T buffer, 4 µL 10 mM dNTP mix (10 mM each for ATP, GTP, CTP, and UTP), 4 µL 10 mM dNTP mix (10 mM each for dATP, dGTP, dCTP, and dTTP), 1 µL RNasin (25 U/µL), 2 µL T7 RNA polymerase (80 U/µL); prepare just before use.
7. Poly(dT)$_{24}$ primer: dephosphorylated 5′-dTTTTTTTTTTT TTTTTTTTTTT TTTT-3′ (100 pmol/µL).
8. AMV reverse transcriptase (50 U/µL).
9. Incubation chamber: 65°C, 42°C, 50°C, 94°C, and 68°C.
10. Incubation mixer: 37°C, 100g vortex for 30 s between every 30 min interval.

3. Methods

3.1. Immobilization of a Poly(A⁺) RNA Library by Oligo-(dT)-Linked Beads

The subtraction force is provided by the homologous affinity of 20 µg RNA-PCR-derived RNA *(1)* or at least 200 µg total RNA to a compared first-strand cDNA library. Poly(A⁺) RNA is selectively bound by beads through poly(dT) ligand, which contains about 20–30 deoxythymidylate oligonucleotides. Many kinds of commercial oligo-(dT)-linked beads are available for mRNA purification, including cellulose, Sephadex, latex resin, and so on. They are all suitable for the subtractive cloning of differential expressed genes as shown here. For demonstration purposes, we use oligo-(dT)-latex resin as an example because of its high binding efficiency.

1. Hybridization of poly(A$^+$) RNA to oligo-(dT)-latex: Suspend 20 µg mRNA in 40 µL of washing buffer, mix well with 40 µL oligo-(dT)$_{30}$-latex beads, heat to 70°C for 5 min to minimize secondary structure, cool to 54°C for 15 min for primer hybridization, and then cool on ice. Occasionally, mix the reaction every 3 min to prevent precipitation of beads. The above procedure can be repeated once for better binding coverage.
2. Covalent immobilization: Add 100 µL of crosslinking buffer and 10 µL of crosslinking agent into the reaction and mix well. Incubate the reaction at 54°C for 10 min.
3. Washing and collecting RNA-bound beads: Spin the reaction 5 min at 14,000g and discard the suspension solution. Add 200 µL of washing buffer into the microtube and mix well with the bead precipitate. Spin 5 min at 14,000g and discard the suspension solution. Add 200 µL of subtractive hybridization buffer into the microtube and mix well with the beads. Spin 5 min at 14,000g and discard the suspension solution. Again, add 40 µL of subtractive hybridization buffer into the microtube and mix well with the beads. Separate the 40 µL of the purified RNA-bound beads into two aliquots and store at a –80°C freezer or perform the next step immediately.

3.2. Preparation of a cDNA Library

Following the procedure described in Chapter 12, a first-strand cDNA library is prepared by reverse transcription and can be further amplified by RNA-PCR for screening differential genes after subtractive cloning. The starting cDNA library is generated by reverse transcription of 0.1 µg total RNA. Poly(A$^+$) RNA is selected using poly(dT) primers, which contains about 20–26 deoxythymidylate oligonucleotides.

1. Primer annealing: Suspend RNA in 5 µL of DEPC-treated water, mix well with 1 µL poly(dT)$_{24}$ primer, heat to 65°C for 5 min for minimizing secondary structure, cool to 50°C for 1 min for primer hybridization, and then cool on ice.
2. First-strand cDNA synthesis: Add 14 µL of first reverse transcriptase mix and heat to 42°C for 50 min. Add another 1 µL of reverse transcriptase and mix. Continue to incubate the reaction at 42°C for 30 min, heat to 50°C for 10 min, and then cool on ice. The RNA is still attached noncovalently to the cDNA.
3. Denaturation: Heat the reaction at 94°C for 3 min and then cool on ice immediately.
4. Primer removal and buffer exchange: Load the reaction into a purification spin column, spin 10 min at 14,000g, and discard the flowthrough. Add 200 µL of DEPC-treated ddH$_2$O into the spin column to wash the cDNA, spin 10 min at 14,000g, and discard the flowthrough. Add 20 µL of subtractive hybridization

buffer into the spin column to dissolve the cDNA, place the spin column upside down in a new collecting microtube, and spin 3 min at 3000g. Store the 20 µL of the purified cDNA in a –20°C freezer or perform the next step immediately.

3.3. Hybridization of the cDNA Library to the Immobilized Poly(A⁺) RNAs

When a cDNA sequence is homologous to a RNA sequence on the bead, the homolog will be trapped by the bead and selected out by centrifugation. The cDNAs remaining in the suspension solution after subtraction hybridization are the differential gene products of interest.

1. Subtractive hybridization: Add 20 µL of the above cDNA preparation into 20 µL of the purified RNA-bound beads and mix well. Incubate the reaction at 65°C for 30 min and occasionally mix the reaction every 3 min for more completely subtractive coverage.
2. Sample collection: Spin the reaction 5 min at 14,000g and collect the suspension solution.
3. Subtractive hybridization: Add the above suspension solution into 20 µL of the purified RNA-bound beads and mix well. Incubate the reaction at 65°C for 30 min and occasionally mix the reaction every 3 min for a more complete subtractive coverage.
4. Sample collection: Spin the reaction 5 min at 14,000g and collect the suspension solution. Load the suspension solution into a purification spin column, spin 10 min at 14,000g, and discard the flowthrough. Add 200 µL of DEPC-treated ddH$_2$O into the spin column to wash the cDNA, spin 10 min at 14,000g, and discard the flowthrough. Add 20 µL of DEPC-treated ddH$_2$O into the spin column to dissolve the cDNA, place the spin column upside down in a new collecting microtube, and spin 3 min at 3000g. Store the 20 µL of the purified cDNA in a –20°C freezer or perform the next step immediately.

3.4. Amplification of Nonhybridized cDNAs Using RNA-PCR

The subtracted cDNA is then tailed by poly(dC)-oligonucleotide and flanked by poly(dT)-oligonucleotide in its 3′ and 5′ termini, respectively. These homopolymeric termini not only maintain the full-length conformation of the cDNA but also serve as binding regions for promoter-linked oligo-(dG) and poly(dT) primers (*see* Chapter 12). The final differential cDNAs are therefore displayed by RNA-PCR amplification.

1. TdT tailing reaction: Add 10 µL of terminal transferase reaction mix to the purified cDNA and mix well. Incubate the reaction at 37°C for 30 min and occasionally mix the reaction every 5 min for better tailing coverage. Heat to 94°C for 1 min to stop the tailing reaction.
2. Buffer exchange: Load the reaction into a purification spin column, spin 10 min

at 14,000g, and discard the flowthrough. Add 200 μL of DEPC-treated ddH$_2$O into the spin column to wash the dC-tailed cDNA, spin 10 min at 14,000g, and discard the flowthrough. Add 20 μL of DEPC-treated ddH$_2$O into the spin column to dissolve the dC-tailed cDNA, place the spin column upside down in a new collecting microtube, and spin 3 min at 3000g. Perform the next step immediately.

3. Primer annealing: Add 1 μL of oligo-(dG)$_{10}$N-T7 RNA promoter primer mix into 20 μl of the dC-tailed cDNA and mix well. Heat to 94°C for 3 min, cool to 50°C for 10 min, and then cool on ice.

4. cDNA double-stranding: Add 9 μL of cDNA double-stranding reaction mix to the 21 μL reaction, mix well, and then incubate the reaction at 68°C for 10 min.

5. Primer removal: Load the reaction into a purification spin column, spin 10 min at 14,000g, and discard the flowthrough. Add 200 μL of DEPC-treated ddH$_2$O into the spin column to wash the double-stranded cDNA, spin 10 min at 14,000g, and discard the flowthrough. Add 20 μL of DEPC-treated ddH$_2$O into the spin column to dissolve the double-stranded cDNA, place the spin column upside down in a new collecting microtube, and spin 3 min at 3000g. Perform the next step immediately.

6. In vitro transcription reaction: Add 15 μL of RNA-PCR reaction mix to the 20 μL purified cDNA and mix well. Incubate the reaction at 37°C for 2 h and occasionally mix the reaction every 30 min.

7. Primer annealing: Add 2 μL of poly(dT)$_{24}$ primer into the reaction and mix well. Heat to 65°C for 5 min, cool to 50°C for 1 min, and then cool on ice.

8. Reverse transcription: Add 4 μL of AMV reverse transcriptase into the reaction and mix well. Heat to 42°C for 55 min and then 52°C for 5 min. Add another 1 μL of reverse transcriptase and mix well. Continue to incubate the reaction at 42°C for 30 min, heat to 94°C for 3 min, and then cool on ice.

9. Primer removal: Load the reaction into a purification spin column, spin 10 min at 14,000g and discard the flowthrough. Add 200 μL of DEPC-treated ddH$_2$O into the spin column to wash the double-stranded cDNA, spin 10 min at 14,000g, and discard the flowthrough. Add 20 μL of DEPC-treated ddH$_2$O into the spin column to dissolve the double-stranded cDNA, place the spin column upside down in a new collecting microtube, and spin 3 min at 3000g. Store the 20 μL of the purified cDNA in a –20°C freezer or perform the next step immediately.

10. Cycling amplification: Repeat **steps 1–7** one more time.

11. Differential gene assessment: 3% agarose gel electrophoresis.

References

1. Hubank, M. and Schatz, D. G. (1994) Identifying differences in mRNA expression by representational difference analysis of cDNA *Nucleic Acids Res.* **22,** 5640–5648.

2. Chien, Y., Becker, D. M., Lindsten, T., Okamura, M., Cohen, D. I., and Davis, M. M. (1984) A third type of murine T-cell receptor gene. *Nature* **31,** 31–35.

3. Heilig, J. S., Glimcher, L. H., Kranz, D. M., Clayton, L. K., Greenstein, J. L., Saito,

H., et al. (1985) Expression of the T-cell-specific gamma gene is unnecessary in T cells recognizing class II MHC determinants. *Nature* **317,** 68–70.

4. Lin, S. L., Chuong, C. M, Widelitz, R. B., and Ying, S. Y. (1999) *In vivo* analysis of cancerous gene expression by RNA–polymerase chain reaction. *Nucleic Acid Res.* **27,** 4585–4589.

23

Strategy for Construction of a cDNA Encoding a Repetitive Amino Acid Sequence

Masahiro Asada and Toru Imamura

1. Introduction

The polymerase chain reaction (PCR) is a powerful method permitting generation of almost any desired cDNA sequence. Construction of cDNA fragments that encode repetitive amino acid sequences by PCR has proven problematic, however, as repetitive nucleotide primers tend to amplify undesired fragments (because of their misannealing with the template). To overcome this problem, Dombrowski and Wright *(1)* developed a solid-phase gene-assembly protocol, but because this technique requires preparation of a solid-phase reaction system, it cannot be routinely performed in most laboratories. Here, we describe a new strategy that makes use of standard PCR protocols, yet enables assembly of cDNAs encoding repetitive amino acid sequences *(2)*.

By strategically varying the third nucleotide of each codon, we were able to construct successfully a cDNA that encoded 10 repeats of the peptide AlaThrProAlaPro, a motif for *O*-glycosylation *(3)*. This strategy is not limited to the introduction of clustered *O*-glycans into proteins, but it should also be useful for the creation of other cDNAs encoding repetitive short peptides.

2. Materials

1. DNA polymerase with adequate buffers: *TaKaRa Taq* DNA polymerase (Takara Shuzo, Siga, Japan) was used in the experiments described in this chapter. More recently, we have routinely used *KOD-plus-* DNA polymerase (Toyobo, Osaka, Japan), which works well for the construction of artificial DNAs rich in repetitive sequences.
2. Oligonucleotide primers: Custom oligonucleotide primers were synthesized by Espec Oligo Service Corp. (Tsukuba, Japan), and purified using a C-18 cartridge.

From: *Methods in Molecular Biology, vol. 221: Generation of cDNA Libraries: Methods and Protocols*
Edited by: S.-Y. Ying © Humana Press Inc., Totowa, NJ

3. Thermal cycling machine: Any thermal cycler designed for PCR can be used. However, because the optimal conditions for PCR vary among thermal cyclers, we strongly recommend that the best thermal cycling conditions for the chosen are ascertained. All of the reactions described here were performed using a TaKaRa PCR Thermal Cycler MP (Takara Shuzo).

4. Reagents and equipment for agarose gel electrophoresis: A mupid-2 mini electrophoresis unit (Cosmo Bio, Tokyo, Japan) was used for agarose gel electrophoresis. Agarose (A-6013) was purchased from Sigma–Aldrich Co. (St. Louis, MO), and molecular-weight markers for electrophoresis was purchased from Invitrogen Corp. (Carlsbad, CA).

5. Reagents and kit to extract DNA from agarose gel: The DNA fragment separated by agarose gel electrophoresis was extracted using ultrafree-MC UFC3-0DV-25 (Millipore Corp., Bedford, MA), the manufacturer's instructions (application sheet Vol. L/S 5) being followed exactly.

6. Vector DNA.

7. A complementary DNA (cDNA) encoding the target protein, into which the repetitive amino acid sequence should be introduced.

8. *Escherichia coli* strains and reagents to cultivate the bacteria: DH5α was used for all the experiments described in this chapter. For the cultivation of the bacteria, Luria-Bertani's broth (LB), LB-agar, and ampicillin are also required.

9. Restriction enzymes.

10. Reagents for DNA ligation: All DNA-ligation reactions described here were performed by means of a DNA ligation kit, ver. 2 (Takara, Shiga, Japan). Recently, we found that a Quick Ligation Kit (New England Biolabs, Beverly, MA) also gave good results for our purposes.

3. Methods

3.1. General Strategy for Creation of an Artificial DNA

The strategy described in this chapter is based on the overlap extension protocol *(4)*. As shown schematically in **Fig. 1**, PCR is performed without template DNA, using primers A and B to generate the 5′ half of the desired sequence. In the same way, the 3′ half is generated by PCR using primers C and D. Because these products overlap each other in sequence at the region common to primers B and C, mixing them can cause the DNA fragments to anneal each other and function as a template for the second-round PCR. Thus, a mixture of the first-round PCR products is used as a template, and the second-round PCR is performed using primers E and F. As the result, a DNA fragment with the desired repetitive sequence is generated, as shown in **Fig. 1**. To introduce this DNA fragment into the target cDNA of interest, we then use it as a template for the subsequent overlap extension PCR.

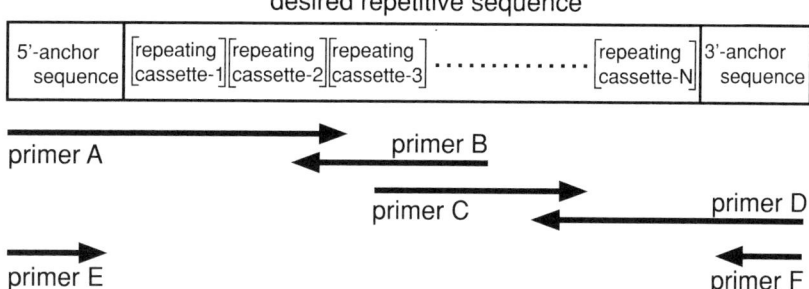

Fig. 1. Outline of the strategy used to construct the DNA encoding a repetitive amino acid sequence. The repetitive sequence comprises an N-times repeat of a peptide-encoding cassette created between anchor sequences. Primers A and B are designed for the 5′ half of the desired DNA and primers C and D for the 3′ half. Primers A and B, or C and D, share complimentary 3′ ends that enable generation of primer dimers.

The products of the first-round PCR (i.e., the 5′ half and 3′ half of the desired DNA) share the same sequence where primers B and C overlap each other. In a mixture, the first-round PCR products therefore anneal each other and can serve as a template for the second-round PCR. In the second-round PCR, primers E and F are used to generate a full-length desired repetitive DNA fragment.

3.2. Designing a cDNA for the Desired Repetitive Amino Acid Sequence

The most critical point in the generation of a cDNA encoding a repetitive amino acid sequence by PCR is the avoidance of undesired misannealing during the PCR process. To avoid misannealing, the third nucleotide of each codon encoding an amino acid should be degenerated as much as possible. In the case of the sequence of a motif for *O*-glycosylation, the amino acids comprising this motif (i.e., alanine [Ala: A], proline [Pro: P], and threonine [Thr: T]) are encoded by codons GCX, CCX, and ACX, respectively, where X is any nucleotide. We therefore designed four nucleotide cassettes that each encoded ATPAP and differed from one another only in the last nucleotide in each codon (*see* **Fig. 2A**).

To avoid misannealing between the cassettes, more than three cassettes of the ATPAP unit were synthesized in a single oligonucleotide format (*see* **Fig. 2B**). In designing the desired cDNA sequence for first-round PCR, the cassettes should not be placed in the same order in the 5′ half and 3′ half (*see* **Note 1**).

At the two ends of the DNA encoding the repetitive amino acid sequence, anchor sequences can be added, and the result would serve as a specific

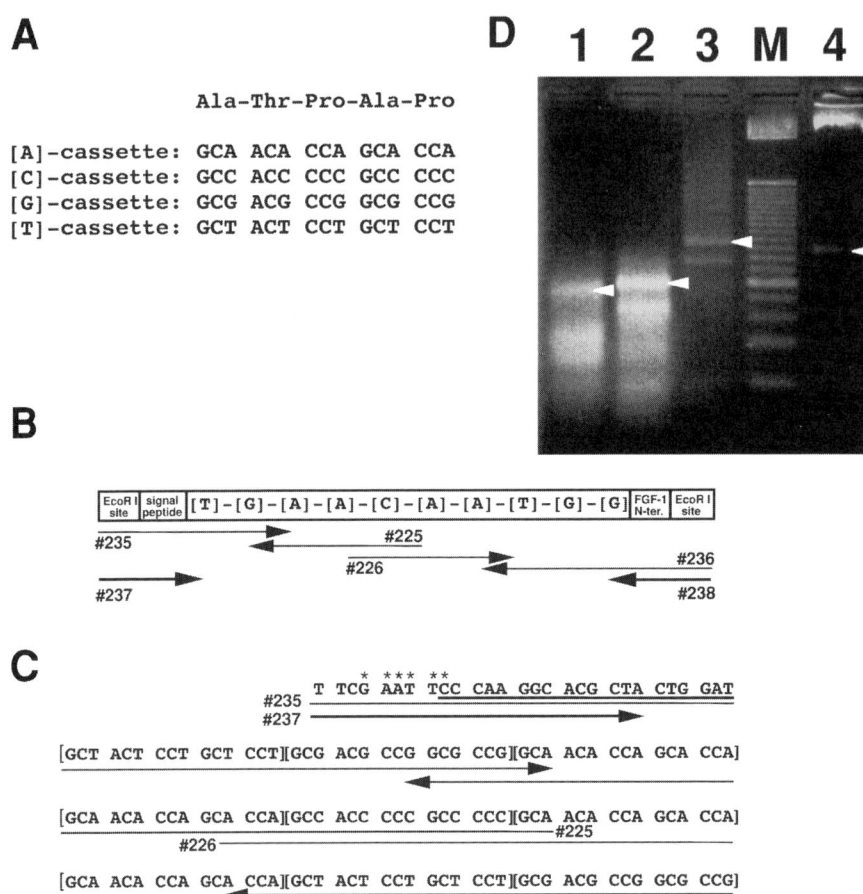

A

Ala-Thr-Pro-Ala-Pro

[A]-cassette: GCA ACA CCA GCA CCA
[C]-cassette: GCC ACC CCC GCC CCC
[G]-cassette: GCG ACG CCG GCG CCG
[T]-cassette: GCT ACT CCT GCT CCT

B

| EcoR I site | signal peptide | [T]-[G]-[A]-[A]-[C]-[A]-[A]-[T]-[G]-[G] | FGF-1 N-ter. | EcoR I site |

#235
#225
#226
#236
#237
#238

C

```
                      *  ***  **
                  T TCG AAT TCC CAA GGC ACG CTA CTG GAT
        #235  ─────────────────────────────────────────
        #237  ─────────

[GCT ACT CCT GCT CCT][GCG ACG CCG GCG CCG][GCA ACA CCA GCA CCA]
                              ◄──────────►

[GCA ACA CCA GCA CCA][GCC ACC CCC GCC CCC][GCA ACA CCA GCA CCA]
     #226 ─────────────────────────────────────────#225

[GCA ACA CCA GCA CCA][GCT ACT CCT GCT CCT][GCG ACG CCG GCG CCG]
                    ◄──────────►

                                          *  ***  **
[GCG ACG CCG GCG CCG] GCT AAT TAC AAG AAG CCC AAG AAT TCC TT
     ◄────────────────────────────────────────────── #236
                                                       #238
```

D 1 2 3 M 4

Fig. 2. Construction of a cDNA encoding a repetitive Ala-Thr-Pro-Ala-Pro sequence. (**A**) Four oligonucleotide cassettes, each encoding the pentapeptide *O*-glycosylation unit Ala-Thr-Pro-Ala-Pro. (**B**) Schematic diagram of the construction of a cDNA encoding the [Ala-Thr-Pro-Ala-Pro]$_{10}$ cassette, which is to be inserted between the signal peptide and the N-terminus of FGF-1. Each *O*-glycosylation unit is indicated by a bracket containing the symbol of the last nucleotide in each codon, as shown in panel **A**. PCR protocols were initially carried out separately using oligonucleotides #235 and #225, or #226 and #236, without a template. A mixture of the two reaction products was then subjected to PCR using primers #237 and #238. (**C**) Alignment of the primers with the completed sequence. The nucleotide sequences encoding the signal peptide and the FGF-1 N-terminus are underlined. The *Eco*RI recognition sites are indicated by asterisks. Primers are shown by arrows. (**D**) Agarose gel (4%) electrophoresis of the PCR products and the enzyme digest of the final clone. Lane 1, the first PCR product obtained using #235 and #225 (106 bp); lane 2, the first PCR product obtained using #226 and #236 (123 bp); lane 3, the second PCR product

template for the subsequent overlap extension PCR, in which the generated repetitive sequence is introduced into the target sequence of interest. In general, each anchor sequence should be around 20-mer, to prevent the primers for the second-round PCR from annealing with the repetitive sequence.

3.3. Primer Design for PCR

The design of the PCR primers is the key to the success of this protocol, especially for the avoidance of misannealing. In the first-round PCR, no template is added to the initial reaction mixture; therefore, the reaction products are generated as a so-called primer dimer. To generate the primer dimer as intended, the 3′ end of each primer is very important. To encode the desired repetitive amino acid sequence, the first and second nucleotides of the codon appear many times. Thus, the 3′ end nucleotide in each primer should match with the third nucleotide in the codon to ensure correct annealing of the primers. For effective annealing, the primers used for the first-round PCR should be designed to overlap about 8–10 nucleotides at their 3′ ends.

To enable effective annealing in the second-round PCR, the 5′ half and 3′ half of the entire repetitive sequence should overlap each other by about 20 nucleotides. As in the case of the first PCR, it is preferable that the termini to be elongated (i.e., the 3′-terminal nucleotide of the 5′ half and the 5′ terminal nucleotide of the 3′ half) should correspond to the third nucleotide in the codon (i.e., the 5′-end nucleotide of primer B and that of primer C should correspond to the third nucleotide in the codon).

The design of the primers for the second-round PCR is relatively easy. To avoid their misannealing with the template repetitive sequence, the greatest part of these primers is designed to anneal with the anchor sequences at the two ends of the generating repetitive sequence. This ensures the specificity of the primers used for the second-round PCR. When introducing a repetitive amino acid sequence into the N-terminus of the target protein, it is an effective approach to include the 5′-terminal nucleotide sequence of the respective cDNA as the 3′-anchor sequence. If the repetitive sequence is to be introduced at the C-terminus, the 5′-anchor sequence can be that of the 3′-terminal sequence of the target protein.

An example of the construction of the DNA encoding the repetitive ATPAP sequence for clustered *O*-glycosylation, including the exact nucleotide sequence and the primers used, is illustrated in **Fig. 2C**.

Fig. 2. *(continued)* (207 bp); lane M, 25 bp ladder markers (Invitrogen Corp.); lane 4, *Eco*RI digest of the clone N-[ATPAP]$_{10}$-secFGF/pBS/#9 (195 bp). Arrowheads indicate the DNA bands having the expected sizes. (From **ref. 2** with permission.)

3.4. First-Round PCR

Without a template DNA, the two combinations of forward and reverse primers (primers A and B, or primers C and D, 8 pmol/µL each) were mixed individually with dNTP (0.2 m*M*) and subjected to PCR using adequate DNA polymerase. The choice of DNA polymerase is a critical issue for this protocol (*see* **Note 2**); therefore, it is recommended that several kinds of polymerase be examined to see which gives the best results.

The thermal cycling conditions are also important for satisfactory results. The annealing temperature is critical to avoid undesired byproducts and to obtain a high yield of the expected products. The extension time is not a problem, as the desired sequence is a short DNA. It is recommended that pilot experiments should be performed to determine the optimal conditions for amplification. In the case of the construction of a DNA fragment encoding ATPAP repeats for clustered *O*-glycosylation, the amplification protocol consisted of 30 cycles of 94°C for 30 s, 68°C for 30 s, and 72°C for 30 s.

After resolving the PCR product by agarose gel electrophoresis, the distinct bands of the 5′- and 3′ halves were excised (**Fig. 2D**, lanes 1 and 2; *see* **Note 3**). The gel pieces were frozen in an ultrafree-MC cup for 10 min and then thawed at 37°C. The extracted DNA was collected by centrifugation (11,500*g*, 10 min, at 4°C), purified by ethanol precipitation, and used for the subsequent experiments.

3.5. Second-Round PCR

The DNAs extracted as in **Subheading 3.4.** were mixed and used as the template for the second-round PCR. The quantity of the template is also important for specific amplification in the second-round PCR (*see* **Note 4**). In **Fig. 1**, primers E and F are used for this reaction (8 pmol/µL each).

The thermal cycling conditions should be optimized as in **Subheading 3.4.** In the case of the construction of a DNA fragment encoding ATPAP repeats for clustered *O*-glycosylation, the amplification protocol of the second PCR consisted of 30 cycles of 94°C for 30 s, 60°C for 30 s, and 72°C for 30 s. The reaction product was resolved by agarose gel electrophoresis, and the target DNA fragment was isolated (**Fig. 2D**, lane 3).

3.6. Cloning of the PCR Product into a Cloning Vector

Any protocol can be used to clone the PCR product into a cloning vector. In our experiment, as *Eco*RI recognition sites were introduced by the second-round PCR as anchor sequences, the PCR product was subcloned into a pBluscript® II (KS+) vector (Stratagene) at the *Eco*RI site. The DNA Ligation Kit, ver. 2 (Takara) is useful and effective for ligation, and the resultant ligation mixture was used to transform DH5α *E. coli* cells (*see* **Note 5**).

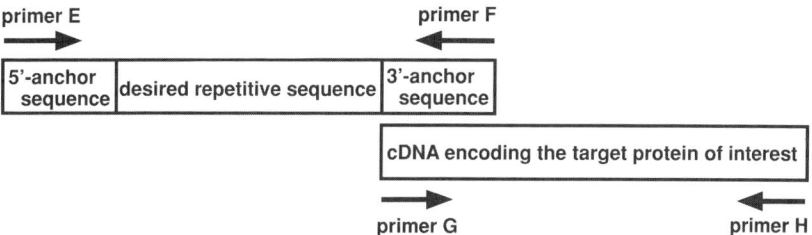

Fig. 3. Introduction of the desired repetitive sequence into the cDNA of interest. As in the case of the second-round PCR, the PCR products generated by primers G and H and primers E and F can anneal at the respective ends of the fragment. Thus, using mixtures of these two reaction products as a template and primers G and F for the third-round PCR, a cDNA is obtained that encodes both the repetitive amino acid sequence and the protein of interest.

Positive clones were initially identified on the basis of their ampicillin resistance and white color. The clones were then screened by direct PCR using vector primers to reveal the size of the inserted sequence, and several promising clones with inserts of the expected size were obtained. The size of the insert was further verified by digestion with *Eco*RI (**Fig. 2D**, lane 4), and the sequence was verified by direct sequencing.

3.7. Introducing the Repetitive Amino Acid Sequence into the Protein of Interest

The resultant DNA encoding the repetitive amino acid sequence is finally linked with a cDNA encoding a protein of interest. This process is also performed using the overlap extension protocol (*see* **Fig. 3**). Because the second-round PCR product contains an anchor sequence sharing a common sequence with the cDNA of the protein of interest, the PCR product obtained by the use of primers G and H can anneal with the PCR product obtained using primers E and F. Therefore, a mixture of the products of each PCR works as a template for the third-round PCR to yield a cDNA encoding the target protein of interest with the desired repetitive amino acid sequence.

In our example, the DNA encoding a cluster of *O*-glycans can be successfully linked to a cDNA encoding secretable fibroblast growth factor (FGF) polypeptide (*5*). The generated cDNA had the expected size and the correct sequence (*6*).

This strategy is an effective way of introducing ATPAP units into any protein of interest. Furthermore, with appropriate modifications, it should be possible to generate cDNAs encoding other repetitive short peptides.

4. Notes

1. In designing the DNA sequence for the desired repetitive amino acids, theoretically the nucleotides designated X (the third nucleotide in each codon) in each cassette could be arranged in a wide variety of ways to eliminate the possibility of misannealing. However, in practice, it is not easy to design the sequence of the nucleotides so as to eliminate emergence of the same sequence. Therefore, we fixed the third nucleotide in each cassette and just shuffled the cassettes to randomize the order of appearance.

2. The choice of DNA polymerase is critical for the success of this protocol. Generally, it is considered that an α-type proofreading polymerse (such as *Pfu* polymerse, Vent polymerse, etc.) is preferable for DNA construction. However, for the present protocol, we believe that the traditional *Taq* polymerase (Pol I-type) is preferable. The reason is that the 3′→5′ exonuclease activity of such a proof-reading polymerase can eliminate mismatched nucleotides if a misannealing of the primers with an undesired region takes place. This leads to the occurrence of mispriming, a consequent amplification of a variety of products differing in size, and a decreased yield of the desired DNA.

    Recently, hybrid- or mixed-type DNA polymerases have become commercially available. Such polymerases, such as *KOD-plus-* polymerase, proved to give good results in our experiments.

3. It is frequently observed that the PCR product appears as smeared or multiple bands on an electrophoresis gel. This is likely to be the result of an unintended amplification reaction and can be ignored. If a distinct band with the expected size can be identified, the DNA extracted from the band is useful for the next process. In our example, as shown in **Fig. 2D** (lanes 1 and 2), several bands appeared in the first PCR product. However, excising only the expected bands gave successful results in subsequent experiments.

4. The amount of the template DNA necessary for the second-round PCR varies among experiments. In our experiments, using one-twentieth to one-fifth of the extract of the first-round PCR product as a template gives a satisfactory results. Using too much template DNA increases the chances of misannealing and of undesired byproducts.

5. Various TA cloning kits are available from a number of companies, such as Invitrogen Corp. and Qiagen Inc. The PCR product, especially the one generated by a Pol I-type DNA polymerase, is easy to clone into one of these vectors by following the standard protocol suggested.

References

1. Dombrowski, K. E. and Wright, S. E. (1992) Construction of a multiple mucin tandem repeat with a mutation in the tumor-specific epitope by a solid-phase gene assembly protocol. *Nucleic Acids Res.* **20,** 6743–6744.
2. Asada, M., Yoneda, A., Oda, Y., and Imamura, T. (2000) Construction of a cDNA encoding a repetitive amino acid sequence. *BioTechniques* **29,** 978–981.

3. Yoshida, A., Suzuki, M., Ikenaga, H., and Takeuchi, M. (1997) Discovery of the shortest sequence motif for high level mucin-type *O*-glycosylation. *J. Biol. Chem.* **272,** 16,884–16,888.

4. Horton, R. M., Cai, Z. L., Ho, S. N., and Pease, L. R. (1990) Gene splicing by overlap extension: tailor-made genes using the polymerase chain reaction. *BioTechniques* **8,** 528–535.

5. Yoneda, A., Asada, M., Suzuki, M., and Imamura, T. (1999) Introduction of an *N*-glycosylation cassette into proteins at random sites: expression of neoglycosylated FGF. *BioTechniques* **27,** 576–582.

6. Yoneda, A., Asada, M., Yamamoto, S., Oki, J., Oda, Y., Ota, K., et al. (2001) Engineering neoglycoproteins with multiple *O*-glycans using repetitive pentapeptide glycosylation units. *Glycoconj. J.* **16,** 321–326.

24

Preparing Lambda Libraries for Expression of Proteins in Prokaryotes or Eukaryotes

Rebecca L. Mullinax and Joseph A. Sorge

1. Introduction

Complementary DNA (cDNA) libraries represent the information encoded in the messenger RNA (mRNA) of a particular tissue or organism. RNA molecules are exceptionally labile and difficult to amplify in their natural form. For this reason, a cDNA library is created by converting the information encoded by the RNA into a stable DNA duplex (cDNA) and then inserting it into a self-replicating lambda or plasmid vector. Individual cDNA representing the original genetic information can be isolated from the cDNA library and examined with relative ease.

Lambda vectors are commonly used to generate high-quality cDNA libraries. For example, the Uni-ZAP® XR vector *(1)* and Lambda ZAP-CMV® XR vector *(2)* systems combine the high efficiency of lambda library construction and the convenience of a plasmid system for expression of proteins in prokaryotes and eukaryotes, respectively.

2. Materials

1. 3 *M* NaOAc: (per 100 mL), 40.82 g NaOAc • 3H$_2$O, and deionized water up to 80 mL. Adjust pH to 5.2 using glacial acetic acid. Adjust volume to 100 mL with additional deionized dH$_2$O. Filter-sterilize or autoclave.
2. 70% (v/v) Ethanol: (per 100 mL) 70 mL molecular-biology-grade ethanol and 30 mL deionized water.
3. ZAP-cDNA® synthesis kit (Stratagene, La Jolla, CA).
4. Uni-ZAP® XR Vector (Stratagene, La Jolla, CA).
5. Lambda ZAP-CMV XR Vector (Stratagene, La Jolla, CA).

From: *Methods in Molecular Biology, vol. 221: Generation of cDNA Libraries: Methods and Protocols*
Edited by: S.-Y. Ying © Humana Press Inc., Totowa, NJ

6. [α-^{32}P]-labeled deoxynucleotide (800 Ci/mmol) ([^{32}P]dATP, [^{32}P]dGTP, or [^{32}P]dTTP.

7. 10X Alkaline agarose: (per 50 mL) 3 mL of 5.0 M NaOH, 2 mL of 0.5 M EDTA, and 45 mL deionized H_2O.

8. Saturated bromophenol blue (BPB): Add a small amount of bromophenol blue crystals to 200 μL water and vortex. Centrifuge briefly and look for the presence of an orange pellet. If a pellet is seen, the solution is saturated. If not, add more crystals and repeat the procedure.

9. Alkaline agarose 2X loading buffer: 200 μL glycerol, 750 μL water, 46 μL saturated BPB, and 5 μL of 5 M NaOH.

10. 10X STE buffer: 1 M NaCl, 200 mM Tris-HCl (pH 7.5), and 100 mM EDTA.

11. 1 mM MgSO$_4$: (per liter) Add 203.3 g MgSO$_4$ • 7 H_2O to 1 L deionized H_2O. Autoclave.

12. 0.5 M isopropyl-1-thio-β-D-galactopyranoside (IPTG): (per 10 mL) Dissolve 1.19 g in 10 mL deionized H_2O. Filter sterilize. Store in 1-mL aliquots at –20°C.

13. X-gal: (per 10 mL) Dissolve 2.5 g in 10 mL dimethyl formamide. Store in the dark at –20°C.

14. SM buffer: (per liter) 5.8 g NaCl, 2.0 g MgSO$_4$ • 7H_2O, 50.0 mL of 1 M Tris-HCl, pH 7.5, 5.0 mL of 2% (w/v) gelatin, and deionized H_2O to a final volume of 1 L.

15. Falcon® 2054 and 2059 polypropylene tubes.

16. LB broth: (per liter) 10 g NaCl, 10 g tryptone, 5 g yeast extract, and deionized H_2O to a final volume of 1 L. Adjust to pH 7.0 with 5 N NaOH. Autoclave.

17. LB-Tetracycline broth: (per liter) Prepare 1 L LB broth. Autoclave. Cool to 55°C and add 1.5 mL of 10 mg/mL filter-sterilized tetracycline. Store broth in the dark, as tetracycline is light sensitive.

18. LB-Tetracycline broth supplemented with 10 mM MgSO$_4$ and 0.2% (w/v) maltose: (per liter). Prepare 1 L LB-tetracycline broth. Add 10 mL of 1 M MgSO$_4$ and 2 g maltose prior to autoclaving.

19. LB-Tetracycline agar: (per liter) Prepare 1 L of LB broth. Add 15 g agar. Autoclave. Cool to 55°C and add 1.5 mL 10 mg/mL filter-sterilized tetracycline. Pour into Petri dishes (approx 25 mL/100-mm plate). Store plates in the dark, as tetracycline is light sensitive.

20. NZY broth: (per liter) 5 g NaCl, 2 g MgSO$_4$ • 7H_2O, 5 g yeast extract, 10 g NZ amine (casein hydrolysate), and deionized H2O to a final volume of 1 L. Adjust the pH to 7.5 with NaOH. Autoclave.

21. LB-Ampicillin agar plates: (per liter) Prepare 1 L of LB broth. Add 15 g agar. Autoclave. Cool to 55°C and add 1.0 mL of 50 mg/mL filter-sterilized ampicillin.

22. LB-Kanamycin agar plates: (per liter) Prepare 1 L of LB broth. Add 15 g agar. Autoclave. Cool to 55°C and add 1.0 mL of 50 mg/mL filter-sterilized kanamycin.

23. NZY top agar: (per liter) Prepare 1 L NZY broth. Add 7 g agarose. Autoclave.

24. NZY agar plates: (per liter) Prepare 1 L NZY broth. Add 15 g agarose. Autoclave. Pour into Petri dishes (approx 80 mL/150-mm plate).

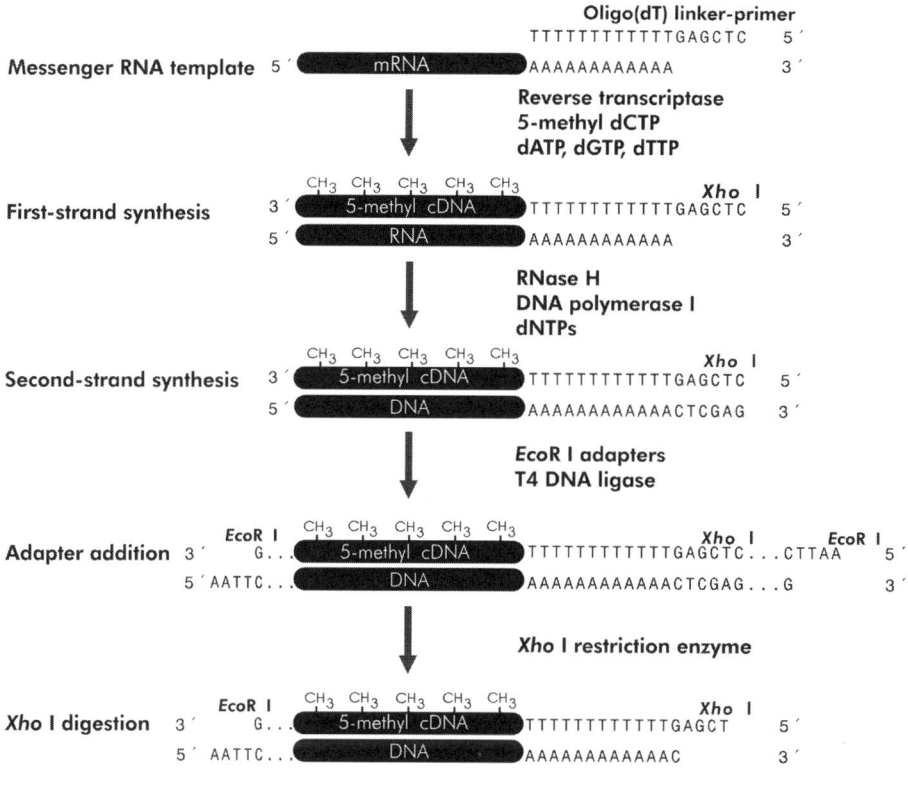

Fig. 1. cDNA synthesis.

3. Methods

The following includes methods for (1) converting mRNA to double-stranded cDNA having *Eco*RI and *Xho*I-compatible ends, (2) directionally ligating the cDNA to a vector, (3) analyzing recombinants to determine library quality, and (4) screening recombinants. Additional information regarding these techniques that is beyond the scope of this chapter can be found in **ref. 3**.

3.1. cDNA Synthesis

The ZAP-cDNA® synthesis kit converts poly(A) mRNA into double-stranded cDNA having *Eco*RI and *Xho*I-compatible ends *(4)* (*see* **Fig. 1**). In the first-strand synthesis reaction, mRNA is primed with a hybrid oligo-(dT) linker–primer that contains an *Xho*I restriction site and is reverse-transcribed using StrataScript reverse transcriptase and 5-methyl dCTP. Incorporation of 5-methyl dCTP during first-strand synthesis results in single-stranded hemimethylated cDNA, which is protected from digestion with the restriction enzyme *Xho*I. Therefore, only the unmethylated *Xho*I site within the linker–primer is cleaved when the synthesized cDNA is incubated with *Xho*I.

First-strand cDNA is converted to double-stranded cDNA during a second-strand synthesis reaction. In this reaction, RNase H nicks the RNA bound to the first-strand cDNA to produce a multitude of fragments that serve as primers for DNA polymerase I. DNA polymerase I "nick-translates" these RNA fragments into second-strand cDNA. The second-strand nucleotide mixture has been supplemented with dCTP to reduce the probability of incorporating 5-methyl dCTP into the second strand, ensuring that the restriction site in the linker–primers will be digested with *Xho*I.

3.1.1. First-Strand cDNA Synthesis

1. For each reaction, combine (in this order) 5 µL of 10X first-strand buffer, 3 µL first-strand methyl nucleotide mixture, 2 µL linker–primer (1.4 µg/µL), X µL diethyl pyrocarbonate (DEPC)-treated water (*see* **Notes 1** and **2**), and 1 µL RNase block ribonuclease inhibitor (40 U/µL) in an RNase-free microcentrifuge tube. Mix.
2. Add X µL poly(A)$^+$ RNA (5 µg) (*see* **Notes 1** and **2**). Mix gently and incubate at room temperature for 10 min.
3. Add 1.5 µL StrataScript RT (50 U/µL). Mix gently and spin briefly. For a first-strand synthesis control reaction, transfer 5 µL of each first-strand synthesis reaction to a separate tube containing 0.5 µL of [α-^{32}P]-dNTP (800 Ci/mmol). Incubate at 42°C for 1 h.
4. Place the nonradioactive first-strand synthesis reaction on ice. Store the radioactive first-strand synthesis control reaction at –20°C for later analysis (*see* **Subheading 3.1.3.2.**).

3.1.2. Second-Strand cDNA Synthesis

1. Add 20 µL of 10X second-strand buffer, 6 µL second-strand dNTP mixture, 114 µL sterile distilled water, and 2 µL of [α-^{32}P]-dNTP (800 Ci/mmol) to the 45-µL first-strand synthesis reaction.
2. Add 2 µL of RNase H (1.5 U/µL) and 11 µL of DNA polymerase I (9.0 U/µL) (*see* **Note 3**).
3. Gently vortex, spin briefly, and incubate for 2.5 h at 16°C.
4. Place on ice.

3.1.3. Generating cDNA with EcoRI- and XhoI-Compatible Ends

Following second-strand synthesis, *Eco*RI- and *Xho*I-compatible ends are created on the ends of the cDNA. In this process, the uneven termini are incubated with cloned *Pfu* DNA polymerase to generate blunt ends. The first-

and second-strand cDNA are then analyzed by denaturing gel electrophoresis to determine the size range of the cDNA and if a secondary structure is present. Following analysis, the cDNA is ligated to *Eco*RI adapters. These adapters are composed of 10- and 14-mer oligonucleotides that anneal to each other to generate a blunt end and a *Eco*RI-compatible end. The 10-mer oligonucleotide is phosphorylated at the 5′ end, which allows it to ligate to the blunt termini on the double-stranded cDNA and to other adapters. The 14-mer oligonucleotide is dephosphorylated at the 5′ end to prevent it from ligating to other adapters. After adapter ligation is complete and the ligase has been heat inactivated, the 14-mer oligonucleotide is phosphorylated to enable its ligation to the termini of a dephosphorylated vector. The cDNA is then size fractionated to select for cDNA of the desired length.

3.1.3.1. BLUNTING cDNA TERMINI

1. Add 23 µL of blunting dNTP mix and 2 µL of cloned *Pfu* DNA polymerase (2.5 U/µL) to the second-strand synthesis reaction. Vortex, spin briefly, and incubate at 72°C for 30 min.
2. Add 200 µL of phenol–chloroform (1 : 1 [v/v]) and vortex. Spin at maximum speed for 2 min at room temperature. Transfer the upper aqueous layer to a new tube. Avoid removing any interface that may be present.
3. Add an equal volume of chloroform and vortex. Spin at maximum speed for 2 min at room temperature and transfer the upper aqueous layer to a new tube.
4. Add 20 µL of 3 *M* sodium acetate and 400 µL of 100% ethanol and vortex. Place at –20°C overnight.
5. Spin at maximum speed for 60 min at 4°C.
6. Carefully remove and discard the radioactive supernatant. Do not disturb the large white pellet that accumulates near the bottom of the microcentrifuge tube and may taper up along the side of the tube.
7. Gently wash the pellet by adding 500 µL of 70% (v/v) ethanol to the side of the tube away from the pellet. Do not mix or vortex. Spin at maximum speed for 2 min at room temperature with the pellet toward the outside of the centrifuge.
8. Carefully aspirate the ethanol wash and dry the pellet by vacuum centrifugation.
9. Resuspend the pellet in 9 µL of the *Eco*RI adapters and incubate at 4°C for at least 30 min.
10. Transfer 1 µL of this second-strand synthesis reaction to a separate tube and save it as the second-strand synthesis control reaction. Store at –20°C.

3.1.3.2. ANALYSIS OF FIRST- AND SECOND-STRAND SYNTHESIS REACTIONS

The size range and the presence of secondary structure in the cDNA is determined by analyzing the first- and second-strand synthesis control reactions by alkaline agarose gel electrophoresis. The size range is determined by comparing the synthesized cDNA to molecular-weight DNA markers.

The presence of a secondary structure or "hairpinning" is determined by denaturation of the cDNA by the alkaline conditions. Hairpinning can occur in either the first- or second-strand reactions when the newly polymerized strand "snaps back" on itself and forms an antiparallel double helix.

1. Melt 0.5 g of agarose in 45 mL of water and cool to 55°C.
2. Add 5 mL of 10X alkaline buffer, swirl to mix, and pour immediately onto the glass slide with minigel comb (*see* **Notes 4** and **5**).
3. Combine the sample with an equal volume of alkaline agarose 2X loading buffer and vortex.
4. Run the gel with 1X alkaline buffer at 100 mA and monitor the system for heat. If the apparatus becomes warmer than 37°C, reduce the amperage. The migration of the BPB in alkaline agarose is similar to the migration in regular agarose and should be run to at least one-half or three-quarters distance of the gel. The alkali condition causes the blue dye to fade.
5. Expose the gel to X-ray film to visualize the radiolabeled first- and second-strand cDNA. The test cDNA sample will run as a tight band at 1.8 kb and will show distinctly different intensity between the first and second strands. This is the result of the relative ratio of α-^{32}P to the amount of NTP in the first- or second-strand reaction. Normally, the second strand will be only one-tenth to one-twentieth the intensity of the first-strand band.

3.1.3.3. *Eco*RI Adapter Ligation and *Xho*I Digestion

Following ligation of the *Eco*RI adapters, the cDNA is digested with *Xho*I, which releases the *Eco*RI adapter and a portion of the linker–primer from the 3′ end of the cDNA. This results in double-stranded cDNA having *Eco*RI- and *Xho*I-compatible ends and part of the linker–primer and *Eco*RI adapter.

1. Add 1 µL of 10X ligase buffer, 1 µL of 10 m*M* rATP, and 1 µL T4 DNA ligase (4 U/µL) to the blunted second-strand cDNA and the *Eco*RI adapters. Mix gently and spin briefly.
2. Incubate overnight at 8°C or at 4°C for 2 d.
3. Heat at 70°C for 30 min.
4. Spin briefly. Cool at room temperature for 5 min.
5. Add 1 µL of 10X ligase buffer, 2 µL of 10 m*M* rATP, 6 µL sterile water, and 1 µL T4 polynucleotide kinase (10 U/µL). Mix gently and incubate for 30 min at 37°C.
6. Incubate at 70°C for 30 min. Spin briefly and cool at room temperature for 5 min.
7. Add 28 µL of *Xho*I buffer supplement and 3 µL *Xho*I (40 U/µL). Incubate for 1.5 h at 37°C.
8. Add 5 µL of 10X STE buffer and 125 µL of 100% (v/v) ethanol and vortex. Precipitate overnight at –20°C.

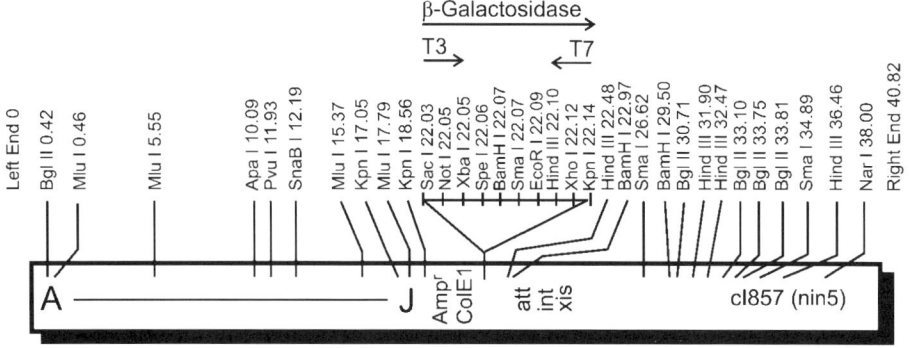

Fig. 2. Map of the Uni-ZAP XR lambda vector.

9. Spin at maximum speed for 60 min at 4°C. Discard the supernatant and dry the pellet completely.

3.1.3.4. SIZE FRACTIONATION OF CDNA

The sample is size fractionated to select for cDNA that are >400 bp. Suitable methods include a drip column containing Sepharose CL-2B gel filtration medium (*see* **Note 6**) and resin-packed microcentrifuge columns (*see* **Note 7**). Size fractionation can be monitored using a handheld Geiger counter because of the incorporation of radioactivity during first- and second-strand synthesis. Phenol–chloroform extract, ethanol precipitate and analyze collected fractions by nondenaturing agarose or acrylamide gel electrophoresis. The analysis assesses the effectiveness of the size fractionation and identifies the fractions that will be used for ligation. The fractions with the highest amount of cDNA of the desired size range are used for ligation.

3.2. Ligating cDNA to a Lambda Vector

The use of lambda vectors results in high-quality cDNA libraries having high numbers of members in the primary library and a low percentage of vector without insert.

3.2.1. Uni-ZAP XR Vector System

The Uni-ZAP XR vector (*see* **Fig. 2**) has *Eco*RI- and *Xho*I-compatible ends and will accommodate DNA inserts from 0 to 10 kb in length. The Uni-ZAP XR vector can be screened with either DNA probes or antibody probes and allows in vivo excision of the ampicillin-resistant pBluescript® phagemid (*see* **Fig. 3**), allowing the insert to be characterized in a plasmid system. The polylinker of the pBluescript phagemid has 21 unique cloning sites (multiple

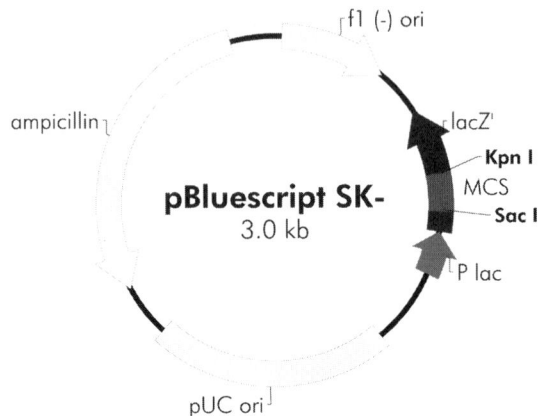

Fig. 3. Circular map of the pBluescript SK(–) phagemid vector (GenBank accession no. X52324).

cloning site [MCS]) flanked by T3 and T7 promoters and a choice of 6 different primer sites for DNA sequencing. The *lacZ* promoter (Plac) is used to drive expression of fusion proteins (cDNA–lacZ') suitable for immunoscreening, Western blot analysis, or protein purification.

3.2.2. ZAP-CMV XR Vector System

The ZAP-CMV Express vector system combines the high efficiency of lambda library construction and the convenience of a plasmid system for expression of proteins in eukaryotes. The ZAP-CMV XR vector has *Eco*RI- and *Xho*I-compatible ends and will accommodate DNA inserts up to 6.5 kb in length (*see* **Fig. 4**). Inserts cloned into the Lambda ZAP-CMV XR vector can be excised out of the phage in the form of the kanamycin-resistant pCMV-Script® EX phagemid vector (*see* **Fig. 5**) by the same excision mechanism used with the Lambda ZAP vectors *(1)*. The polylinker of pCMV-Script EX phagemid has 15 unique cloning sites flanked by T3 and T7 promoters and has three primer sites for DNA sequencing. Selection of clones in prokayotic cells is by expression of the neomycin- and kanamycin-resistance gene (neo/kan) driven by the β-lactamase promoter (P bla) to render transfectants resistant to kanamycin.

The unique ability to quickly and efficiently convert the lambda library to a phagemid library allows for screening of either plaques or colonies for the desired cDNA insert. Additionally, the library can be transfected into mammalian cells for functional screening. Eukaryotic expression of inserts in pCMV-Script EX is driven by the cytomegalovirus (CMV) immediate early (IE)

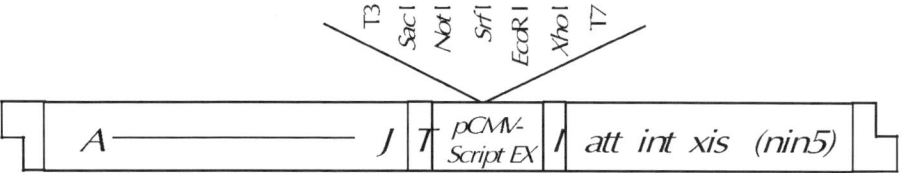

Fig. 4. Map of the Lambda ZAP-CMV vector.

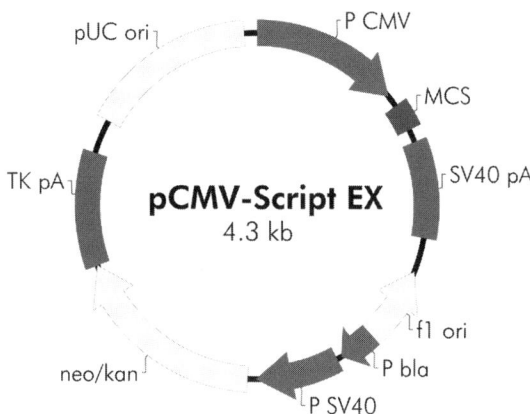

Fig. 5. Circular map of the pCMV-Script EX phagemid vector.

promoter (P CMV) with the SV40 transcription terminator and polyadenylation signal (SV40, pA) (*see* **Fig. 5**). Stable selection of clones in eukaryotic cells is by expression of the neomycin- and kanamycin-resistance gene (neo/kan) with TK transcription polyadenylation signals (TK, pA) driven by the SV40 early promoter (P SV40) to render transfectants resistant to G418 (geneticin). The pCMV-Script EX phagemid within the Lambda ZAP-CMV vector does not contain an ATG initiation codon. Clones must contain their own ATG initiation codon to be expressed.

3.2.3. cDNA and Lambda Vector Ligation

Lambda XR vector refers to either the Uni-ZAP® XR vector or the Lambda ZAP®-CMV XR vector. The lambda XR vectors are digested with *Eco*RI and *Xho*I and dephosphorylated with calf intenstinal alkaline phosphatase (CIAP).

1. Prepare a control ligation by combining 1.0 µL of the lambda XR vector (1 µg), 1.6 µL of the test insert (0.4 µg), 0.5 µL of 10X ligase buffer, 0.5 µL

of 10 m*M* rATP (pH 7.5), 0.9 µL of water, and 0.5 µL of T4 DNA ligase (4 U/µL) (*see* **Note 8**).

2. Prepare a sample ligation by combining *X* µL of resuspended cDNA (approx 100 ng), 0.5 µL of 10X ligase buffer, 0.5 µL of 10 m*M* rATP (pH 7.5), 1.0 µL of the lambda XR vector (1 µg/µL), *X* µL of water for a final volume of 4.5 µL, and 0.5 µL of T4 DNA ligase (4 U/µL) (*see* **Note 8**).

3. Mix gently and incubate overnight at 12°C or for up to 2 d at 4°C.

3.2.4. Packaging Ligated cDNA and Lambda Vector

Packaging extracts are used to assemble the recombinant lambda phage DNA and phage proteins into phage particles. Gigapack III Gold packaging extract *(5–8)* is recommended because of its high packaging efficiency, resulting in an increase in the size and representation of the gene library, and because it is restriction minus (HsdR⁻ McrA⁻ McrBC⁻ McrF⁻ Mrr⁻). Other commercially available packaging extracts can restrict hemimethylated cDNA, therefore producing low-titer libraries.

Package 1 µL of each ligation, including the control ligations, using Gigapack III Gold packaging extract. A good representational primary library size consists of approx 1×10^6 clones. If a low number of plaque-forming units results from packaging the 1-µL ligation, package 2–3 µL of the remaining ligation mixture in one packaging reaction.

1. Streak the XL1-Blue MRF' bacterial glycerol stock onto LB-tetracycline plates and incubate overnight at 37°C.

2. Inoculate LB-tetracycline, supplemented with 10 m*M* MgSO$_4$ and 0.2% (w/v) maltose, with a single colony. Incubate at 37°C with shaking for 4–6 h (do not grow past an optical density [OD$_{600}$] of 1.0). Alternatively, grow overnight at 30°C, with shaking (*see* **Note 9**).

3. Spin the cells at 500*g* for 10 min and discard the supernatant.

4. Gently resuspend the cells in half the original volume with sterile 10 m*M* MgSO$_4$.

5. Dilute the cells to an OD$_{600}$ of 0.5 with additional sterile 10 m*M* MgSO$_4$.

6. Remove the appropriate number of packaging extract vials from the –80°C freezer and place on dry ice.

7. Quickly thaw a packaging extract by holding the tube between your fingers until the contents of the tube just begins to thaw.

8. Add the experimental DNA immediately (1–4 µL containing 0.1–1.0 µg of ligated DNA) to the packaging extract.

9. Stir the reaction mixture with a pipet tip to mix well. Gentle pipetting is allowable, provided that air bubbles are not introduced.

10. Spin the tube quickly (for 3–5 s), if desired, to ensure that all contents are at the bottom of the tube.

11. Repeat **steps 7–10** with any additional samples.

12. Incubate the tube at room temperature (22°C) for 2 h. Do not exceed 2 h (*see* **Note 10**).
13. Add 500 µL of SM buffer and 20 µL of chloroform. Vortex and spin the tube briefly.
14. Remove the supernatant containing phage particles and transfer to a fresh tube.
15. Add 20 µL of chloroform and vortex.
16. The supernatant containing the phage is ready for titering. Store the supernatant at 4°C. The phage titer will decrease upon storage, therefore, amplification of the primary library within a few days is highly recommended.

3.2.5. Determining Library Quality

The quality of a cDNA library is typically expressed as the number of clones in the primary library and the percentage of vector with insert. The number of clones in the primary library can easily be determined by titering the packaged phage. The percentage of vector with insert can be determined by several methods, including α-complementation *(1,3)* polymerase chain reaction (PCR) analysis of individual plaques or colonies using vector-specific primers, and restriction digestion or nucleotide sequencing of miniprep DNA isolated from excised clones.

1. Prepare the XL1-Blue MRF' strain as outlined in **steps 1–5** of **Subheading 3.2.4.**
2. Prepare a 1:10 dilution of the primary library in SM buffer.
3. Add 1 µL of the primary library to 200 µL of host cells in a Falcon 2054 tube. Add 1 µL of the 1:10 dilution of the primary library to a separate aliquot of 200 µL of host cells. Incubate at 37°C for 15 min.
4. Add 15 µL of 0.5 *M* IPTG (in water) and 50 µL of 250 mg/mL X-gal (in dimethyl formamide [DMF]) to 2–3 mL of NZY top agar, melted and cooled to approx 48°C (*see* **Note 11**).
5. Add top agar mixture to phage and cells. Mix well by inversion. Pour immediately on NZY agar plates and allow plates to set for 10 min or until top agar is solidified.
6. Invert plates and incubate plates at 37°C. Plaques are visible after incubation for 6–8 h, although color detection may require overnight incubation. Background plaques (vector without cDNA insert) are blue and should be $<1 \times 10^5$ pfu/µg of lambda vector, whereas recombinant plaques (vector with cDNA insert) will be white or clear and should be 10- to 100-fold above the background (*see* **Note 12**).
7. Package remaining ligated cDNA as described in **Subheading 3.2.4.**

3.2.6. Amplification of Primary Library

It is usually desirable to amplify libraries prepared in lambda vectors to make a large, stable quantity of a high-titer stock of the library. More than one round of amplification is not recommended, because slower growing clones may be significantly underrepresented.

1. Prepare the XL1-Blue MRF' strain as outlined in **steps 1–5** of **Subheading 3.2.4.**
2. Combine aliquots of the packaged mixture containing approx 5×10^4 pfu of bacteriophage with 600 µL of host cells at an OD_{600} of 0.5 in Falcon 2059 polypropylene tubes. To amplify 1×10^6 plaques, use a total of 20 aliquots (each aliquot contains 5×10^4 plaques/150-mm plate). Do not add more than 300 µL of phage/600 µL of cells. Incubate for 15 min at 37°C.
3. Mix 6.5 mL of NZY top agar, melted and cooled to approx 48°C, with each aliquot of infected bacteria and pour evenly onto a freshly poured 150-mm NZY agar plate.
4. After the NZY top agar solidifies, invert and incubate the plates at 37°C for 6–8 h. Do not allow the plaques to grow larger than 1–2 mm. On completion, the plaques should be touching.
5. Overlay the plates with approx 8–10 mL SM buffer. Store the plates at 4°C overnight (with gentle rocking, if possible).
6. Recover the bacteriophage suspension from each plate by tilting the plates and removing the bacteriophage suspension with a 10-mL pipet. Pool the bacteriophage suspensions into a sterile polypropylene container. Rinse the plates with an additional 2 mL SM buffer and pool. Add chloroform to a 5% (v/v) final concentration. Vortex and incubate for 15 min at room temperature.
7. Remove the cell debris by centrifugation for 10 min at 500*g*.
8. Transfer the supernatant to a sterile polypropylene container. If the supernatant appears cloudy or has a high amount of cell debris, repeat the chloroform extraction. If the supernatant is clear, add chloroform to a 0.3% (v/v) final concentration and store at 4°C. Preparing and storing aliquots of the amplified library in 7% (v/v) dimethyl sulfoxide (DMSO) at –80°C as soon as possible is highly recommended.

3.2.7. Titering the Amplified Library

Assume the amplified library is approx 10^9–10^{11} pfu/mL.

1. Prepare the XL1-Blue MRF' strain as outlined in **steps 1–5** of **Subheading 3.2.4.**
2. Prepare the following dilutions of the amplified phage stock in SM buffer: 1:10,000, 1:100,000, 1:1,000,000.
3. Add 1 µL of the diluted phage stock to 200 µL of diluted host cells in a Falcon 2054 tube. Incubate at 37°C for 15 min.
4. Continue as outlined in **steps 4–6** of **Subheading 3.2.5.**
5. Store aliquots of the amplified library in 7% (v/v) DMSO at –80°C. When needed, remove an aliquot from –80°C and store unused portion at 4°C. Avoid freeze thawing of aliquots.

3.3. Identifying Desired cDNA Inserts in Lambda Library

The lambda library can be screened to identify clones containing any desired cDNA insert using standard methods *(3)*. The most commonly used method is to hybridize labeled oligonucleotide or double-stranded probes having homology to the desired cDNA insert to lambda DNA bound to membranes. Alternatively, in an immunoscreening assay, proteins expressed by the cDNA clones are bound to membranes and immunoreacted with an antibody that specifically binds to the desired protein. Following identification of the desired clone, perform in vivo excision to convert the lambda vector to a phagemid vector for further analysis.

3.3.1. In Vivo Excision

In vivo excision rapidly and easily converts a lambda clone to a phagemid clone *(1,9)*. Clones can be excised individually or in a mass excision.

3.3.1.1. SINGLE-CLONE EXCISION

1. Core the plaque from the agar plate and transfer to a sterile microcentrifuge tube containing 500 µL SM buffer and 20 µL chloroform. Vortex and incubate for 1–2 h at room temperature or overnight at 4°C. (This phage stock is stable for up to 6 mo at 4°C.)
2. Grow separate overnight cultures of XL1-Blue MRF′ cells, supplemented with 0.2% (w/v) maltose and 10 mM MgSO$_4$, and XLOLR cells in NZY broth at 30°C.
3. Gently spin down the XL1-Blue MRF′ and XLOLR cells (1000g). Resuspend the XL1-Blue MRF′ and XLOLR cells at an OD$_{600}$ of 1.0 in 10 mM MgSO$_4$ (8×10^8 cells/mL).
4. Combine 200 µL XL1-Blue MRF′ cells at an OD$_{600}$ of 1.0, 250 µL of phage stock (containing $>1 \times 10^5$ phage particles) and 1 µL of the ExAssist helper phage ($>1 \times 10^6$ pfu/µL) in a Falcon 2059 polypropylene tube. Incubate at 37°C for 15 min.
5. Add 3 mL of NZY broth and incubate for 2.5–3 h or overnight at 37°C with shaking.
6. Heat at 65–70°C for 20 min and spin the tube at 1000g for 15 min.
7. Decant the supernatant into a sterile Falcon 2059 polypropylene tube. This stock contains the excised phagemid vector packaged as filamentous phage particles. (This stock may be stored at 4°C for 1–2 mo.)
8. To plate the excised phagemids, add 200 µL of freshly grown XLOLR cells from **step 3** (OD$_{600}$ = 1.0) to two Falcon 2059 polypropylene tubes. Add 100 µL of the phage supernatant to one tube and 10 µL of the phage supernatant to the other tube. Incubate at 37°C for 15 min.
9. Add 300 µL of NZY broth and incubate at 37°C for 45 min.
10. Plate 200 µL of each sample on LB-ampicillin agar plates (50 µg/mL) (pBluescript SK- clones) or LB-kanamycin agar plates (50 µg/mL) (pCMV-Script clones) and incubate overnight at 37°C. If isolated single colonies are not present, replate a higher dilution of the cell mixture.

3.3.1.2. Mass Excision

1. Prepare host cells as outlined in **steps 2** and **3** of **Subheading 3.3.1.1.**
2. In a 50-mL conical tube, combine a portion of the amplified lambda bacteriophage library with XL1-Blue MRF′ cells at a multiplicity of infection (MOI) of 1:10 lambda phage-to-cell ratio. Excise 10- to 100-fold more lambda phage than the size of the primary library to ensure statistical representation of the excised clones. Add ExAssist helper phage at a 10:1 helper phage-to-cells ratio to ensure that every cell is coinfected with lambda phage and helper phage. For example, for a library having 10^6 primary clones, combine 10^7 pfu of the lambda phage (i.e., 10-fold above the primary library size), 10^8 XL1-Blue MRF′ cells (1:10 lambda phage-to-cell ratio, noting that an OD_{600} of 0.3 corresponds to 2.5×10^8 cells/mL) and 10^9 pfu of ExAssist helper phage (10:1 helper phage-to-cells ratio).
3. Incubate at 37°C for 15 min.
4. Add 20 mL of LB broth and incubate for 2.5–3 h at 37°C, with shaking. Incubating longer than 3 h may alter the clonal representation. The turbidity of the media is not indicative of the success of the excision.
5. Incubate at 65–70°C for 20 min.
6. Spin down the debris at 1000*g* for 10 min and decant the supernatant into a sterile conical tube.
7. To titer the excised phagemids, combine 1 μL of this supernatant with 200 μL XLOLR cells from **step 1** in a 1.5-mL microcentrifuge tube. Incubate at 37°C for 15 min.
8. Plate 100 μL of the cell mixture onto LB-ampicillin agar plates (50 μg/mL) (pBluescript SK- clones) or LB-kanamycin agar plates (50 μg/mL) (pCMV-Express clones) and incubate overnight at 37°C. If isolated single colonies are not present, replate a higher dilution of the cell mixture.

At this stage, individual colonies may be selected for plasmid preps, or the cell mixture may be plated directly onto filters for colony screening.

3.3.2. Functional Screening of pCMV-Express Libraries in Eukaryotic Cells

Screening libraries in eukaryotic cells is an effective way of identifying clones otherwise nonidentifiable in prokaryotic screening systems. The screening technique used will depend on the clone of interest and on the type of assay available. An appropriate cell line for screening must be obtained, and an assay or reagent capable of identifying the cell or cells expressing the desired target protein must be developed. Panning assays and functional analysis of clone pools are two potential screening techniques.

3.3.2.1. Panning Assay

Clone identification by "panning" requires the transfection of a library into a cell line deficient in the desired surface protein. When the clone of interest is translated and expressed on the surface of eukaryotic cells, the translated protein product is made accessible to an antibody, ligand, or receptor coupled either directly or indirectly to a solid-phase matrix. Cells expressing the appropriate insert are selected by binding to an affinity matrix. Either transient or stable transfection protocols can be used.

3.3.2.2. FUNCTIONAL ASSAY

Functional assay screening can also be performed on either transiently or stably transfected cells. Transient expression will likely require subdividing the amplified library into smaller pools of clones to prevent the dilution of a positive cell signal with an excess of negative cell signals. Each clone pool is amplified separately and transfected into the eukaryotic cells. The transfected cells are then tested for the expression of the desired clone. Once a pool is identified as containing the clone of interest, it is subdivided into smaller pools for a second round of prokaryotic amplification, eukaryotic transfection, and screening. After several rounds of enriching for the desired clone, a single clone can be isolated. The initial pool size is determined according to the sensitivity of the available assay so that a single clone within the pool is still theoretically detectable in the transfected cells. For example, if a positive assay signal is 1000-fold above background, pools containing 500–1000 members should still give a signal above background. The sensitivity of the assay dictates the initial size of the pools, as well as the number of pools required to screen. If stable transformants are created using G418 selection, pools of stable clones can be assayed. This simplifies the identification of isolated positive eukaryotic clones, because the eukaryotic colonies can be picked or diluted in microtiter tissue culture plates.

3.3.3. Recovery of Clone from Eukaryotic Cells

After a clone has been identified within the eukaryotic cells, the clone can be retrieved from the transfected cells by several methods. Plasmid DNA within the tissue culture cells can be collected using the Hirt, Birnboim, and Doly procedures *(10,11)* and then transferred into *E. coli* cells for amplification and plasmid DNA preparations. Screening libraries in COS cells *(12)* where the presence of the SV40 T-antigen increases the copy number of phagemids containing the SV40 origin of replication is also recommended. This results in a higher episomal copy number, which may help in the retrieval of the plasmids. Inserts can also be isolated by PCR amplification of the tissue culture cells using T3/T7 primer sets. The resulting PCR fragment can be digested using restriction sites flanking the insert and then recloned into pCMV-Script EX phagemid DNA for further analysis.

4. Notes

1. The quality and quantity of the mRNA used is critical to the construction of a large, representative cDNA library. Stratagene's RNA Isolation Kit uses the guanidine isothiocyanate (GITC)–phenol–chloroform extraction method, which quickly produces large amounts of undegraded RNA. It is imperative to protect the RNA from any contaminating RNases until the first-strand cDNA synthesis is complete. Wear fresh gloves, use newly autoclaved pipet tips, and avoid using pipet tips or microcentrifuge tubes that have been handled without gloves. Ribonuclease A *cannot* be destroyed by normal autoclaving alone. Baking or DEPC treatment is recommended.

2. Calculate the amount of DEPC-treated water to add to the reaction so the total volume of DEPC-treated water and 5 μg poly(A) mRNA is 37.5 μL. For the control reaction, 5 μg of test RNA will be combined with 12.5 μL of DEPC-treated water.

3. All reagents must be <16°C when the DNA polymerase I is added.

4. The easiest method of preparing an alkaline agarose gel is to position a minigel comb over a 5×7.5-cm glass slide with high-tension clips and add 10 mL of molten alkaline agarose near the upper center of the slide. The surface tension of the solution will prevent overflow and produce a small, thin gel that can be exposed without further drying. Do not allow the teeth of the comb to overlap the edge of the plate or the surface tension may be broken. To improve the resolution, pat the gel dry with several changes of Whatman 3MM paper following electrophoresis and before exposure.

5. If buffer is added before the correct temperature is reached, the agarose may not solidify.

6. A detailed protocol on the preparation and collection of fractions from a size-fractionation column is beyond the scope of this chapter but can be found in the manual for the cDNA Synthesis Kit at www.stratagene.com.

7. Suitable columns include the Spin Column-200 (Sigma, St. Louis, MO).

8. Optimal packaging efficiencies are obtained with lambda DNAs that are concatameric, of high concentration, and free of contaminants. Ligations should be carried out at DNA concentrations of 0.2 μg/μL or greater to favor concatamers. DNA to be packaged should be relatively free from contaminants. DNA may be used directly from ligation reactions in most cases; however, polyethylene glycol (PEG), which is contained in some ligase buffers, has been shown to inhibit packaging. Use the ligase buffer provided in the ZAP-cDNA Synthesis Kit, which does not contain PEG. The volume of DNA added to each extract should be <5 μL.

9. The lower temperature keeps the bacteria from overgrowing, which reduces the number of nonviable cells. Phage can adhere to nonviable cells, resulting in a decreased titer.

10. The highest packaging efficiency occurs between 90 min and 2 h. Efficiency may drop dramatically during extended packaging times.

11. The high concentrations of IPTG and X-gal may result in the formation of a precipitate, which disappears after incubation. Add the IPTG and X-gal separately, with mixing in between additions, to the NZY top agar to minimize the formation of precipitate.
12. If the blue color is difficult to see, incubate the plates at 4°C overnight.

References

1. Short, J. M., Fernandez, J. M., Sorge, J. A., and Huse, W. D. (1988) Lambda ZAP: a bacteriophage lambda expression vector with in vivo excision properties. *Nucleic Acids Res.* **16,** 7583–7600.
2. Lu, Q., Hosfield, T., Dewar, C., Sanchez, T., and Kobrin, M. (1999) A new lambda vector for mammalian expression. *Strategies* **12,** 11–13.
3. Sambrook, J., Fritsch, E. F., and Maniatis, T. (1989) *Molecular Cloning: A Laboratory Manual*, Cold Spring Harbor Laboratory, Cold Spring Harbor, NY.
4. Huse, W. D. and Hanse, C. (1988) cDNA cloning redefined: a rapid, efficient, directional method. *Strategies* **1,** 1–3.
5. Kretz, P. L., Danylchuk, T., Hareld, W., Wells, S., Provost, G. S., and Short, J. A. (1994) Gigapack® III high-efficiency lambda packaging extract with single-tube conveneince. *Strategies* **7,** 44–45.
6. Kohler, S. W., Provost, G. S., Kretz, P. L., Dycaico, M. J., Sorge, J. A., and Short, J. A. (1990) Development of a short-term, in vivo mutagenesis assay: the effects of methylation on the recovery of a lambda phage shuttle vector from transgenic mice. *Nucleic Acids Res.* **18,** 3007–3013.
7. Kretz, P. L., Kohler, S. W., and Short, J. M. (1991) Identification and characterization of a gene responsible for inhibiting propagation of methylated DNA sequences in mcrA mcrB1 *Escherichia coli* strains. *J. Bacteriol.* **173,** 4707–4716.
8. Kretz, P. L., Reid, C. H., Greener, A., and Short, J. M. (1989) Effect of lambda packaging extract mcr restriction activity on DNA cloning. *Nucleic Acids Res.* **17,** 5409.
9. Hay, B. and Short, J. M. (1992) ExAssist™ helper phage and SOLR™ cells for Lambda ZAP® II excisions. *Strategies* **5,** 16–18.
10. Hirt, B. (1967) Selective extraction of polyoma DNA from infected mouse cell cultures. *J. Mol. Biol.* **26,** 365–369.
11. Birnboim, H. C. and Doly, J. (1979) A rapid alkaline extraction procedure for screening recombinant plasmid DNA. *Nucleic Acids Res.* **7,** 1513–1523.
12. Simmons, D. L., Satterthwaite, A. B., Tenen, D. G., and Seed, B. (1992) Molecular cloning of a cDNA encoding CD34, a sialomucin of human hematopoietic stem cells. *J. Immunol.* **148,** 267–271.

25

Peptide Library Construction from RNA-PCR-Derived RNAs

Shi-Lung Lin

1. Introduction

The generation of peptide from messenger RNA (mRNA) provides a convenient source for current proteomic analysis. Intron-free mRNA possessing adenine-uracil-guanine (AUG) start codons can be translated into labeled or unlabeled peptides under a predetermined reticulocyte lysate condition. In conjunction with RNA-polymerase cycling reaction (*see* **Fig. 1**; RNA-PCR), full-length gene transcripts can be unlimitedly amplified for protein/peptide synthesis in vitro (*1*). Many commercialized in vitro translation systems provide a cap nucleotide, which can be added to the 5′ end of the amplified poly(A$^+$) RNAs during the transcription step of RNA-PCR. Totally resembling mRNAs, the capped poly(A$^+$) RNAs can be used to synthesize proteins/peptides with labeling and may help the functional analysis of protein activity if they fold correctly.

We has successfully tested the RNAPCR-derived protein analysis in a prostatic cancer cell line, LNCaP, of which the protein data matched previous findings using mRNA. The antiapoptotic gene family of *bcl-2* has been well known for their ability to increase cancer resistance to multiple anticancer drugs. Previous data has predicted that a mutated form of *bcl-2* may be elevated in the drug-resistant LNCaP cells after androgen retrieval treatment (*2*). Using Northern and Western blots, we confirmed that a truncated form of the *bcl-2* member can be clearly detected in both the mRNA and protein levels, indicating a consistent result between these two methods. Because the folding of synthesized proteins may be different from that of the original one, the detection of such mutated changes will depend on the antibody used. For the

From: *Methods in Molecular Biology, vol. 221: Generation of cDNA Libraries: Methods and Protocols*
Edited by: S.-Y. Ying © Humana Press Inc., Totowa, NJ

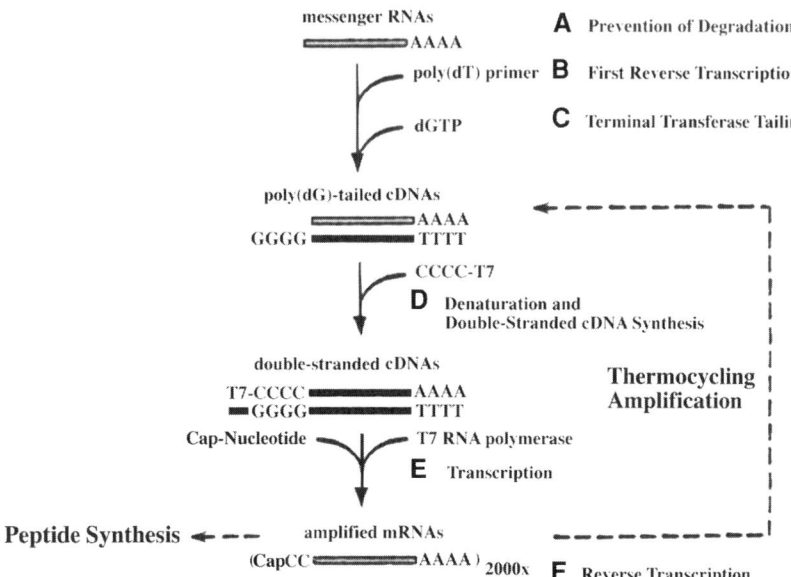

Fig. 1. An illustration of using RNA-PCR procedure for peptide library generation. The mRNA generated in step d can be capped by a methylated nucleotide such as P^1-5′-(7-methyl)-guanosine-P^3-5′-adenosine/guanosine-triphosphate for in vitro translation.

utilization of RNA-PCR-derived proteins in immunohistological analysis, a monoclonal antibody that recognizes denatured peptides is preferable. The same principle may be applied to the design of protein chip analysis.

2. Materials

2.1. Generation of Capped mRNA Construct by RNA-PCR

1. Diethyl pyrocarbonate (DEPC) H_2O: Stir double distilled water (ddH_2O) with 0.1% DEPC for more than 12 h and then autoclave at 120°C under about 1.2 kgf/cm^2 for 20 min, twice.
2. 10X In vitro transcription (IVT) buffer: 400 mM Tris-HCl (pH 8.0) at 25°C, 100 mM MgCl$_2$, 50 mM dithiothreitol (DTT), and 5 mg/mL nuclease-free bovine serum albumin (BSA).
3. T7 RNA polymerase (80 U/μL).
4. IVT reaction mixture: 4 μL DEPC-treated ddH_2O, 4 μL of 10X IVT buffer, 4 μL of 10 mM dNTP mix (10 mM each of rATP, rCTP, and rUTP), 1 μL of 10 mM rGTP, 3 μL of 10 mM P^1-5′-(7-methyl)-guanosine-P^3-5′-adenosine-triphosphate, 2 μL RNasin (25 U/μL), and 2 μL T7 RNA polymerase; prepare just before use.
5. Incubation mixer: 37°C, 100g vortex for 30 s between every 30-min interval.
6. Purification spin column: 100 bp cutoff filter, such as a Microcon-50 centrifugal filter (Amicon, Beverly, MA).

2.2. In Vitro Translation of Capped mRNA to Peptide

1. Reticulocyte lysate (such as Translation kit, type 2, Roche, Indianapolis, IN).
2. Incubation mixer: 30°C, 100g vortex for 30 s between every 10-min interval.
3. Purification spin column: <30 bp cutoff filter, such as a Microcon-30 centrifugal filter (Amicon, Beverly, MA).

3. Methods

3.1. Generation of Capped mRNA Construct by RNA-PCR

The RNA-PCR-derived cDNA contains a RNA promoter in its 5′-end, which can serve as a recognition site for RNA polymerase during an in vitro transcription reaction. The in vitro transcription provides linear amplification up to 2000-fold the amount of starting materials *(1,2)*. The proofreading capability of the RNA polymerase ensures the fidelity of the resulting nucleic acid products. Because the promoter is incorporated in the same orientation of mRNA, the resulting product is sense RNA (mRNA) rather than antisense RNA. A cap structure therefore can be added to the sense RNA for further peptide synthesis *(3)*.

1. In-vitro transcription reaction: Add 2 pmol (in 10 µL DEPC-treated ddH$_2$O) of RNA-PCR-derived double-stranded cDNA to 20 µL of IVT reaction mixture and mix well. Incubate the reaction at 37°C for 2–3 h and occasionally mix the reaction every 30 min for better RNA elongation (*see* **Note 1**).
2. Buffer exchange and sample concentration: Load the reaction into a purification spin column, spin 10 min at 14,000g, and discard the flowthrough (*see* **Note 2**). Add 200 µL of DEPC-treated ddH$_2$O into the spin column to wash the poly(A$^+$) RNA, spin 10 min at 14,000g, and discard the flowthrough. Add 20 µL of DEPC-treated ddH$_2$O into the spin column to dissolve the poly(A$^+$) RNA, place the spin column upside down in a new collecting microtube, and spin 3 min at 3000g. Store the 20 µL of the purified capped mRNA (about 40 µg) in a –80°C freezer or perform the next step immediately.

3.2. In Vitro Translation of Capped mRNA to Peptide

In this method, which is currently unpublished by Lin and Ying et al., the m^{7}G(5′)ppp(5′)A/G-capped poly(A$^+$) RNA is efficiently translated into polypeptide in a reticulocyte extract system *(4)* (*see* **Fig. 2**). Specific peptides, homologs, and their truncated products can be detected by Western blots *(5,6)*. The utilization of the RNA-PCR-derived peptide library in proteomic analysis is expected to be useful.

1. In vitro translation reaction: Add 2 µL of the purified capped mRNA to 30 µL reticulocyte lysate and mix well. Incubate the reaction at 30°C for 30 min and occasionally mix the reaction every 10 min for better elongation.

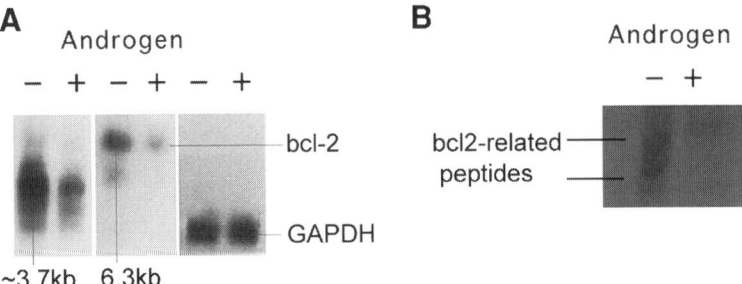

Fig. 2. RNA-PCR-amplified m⁷G(5′)ppp(5′)A/G-capped poly(A⁺) RNAs (10 µL) were added into an in vitro translation reaction (rabbit cell lysate extractions with RNase inhibitors) and incubated at 30°C for 30 min for peptide synthesis. The resulting *bcl-2* RNA and its peptide products were detected only in the LNCaP cells after androgen retrieval on 6% polyacrylamide gel by Northern and Western blots, respectively, showing the immortal and tumorigenic properties of these metastatic prostate cancer cells. It is noted that a truncated form of the mutated *bcl-2* was consistently found using both RNA-PCR-derived mRNAs and peptides.

2. Buffer removal and sample concentration: Load the reaction into a purification spin column, spin 10 min at 14,000*g*, and discard the flowthrough. Add 200 µL of DEPC-treated ddH₂O into the spin column to wash the poly(A⁺) RNA, spin 10 min at 14,000*g*, and discard the flowthrough. Add 20 µL of DEPC-treated ddH₂O into the spin column to dissolve the poly(A⁺) RNA, place the spin column upside down in a new collecting microtube, and spin 3 min at 3000*g*. Store the 20 µL of the purified peptides in a –80°C freezer.

4. Notes

1. The most stable and efficient IVT reaction occurs during first 2 h of incubation at 37°C. The rate of RNA synthesis decreases considerably (40–50%) after 3 h of incubation or below 37°C incubation. A longer reaction may increase yield, but the possibility of degradation by RNase increases. Occasionally, gentle mixing can prevent the stall of crowded RNA polymerases on a template and enhance full-length synthesis. The overall rate of RNA polymerization is maximal between pH 7.7 and 8.3, but it remains about 70% of maximum at pH 7.0 or 9.0. High concentrations of NaCl, KCl, or NH₄Cl above 75 m*M* will inhibit the reaction.
2. Relative Centrifugal Force (RCF) (g) = $(1.12 \times 10^{-5}) \cdot (\text{rpm})^2 \cdot r$, where *r* is the radius in centimeters measured from the center of the rotor to the middle of the spin column and rpm is the speed of the rotor in revolutions per minute.

References

1. Lin, S. L., Chuong, C. M., Widelitz, R. B., and Ying, S. Y. (1999) In vivo analysis of cancerous gene expression by RNA-polymerase chain reaction. *Nucleic Acid Res.* **27**, 4585–4589.

2. Chen, Y. G., Lui, H. M., Lin, S. L., Lee, J. M., and Ying, S. Y. (2002) Regulation of cell proliferation, apoptosis, and carcinogenesis by activin. *Exp. Biol. Med.* **227,** 75–87.

3. Contreras, R., Cheroutre, H., Degrave, W., and Fiers, W. (1982) Simple, efficient *in vitro* synthesis of capped RNA useful for direct expression of cloned eukaryotic genes. *Nucleic Acids Res.* **10,** 6353–6362.

4. Krieg, P. A. and Melton, D. A. (1984) Formation of the 3′ end of histone mRNA by posttranscriptional processing. *Nature* **308,** 203–206.

5. Towbin, H. T., Stachelin, T., and Gordon, J. (1979) Electrophoretic transfer of proteins from acrylamine gels to nitrocellulose sheets: Procedure and some applications. *Proc. Natl. Acad. Sci. USA* **76,** 4350–4354.

6. Burnette, W. N. (1981) Western blotting: electrophoretic transfer of proteins from sodium dodecyl sulfate–polyacrylamide gels to unmodified nitrocellulose and radiographic detection with antibody and radioiodinated protein A. *Anal. Biochem.* **112,** 195–203.

26

Identifying Interacting Proteins in an *Escherichia coli*-Based Two-Hybrid System

Bonnie Wu, Rebecca L. Mullinax, and Joseph A. Sorge

1. Introduction

The BacterioMatch™ two-hybrid system provides a fast, simple, and efficient method for in vivo detection of protein–protein interactions in *Escherichia coli (1,2)*. This system is based on transcriptional activation of a reporter gene. In this system, a protein of interest (the bait) is fused to the full-length bacteriophage λ repressor protein (λcI), which consists of an amino-terminal DNA-binding domain and a carboxyl-terminal dimerization domain. Target proteins are fused to the N-terminal domain of the α-subunit of RNA polymerase (RNAP-α). When expressed in the bacterial reporter strain, the bait is tethered to the λ operator sequence upstream of the reporter promoter through the DNA-binding domain of λcI. When the bait and target proteins interact, they recruit and stabilize the binding of RNA polymerase at the promoter and activate transcription of the β-lactamase (Amp^r) and β-galactosidase (*lacZ*) reporter genes (*see* **Fig. 1**). Expression from the reporter promoter is correlated with the interaction affinity of the bait and target. A stronger interaction between bait and target proteins results in higher levels of gene expression of the reporter genes, as indicated by higher levels of carbenicillin-resistance *(3)* and β-galactosidase activity in the reporter strain. The BacterioMatch system detects an interaction between proteins with an equilibrium dissociation constant in the high nanomolar range.

The BacterioMatch two-hybrid system has many significant advantages over yeast two-hybrid systems. These advantages include the ability to screen larger libraries and the relative ease and quickness when using an *E. coli*-based rather than a yeast-based system. Furthermore, using *E. coli* for two-hybrid screening reduces the chance that the host expresses a eukaryotic homolog

From: *Methods in Molecular Biology, vol. 221: Generation of cDNA Libraries: Methods and Protocols*
Edited by: S.-Y. Ying © Humana Press Inc., Totowa, NJ

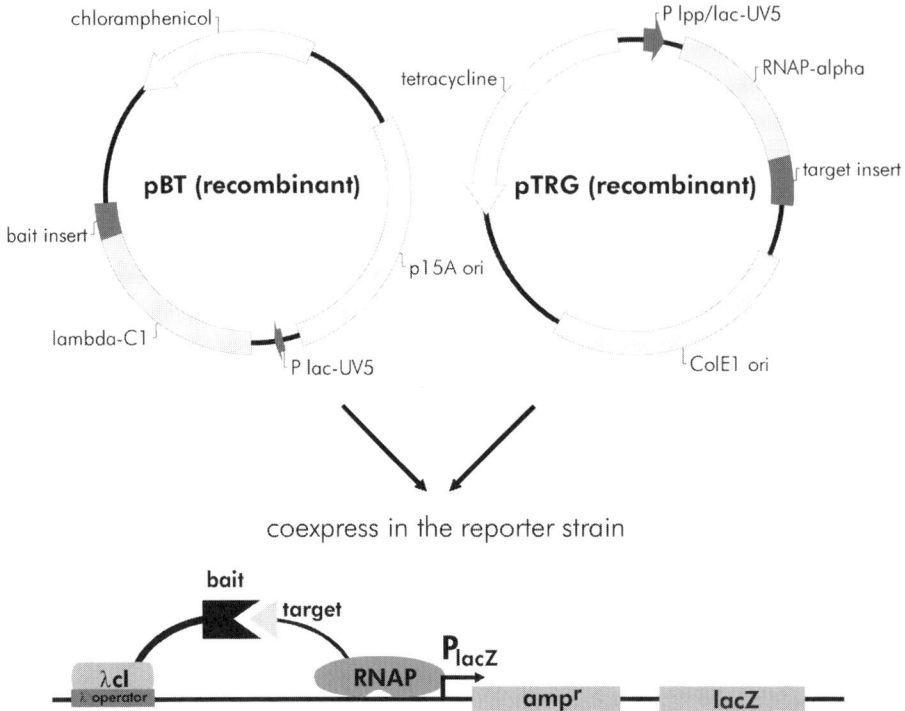

Fig. 1. Interaction of the bait and target proteins activates transcription of the Amp[r] and *lacZ* reporter genes.

of one of the interacting proteins. For example, some eukaryotic regulatory proteins, such as cell cycle checkpoint proteins and signal transduction pathway proteins, may be toxic in yeast because they interfere with the function of yeast homologues. For the same reason, use of a bacterial system could also result in fewer false positives observed. However, the disadvantages over using a yeast-based system are that some heterologous proteins could be toxic to an *E. coli* host and bacteria may lack the ability to perform posttranslational modifications. Because of the tendency of both the λ repressor protein and the N-terminal domain of the α-subunit of RNA polymerase to dimerize, the system might not be optimal for the analysis of proteins that self-associate unless their interaction with other proteins depends on the oligomerization.

Following the identification of target proteins that interact with a bait protein, the system is useful in identifying protein domains or amino acids that are critical for the interaction. For example, specific mutations, insertions, or deletions that affect the encoded amino acids can be introduced into DNA encoding the target protein, and the mutant target proteins can be assayed for the interaction with the bait protein.

2. Materials

1. 4 M NH$_4$OAc: (per 100 mL) 30.83 g NH$_4$OAc and deionized water up to 100 mL. Sterile filter.
2. 70% (v/v) Ethanol: (per 100 mL) Combine 70 mL molecular-biology-grade ethanol and 30 mL deionized water.
3. TE: 10 mM Tris-HCl (pH 7.5) and 1 mM EDTA. Autoclave.
4. 80% (v/v) Glycerol: (per 100 mL) Combine 80 mL molecular-biology-grade glycerol and 20 mL deionized water. Autoclave.
5. BacterioMatch™ Two-Hybrid System cDNA Library Construction Kit (Stratagene, La Jolla, CA).
6. BacterioMatch two-hybrid system reporter strain competent cells including β-ME (Stratagene, La Jolla, CA).
7. XL1-Blue MRF′ competent cells including β-ME (Stratagene, La Jolla, CA).
8. XL1-Blue MRF′ Kan Library Pack Competent Cells including Library Pack β-ME (Stratagene, La Jolla, CA).
9. NZY broth: (per liter) 5 g NaCl, 2 g MgSO$_4$ • 7H$_2$O, 5 g of yeast extract, 10 g of NZ amine (casein hydrolysate), and deionized H$_2$O to a final volume of 1 L. Adjust the pH to 7.5 with NaOH. Autoclave.
10. 1 M glucose: (per 100 mL) Dissolve 18 g glucose in 100 mL deionized dH$_2$O. Sterile filter.
11. SOC medium: (per liter) 20.0 g tryptone, 5.0 g yeast extract, 0.5 g NaCl, and deionized H$_2$O to a final volume of 1 L. Adjust pH to 7.0 using 5 N NaOH. Autoclave. Add 10 mL 1 M MgCl$_2$ and 10 mL of 1 M MgSO$_4$. Immediately before use, add 2 mL of 1 M glucose per 100 mL medium. Sterile filter.
12. LB-agar: (per liter) 10 g NaCl, 10 g tryptone, 5 g yeast extract, 20 g agar, and deionized H$_2$O to a final volume of 1 L. Adjust pH to 7.0 with 5 N NaOH. Autoclave. Cool to 45°C and pour into Petri dishes.
13. LB-Chloramphenicol agar plates (per liter): Prepare 1 L of LB-agar and autoclave. Cool to 45°C. Add 3.4 mL of 10-mg/mL chloramphenicol (prepared in 100% EtOH). Pour into Petri dishes.
14. LB-Tetracycline agar plates: (per liter) Prepare 1 L of LB-agar and autoclave. Cool to 45°C. Add 1.5 mL of 10-mg/mL tetracycline (prepared in 50% EtOH). Pour into Petri dishes.
15. LB-CTCK agar plates: (per liter) Prepare 1 L of LB-agar, autoclave and cool to 45°C. Add 25 mL of 10 mg/mL filter-sterilized carbenicillin, 1.5 mL of 10 mg/mL tetracycline (prepared in 50% EtOH), 3.4 mL of 10 mg/mL chloramphenicol (prepared in 100% EtOH), and 5.0 mL of 10 mg/mL filter-sterilized kanamycin. Store plates in a dark, cool place or cover plates with foil.
16. LB-TCK agar plates: (per liter) Prepare 1 L of LB-agar, autoclave and cool to 45°C. Add 1.5 mL of 10-mg/mL tetracycline (prepared in 50% EtOH),

3.4 mL of 10-mg/mL chloramphenicol (prepared in 100% EtOH), and 5.0 mL of 10-mg/mL filter-sterilized kanamycin. Store plates in a dark, cool place or cover plates with foil.

17. X-gal indicator plates: (1 L) Prepare 1 L of LB-agar, autoclave, and cool to 45°C. Add 1.5 mL of 10 mg/mL tetracycline (prepared in 50% EtOH), 3.4 mL of 10 mg/mL chloramphenicol (prepared in 100% EtOH), 5.0 mL of 10 mg/mL filter-sterilized kanamycin, 1 mL of an 80-mg/mL stock solution of X-gal (prepared in dimethyl formamide), and 1 mL of a 200-mM stock solution of β-galactosidase inhibitor (phenylethyl β-D-thio galactoside, prepared in dimethyl formamide). Pour into Petri dishes.

18. Falcon® 2059 polypropylene tubes.

3. Methods

The following includes methods for (1) preparing the bait plasmid, (2) preparing a library of target plasmids, (3) screening the library to identify target proteins that interact with the bait protein, and (4) validating the interaction. Additional information regarding these techniques that is beyond the scope of this chapter can be found in **ref. 4**.

3.1. Plasmids

To detect protein–protein interactions in *E. coli*, the bait plasmid expressing a protein of interest (bait) fused to λcl and target plasmids expressing a library of proteins (targets) fused to RNAP-α are prepared.

3.1.1. Target and Bait Plasmid Descriptions

The 3.2-kb pBT bait plasmid (*see* **Fig. 2**) carries a p15A replication origin (p15A ori) and confers chloramphenicol resistance. The plasmid encodes the full-length bacterial phage λcl protein (lambda-C1) under the control of the *lacUV5* promoter (P lac-UV5). The multiple cloning site (MCS) contains several restriction sites to facilitate insertion of DNA encoding the bait protein in the same reading frame as lambda-C1.

The 4.4-kb target plasmid, pTRG (*see* **Fig. 3**), carries a ColE1 replication origin (ColE1 ori) and confers tetracycline resistance (*see* **Note 1**). The plasmid encodes the amino-terminal domain of RNA polymerase α-subunit (RNAP-α) under the control of the tandem promoter *lpp/lacUV5* (P lpp/lac-UV5). The arrangement of *Eco*RI and *Xho*I restriction sites in the MCS is designed to express cDNA inserts generated with Stratagene's cDNA Synthesis Kit (described in detail in Chapter 24).

3.1.2. Control Plasmids

The BacterioMatch two-hybrid system includes the pTRG-Gal11[P] and the pBT-LGF2 positive control plasmids (*see* **Fig. 4**). pBT-LGF2 expresses the

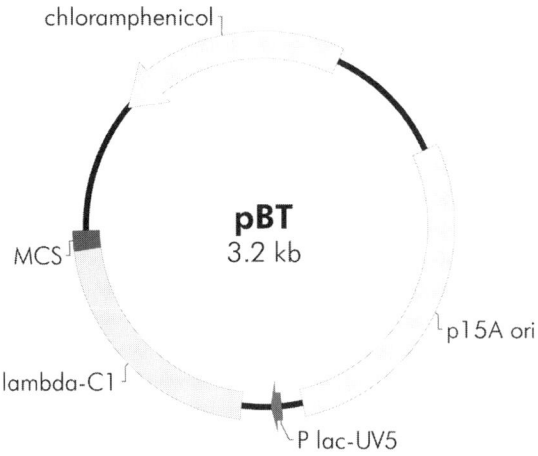

pBT Multiple Cloning Site Region
(sequence shown 2327–2394)

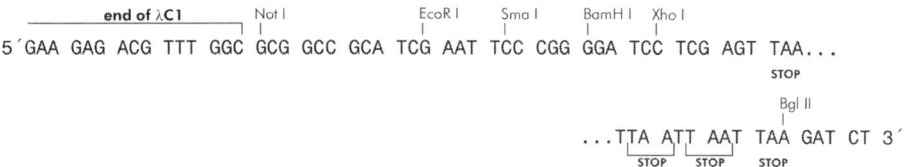

Feature	Position
chloramphenicol resistance ORF	2770–219
p15A origin of replication	581–1493
lac-UV5 promoter	1556–1586
λC1 ORF	1631–2341
pBT forward primer [5' TCCGTTGTGGGGAAAGTTATC 3']	2291–2311
multiple cloning site	2342–2394
pBT reverse primer [5' GGGTAGCCAGCAGCATCC 3']	2419–2436

Fig. 2. The pBT bait plasmid.

dimerization domain (40 amino acids) of the Gal4 transcriptional activator protein. pTRG-Gal11[P] expresses a domain (90 amino acids) of the mutant form of the Gal 11 protein (Gal11[P]). Gal 4 and Gal11[P] have been shown to interact in *E. coli* cells (*2*). Cotransformation of the BacterioMatch two-hybrid system reporter strain competent cells with these plasmids results in the expression and interaction of these proteins. This interaction is detected by growth of the colonies on LB-agar plates supplemented with 250 µg/mL carbenicillin,

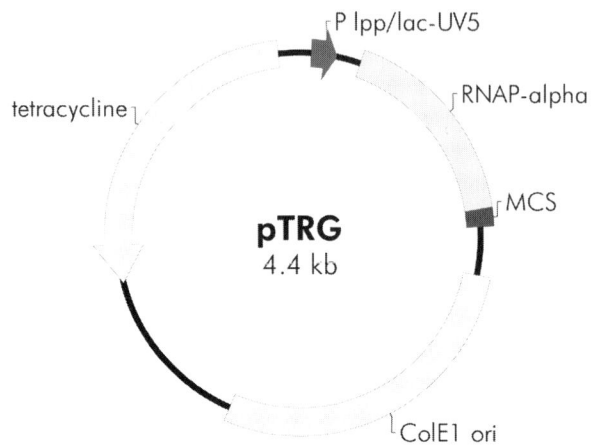

pTRG Multiple Cloning Site Region
(sequence shown 978–1065)

end of RNAPα   BamH I   Not I   EcoR I

5′ AAA CCA GAG GCG GCC GGA TCC GCG GCC GCA AGA ATT CAG TCT GAG CTG GCG...

Xho I   Spe I

...CTC GAG TAA TTA ATT AAT TAA TGA ACT AGT GAG ATC C 3′
            STOP   STOP STOP STOP STOP   STOP STOP

Feature	Position
lpp promoter	47–76
lac-UV5 promoter	119–148
RNAPα ORF	243–992
pTRG forward primer [5′ CAGCCTGAAGTGAAAGAA 3′]	957–974
multiple cloning site	993–1058
pTRG reverse primer [5′ ATTCGTCGCCCGCCATAA 3′]	1102–1119
ColE1 origin of replication	1243–2475
tetracycline resistance ORF	3120–4310

Fig. 3. The pTRG target plasmid.

15 µg/mL tetracycline, 34 µg/mL chloramphenicol, and 50 µg/mL kanamycin (LB-CTCK agar plates).

3.2. Reporter Gene Cassette and Reporter Strain

The reporter gene cassette contains the ampicillin (Amp^r) and the β-galactosidase (*lacZ*) genes on the F′ episome in the BacterioMatch reporter

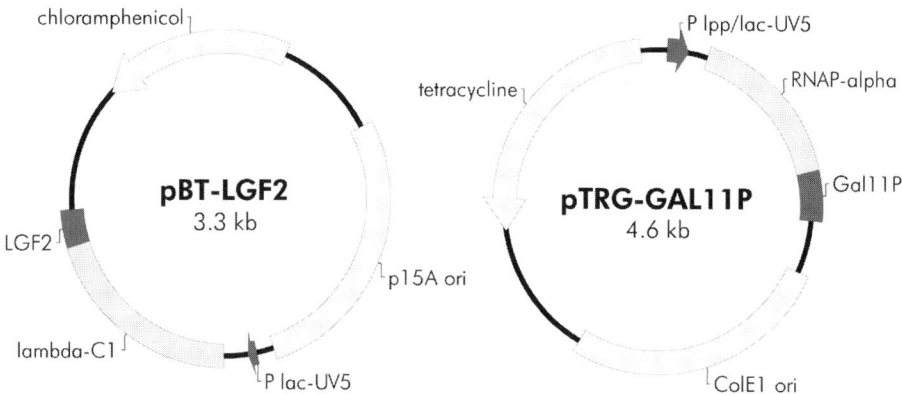

Fig. 4. Circular maps of the control plasmids.

strain (*see* **Fig. 1**). The activatable promoter in the reporter gene cassette is a modified *lac* promoter that contains a single λ operator (O_R2) centered at position –62 in place of the original cAMP receptor protein (CRP)-binding site. This modified lac promoter is not induced by isopropyl-1-thio-β-D-galactopyranoside (IPTG). To enhance reporter gene expression, the Shine–Dalgarno translational signal is also included upstream of the reporter genes.

The BacterioMatch two-hybrid system reporter strain was derived from Stratagene's XL1-Blue MR, resulting in high transformation efficiency in a restriction minus host. The reporter strain has *lac*I^q on the F′ episome to repress expression of the bait and target proteins. Expression of bait and target is enhanced during screening by the addition of IPTG (5–10 μ*M*). In addition, a basal level of expression of the reporter genes occurs in the reporter strain, resulting in a background level of ampicillin resistance. This background growth of the reporter strain is reduced when carbenicillin and not ampicillin is used in the LB-CTCK plates.

3.3. Bait Plasmid Construction

3.3.1. Preparing Insert DNA Encoding Bait Protein

DNA encoding the bait protein is prepared either by restriction digestion or polymerase chain reaction (PCR) amplification. DNA encoding the bait protein must be inserted so that the bait protein is expressed in the same reading frame as the λcI protein. It is advisable that DNA encoding a flexible linker, such as (Gly$_4$Ser)$_3$ be incorporated between DNA encoding the λ repressor protein and the bait protein. The *Not*I, *Eco*RI, *Sma*I, *Bam*HI, *Xho*I, and *Bgl*II sites are unique in the pBT plasmid (*see* **Fig. 2**).

3.3.2. Ligating Insert into the pBT Plasmid

1. Digest 2 µg of the pBT plasmid.
2. Extract with an equal volume of phenol–chloroform until a clear interface is obtained.
3. Repeat the extraction with an equal volume of chloroform.
4. Add an equal volume of 4 M NH$_4$OAc to the aqueous phase of the tube and 2 volumes of 100% (v/v) ethanol at room temperature. Spin in a microcentrifuge at room temperature.
5. Wash the pellet twice with 70% (v/v) ethanol.
6. Resuspend the pellet in a volume of TE buffer to a DNA concentration of approx 0.1 µg/µL.
7. Verify that the plasmid has been digested completely by agarose gel electrophoresis.
8. Calculate the amount of insert required for a molar ratio of insert-to-vector DNA of 2 : 1 to 3 : 1 (*see* **Notes 2** and **3**). Combine 1 µL 10X ligase buffer, 1 µL of 10 mM rATP (pH 7.0), 1 µL digested pBT plasmid DNA (approx 0.1 µg/µL), X µL digested insert DNA, and double-distilled water up to a volume of 9.5 µL. Add 0.5 µL of 4 U/µL T4 DNA ligase. Incubate overnight at 12°C.
9. Transform XL1-Blue MRF′ with the ligation reactions following the manufacturer's protocol.
10. Plate on LB-chloramphenicol agar plates and incubate overnight at 30°C to minimize potential bait protein toxicity.
11. Identify recombinants by PCR analysis using vector-specific (**Fig. 2**) or insert-specific primers or restriction analysis of miniprep DNA (*see* **Note 4**). Determining the nucleotide sequence of the insert and the junctions between the vector and insert is highly recommended.

3.3.3. Testing Bait Plasmid for Expression of Reporter Genes in the Absence of an Interacting Target Protein

The ability of the bait to induce expression of the reporter genes in the absense of target must be determined prior to using the bait to detect protein–protein interactions in the BacterioMatch two-hybrid system. This is easily determined by cotransforming the BacterioMatch two-hybrid system reporter strain with the recombinant pBT (bait plasmid expressing the bait protein) and the nonrecombinant pTRG plasmid (not expressing a target protein). No carbenicillin-resistant colonies should grow on LB-CTCK plates. If growth is observed, the recombinant pBT plasmid is not suitable for use in the BacterioMatch two-hybrid system. It may be possible to selectively alter specific amino acid residues or protein regions of the bait to eliminate this undesireable expression.

1. Thaw the BacterioMatch two-hybrid system reporter strain competent cells on ice. Gently mix cells by tapping.
2. Aliquot 100 µL of the cells into prechilled Falcon 2059 polypropylene tubes.
3. Add 1.7 µL β-ME to each of the aliquots of cells and swirl the contents of the tube gently. Incubate on ice for 10 min.
4. Add 10 ng of the recombinant pBT plasmid and 10 ng of the nonrecombinant pTRG plasmid into the reporter strain competent cells.
5. Swirl the contents of the tubes gently and incubate on ice for 30 min.
6. Heat pulse in a 42°C water bath for 45 s. The duration and temperature of the heat pulse is critical for obtaining the highest efficiencies. Do not exceed 42°C or 45 s.
7. Incubate the tubes on ice for 2 min.
8. Add 0.9 mL of NZY broth (prewarmed to 30°C) to each tube and incubate at 30°C for 1.5 h, with shaking.
9. Plate 1 µL and 10 µL of each transformation onto 100 mm LB-CTCK agar plates. To plate the samples, pipet 100 µL of NZY broth onto the plate, pipet the aliquot of transformed cells into the pool of broth, and spread evenly.
10. Incubate the plates at 30°C for 17–24 h. To reduce the appearance of satellite colonies and/or false positives, do not incubate longer than 24 h.
11. Examine the test plates after incubation. Cotransformants of various sizes may be present. Growth of more than 20 colonies on the plates containing the recombinant pBT cotransformation indicates transcription of the Amp^r reporter gene in the absence of interacting proteins and that the recombinant pBT plasmid is not suitable for use in this system. If fewer than 20 colonies is observed, the recombinant pBT plasmid can be used in protein–protein interaction studies.

3.4. Construction of Target Library in the pTRG Plasmid

3.4.1. Preparing a cDNA Library in the pTRG Plasmid

Details on the preparation of the cDNA inserts encoding target proteins using the reagents in the BacterioMatch Two-Hybrid System cDNA Library Construction Kit can be found in Chapter 24. The cDNA is inserted into the pTRG XR plasmid in a sense orientation (*Eco*RI–*Xho*I) with respect to the p_{GAL1}, which controls expression of the target proteins.

3.4.1.1. TEST LIGATION OF THE TARGET CDNA INTO THE PTRG PLASMID

1. To prepare the sample ligation, add the following components to an RNase/DNase-free microcentrifuge tube: *X* µL of resuspended cDNA (use 3:1 molar ratio of insert to vector [approx 25–35 ng]), 0.6 µL of 10X ligase buffer, 0.6 µL of 10 m*M* rATP (pH 7.5), 30 ng pTRG XR plasmid (0.1 µg/µL), *X* µL distilled water for a final volume of 5.4 µL, and 0.6 µL of T4 DNA ligase (4 U/µL) (*see* **Note 5**).
2. Flick the tube gently to mix and then spin down briefly. Incubate the reaction tubes overnight at 12°C.

3. Thaw the 500-µL aliquot of XL1-Blue MRF′ Kan Library Pack Competent Cells on ice. Gently mix the cells by flicking the tube.

4. Aliquot 100 µL of the cells into prechilled 15-mL Falcon 2059 polypropylene tubes.

5. Add 2 µL of the Library Pack β-ME provided with the kit to each 100-µL aliquot of cells. Swirl the contents of the tube gently.

6. Incubate the cells on ice for 10 min, swirling gently every 2 min. Do not exceed 10 min.

7. Add 1.6 µL of each ligation reaction to separate aliquots of the cells and swirl gently. Performing duplicate transformations of the test ligation is recommended.

8. Incubate on ice for 30 min.

9. Heat pulse in a 42°C water bath for 33 s. The duration and temperature of the heat pulse is critical for obtaining the highest efficiencies. Do not exceed 42°C or 33 s.

10. Incubate on ice for 2 min.

11. Add 0.9 mL of SOC medium (prewarmed to 30°C) to each tube.

12. Incubate at 30°C for 1.5 h, with shaking.

13. Place 100-µL pools of SOC medium on four separate 100-mm LB-tetracycline agar plates.

14. Pipet 2 µL and 5 µL of the cell culture from the each transformation tube into two different pools of SOC medium on the agar plates. Spread the mixture evenly on the plates.

15. Incubate at 30°C for 17–30 h. Incubation at 30°C is necessary to minimize potential toxic effects of the target proteins on the *E. coli* host, thereby ensuring good library representation.

16. Count the number of tetracycline-resistant colonies and calculate total colony-forming units (cfu) for the transformation using the formula

$$\frac{(\text{No. of colonies})\,(\text{Total volume of transformation }[\mu L])}{\text{Volume of transformation plated }(\mu L)} = \text{Total cfu}$$

17. The percentage of vectors with cDNA insert and the average insert size can be determined by either PCR directly from individual colonies with pTRG-specific primers (*see* **Fig. 3**) or by restriction analysis of plasmid DNA isolated from individual colonies.

3.4.1.2. Scaling Up the Ligations and Transformations for pTRG cDNA Libraries

Perform individual ligations to reach the target primary library size desired. The number of ligations necessary may vary with each insert and should be based on efficiencies of the test ligation.

1. Combine *X* µL of cDNA (use 3:1 molar ratio of insert to vector [approx 145–155 ng]), 3.0 µL of 10X ligase buffer, 3.0 µL of 10 m*M* rATP (pH 7.5), 150

ng of pTRG XR plasmid (0.1 µg/µL), X µL of distilled water (to a final volume of 30 µL), and 3.0 µL of T4 DNA ligase (4 U/µL).

2. Flick the tube gently to mix and then spin down briefly. Incubate the reaction tubes overnight at 12°C.

3. Thaw the 500-µL aliquot of XL1-Blue MRF′ Kan Library Pack Competent Cells on ice.

4. Immediately after thawing, gently mix the cells by flicking the tube. Transfer all of the cells into a prechilled 15-mL Falcon 2059 polypropylene tube.

5. Add 10 µL of the Library Pack β-ME provided with the kit to the 500 µL of bacteria. Swirl the contents of the tube gently.

6. Incubate the cells on ice for 10 min, swirling gently every 2 min. Do not exceed 10 min.

7. Add 8 µL of the ligation reaction to the cells and swirl gently. Incubate on ice for 30 min.

8. Heat pulse the tubes in a 42°C water bath for 50 s. The duration and temperature of the heat pulse is critical for obtaining the highest efficiencies. Do not exceed 42°C or 50 s.

9. Incubate the tubes on ice for 2 min.

10. Add 4.5 mL of SOC medium (preheated to 30°C) and transfer all of the liquid to a sterile 50-mL conical tube. Incubate at 30°C for 1.5 h, with shaking. This is the primary library. Store the primary library at 4°C and use to generate an amplified library within 1–2 d.

11. Plate 2 µL and 5 µL of the primary library onto 100-mm LB-tetracycline plates to determine the total number of clones in the primary library as described in **steps 13–16** in **Subheading 3.4.1.1.**

3.4.1.3. Amplification of the Primary Library

1. Plate the remaining primary library on 150-mm LB-tetracycline plates. Plate approximately 250 µL of culture on each plate and spread the liquid evenly over the surface of the agar. This plating volume ensures that the colonies are not overly populated on the plate. If using 25 cm × 25-cm plates, 1 mL of culture per plate can be plated.

2. Incubate at 30°C for 17–30 h.

3. Using a wide, sterile scraper (e.g., cell scraper, cell lifter, or rubber policeman), gently scrape all of the bacteria to one edge of the first 150-mm plate. Lift or scrape the dense paste from the edge into a 50-mL sterile container on ice (*see* **Note 6**).

4. Repeat **step 3** for all additional plates and pool the bacteria from all of the plates into the same sterile container.

5. Add enough SOC medium to the collected bacterial paste to allow pipetting, but keep the volume of added medium as small as possible.

6. Resuspend the harvested bacteria by pipeting repeatedly but gently until the cells are completely resuspended. Calculate the volume of resuspended cells. This is the amplified library.

7. Plate 2 μL and 5 μL of the resuspended amplified library onto LB-tetracycline plates to determine the number of colony-forming units per milliliter (cfu/mL) of the amplified library, as described in **steps 14–16** of **Subheading 3.4.1.1.**

8. Isolate plasmid DNA from one-half of the pooled amplified library using a commercially available maxiprep DNA column or alkaline lysis/CsCl gradient purification method (*see* **Note 7**).

9. Add 0.2 volumes of 80% (v/v) glycerol to the remaining bacterial resuspension and mix thoroughly by pipetting gently or inverting the container.

10. Dispense 1-mL aliquots into sterile 1.5-mL microcentrifuge tubes and store at –80°C.

3.5. Screening Library for Target Proteins That Interact with Bait Protein

3.5.1. Screening for Interacting Proteins

Cotransform 10 ng of the pTRG target cDNA library and 10 ng of the pBT bait plasmid into the BacterioMatch two-hybrid system reporter strain competent cells (*see* **Subheading 3.3.1.1.**) and plate on LB-CTCK agar plates to determine the number of colonies that will be obtained per transformation.

To screen for interacting proteins, repeat the transformation using the required number of cells and plasmid DNA to obtain $(1–2) \times 10^6$ colonies. Do not plate more than 5×10^5 cotransformants per 150-mm LB-CTCK agar plate (*see* **Note 8**). This will require a minimum of 20 plates. Include separate cotransformations of 10 ng of each plasmid as follows: (1) pBG-LGF2 and pTRG-Gal11^P and (2) pTRG and pBT-LGF2. The interaction between expressed Gal11^P and LGF2 proteins in cotransformation 1 should result in growth on LB-CTCK agar plates. The lack of interaction between proteins in cotransformation 2 should result in insignificant growth [≤ 1% when compared to cotransformation (1)]. If interacting bait and target proteins are present in the sample cotransformation, colonies should grow on the LB-CTCK agar plates. Patch putative positive colonies onto LB-TCK agar plates without carbenicillin. These patched plates serve as the primary source of target plasmid DNA for validation analyses. If no colonies are present, no interactions can be detected at the carbenicillin concentration of 250 μg/mL in the LB-CTCK agar plates. Reducing the carbenicillin concentration may enable detection of lower-affinity interactions.

3.5.2. Adjusting Carbenicillin Concentration in LB-CTCK Agar Plates

Expression from the promoter of the reporter cassette is proportional to the strength of the protein–protein interaction between bait and target proteins. Significant growth on 250 μg/mL carbenicillin indicates an interaction that has a $K_d \leq 10^{-7}$ M. Interactions with higher K_d values (weaker interactions) may be detected by decreasing the concentration of carbenicillin in the LB-CTCK agar plates. For example, analyze growth on plates containing 200 µg/mL and 150 µg/mL. Carbenicillin concentrations below 150 µg/mL are not recommended, as this can lead to undesired background growth. If the detection of stronger interactions is desired, growth can be analyzed on LB-CTCK plates containing up to 500 µg/mL carbenicillin.

3.6. Validation of Putative Protein–Protein Interactions

The interaction between the bait and target proteins can be verified by the following two methods. Which method is performed and in what order is determined by the researcher according to his or her individual priority and requirements.

3.6.1. Interaction Validation Using β-Lactamase (Amp^r) Reporter Gene

First, isolate the target and bait plasmids from the reporter strain using protocols that have been modified for isolation of low copy number plasmids. Second, transform XL1-Blue MRF′ Kan with the isolated plasmids and select for cells that have been transformed with the target plasmid by plating on LB-tetracycline agar plates. Identify colonies that have been transformed with only the target plasmid by patching the transformants onto LB-chloramphenicol and LB-tetracycline agar plates. Transformants that are tet^r and cam^S contain only the target plasmid DNA. Third, isolate target plasmid DNA. Fourth, cotransform the reporter strain using the target plasmid and the following: (1) recombinant pBT plasmid, (2) pBT plasmid without insert, and (3) recombinant pBT plasmid expressing any protein that does not interact with the target protein. In addition, transform the reporter strain with only the target plasmid.

Select target plasmids for further analysis that induce the growth of colonies on LB-CTCK agar plates in the presence of the recombinant pBT plasmid. Do not select target plasmids that induce the growth of colonies on LB-CTCK agar plates in the absence of the bait protein.

3.6.2. Interaction Validation Using β-Galactosidase (lacZ) Reporter Gene

A second reporter gene, β-galactosidase, is expressed from the same reporter promoter as the Amp^r gene and provides an additional means of validating the bait and target interaction. To verify interacting target and bait proteins, streak colonies cotransformed with the bait and target plasmids directly onto X-gal indicator plates. As a positive control, streak a colony containing the control plasmids pTRG-Gal11^P and pBT-LGF2, and as a negative control, streak a colony containing the pTRG and pBT-LGF2 plasmids. Incubate at 30°C overnight and examine colonies for the presence of a blue color. It generally takes longer than 17 h for the blue color to develop. However, avoid very long incubations as this can lead to increased background color development. Compare the test colonies to the control colonies to correctly interpret the results.

3.6.3. Further Analysis of Verified Positives

To identify the protein encoded by the target DNA, determine the nucleotide sequence of the target DNA and compare it to nucleotide and protein sequence databases. In addition, the target DNA can be used as a hybridization probe to screen libraries for full-length target DNA clones and for clones with high homology to the target DNA.

4. Notes

1. The pTRG plasmid can be toxic to some cells; therefore, all forms of pTRG (with or without insert present) should be grown at 30°C in a host strain containing the *lacI^q* gene. This will minimize the overall toxicity of the plasmid and will minimize the selection for less toxic pTRG mutants and/or less toxic pTRG recombinants.

2. For ligation, the ideal ratio of insert to vector DNA is variable; however, 2:1 or 3:1 ratios are recommended (insert-to-vector molar ratio). The amount of insert DNA required for a 1:1 ratio is calculated as follows:

$$X \text{ ng of insert} = \frac{(\text{No. of base pairs of insert}) (100 \text{ ng of vector})}{3210 \text{ bp (pBT plasmid)}}$$

3. The preparation of addtional control ligations with vector only and insert only to determine the number of background colonies resulting from each component is highly recommended.

4. The pBT plasmid is present at about 10 copies per cell. When preparing plasmid DNA from host cells harboring this plasmid, use suggested protocol modifications developed for low-copy-number plasmids.

5. A separate positive control ligation using 1.5 µL of XR Amp test insert (15 ng) is also highly recommended.

6. Some researchers prefer to add liquid medium (e.g., SOC medium) to the plate before scraping or prefer to "wash" the plate with liquid medium after scraping. If this is desired, keep the volume of liquid medium added to a minimum, just enough to allow pipetting during the resuspension step.

7. The pTRG plasmid is present at about 20–30 copies per cell. When preparing plasmid DNA from host cells harboring this plasmid, use suggested protocol modifications developed for low-copy-number plasmids.

8. Avoid excessive surface moisture on the LB-CTCK agar plates, which can lead to an increase in the appearance of false positives. However, if the plates are too dry, the growth of colonies will be inhibited.

References

1. Dove, S. L., Joung, J. K., and Hochschild, A. (1997) Activation of prokaryotic transcription through arbitrary protein–protein contacts. *Nature* **386,** 627–630.
2. Dove, S. L. and Hochschild, A. (1998) Conversion of the omega subunit of *Escherichia coli* RNA polymerase into a transcriptional activator or an activation target. *Genes Dev.* **12,** 745–754.
3. Shaywitz, A. J., Dove, S. L., Kornhauser, J. M., Hochschild, A., and Greenberg, M. E. (2000) Magnitude of the CREB-dependent transcriptional response is determined by the strength of the interaction between the kinase-inducible domain of CREB and the KIX domain of CREB-binding protein. *Mol. Cell Biol.* **20,** 9409–9422.
4. Sambrook, J., Fritsch, E. F., and Maniatis, T. (1989) *Molecular Cloning: A Laboratory Manual,* Cold Spring Harbor Laboratory, Cold Spring Harbor, NY.

27

Future Perspectives

Shao-Yao Ying

We have made tremendous strides in our understanding and utilization of the new methodologies in molecular biology over the past 50 yr. With our recent access to complete genome sequences, the function of different transcripts and proteins, the gene products, will soon be determined and, consequently, their functions in cells, tissues, and organisms will be defined. Therefore, gene expression analysis not only provides an important functional genomics technology but also will lead to future research opportunities, diagnostic utilization, targeted drug development, and clinical therapeutic applications. With more defined methodologies, more specific questions in molecular biology can be answered. The study of molecular biology has certainly entered a phase of exponential growth and productivity. In this chapter, general principles will be proposed, and examples will be provided for exploring the future perspectives of the biomedical field, using some novel technologies described in this book.

1. Research Opportunity

The newly developed technologies described in this book are most suitable for defining functional genomics by global approaches. For instance, replication and amplification of mRNA in microdissected single cells facilitates the exploration of structural genomics in homogenous starting materials. This approach circumvents most problems encountered in tissues with multiple heterogeneous loci. Coupling this approach to a RNA microarray technology, large-scale elucidation of functional genomics can be achieved (*1*). This combination provides the fundamental basis for drug discovery (*2*). Furthermore, alterations of genes so identified can be employed in knockout studies to ascertain gene function (*3*). With the environment as a potential factor largely

From: *Methods in Molecular Biology, vol. 221: Generation of cDNA Libraries: Methods and Protocols*
Edited by: S.-Y. Ying © Humana Press Inc., Totowa, NJ

accounting for susceptibility to disease, drug efficacy, and drug side effects, it is pertinent to examine genetic variation as a risk factor. By the same token, studies of dietary components traditionally observed to have beneficial effects in certain diseases may lead to the identification of preventive or protective genes against the disease. However, first and foremost, the generation of full-length mRNA/cDNA libraries is still the major determining step for the study of systematic genomics, transcriptomics, and proteomics.

1.1. Principles

The following systemic approach can be applied to a disease, a condition, a system, or functional variation to better understand the underlying mechanism(s). First, the integration of mRNAs in cells is preserved by fixation, homogeneous single cells are obtained by microdissection, replication and/or amplification of mRNA in cells is performed, the genetic etiology or pathway is determined by microarray or subtractive hybridization analysis, differentially expressed genes so identified are assessed by Northern blot analysis, immuno-cytochemistry, *in situ* hybridization, reverse transcription–polymerase chain reaction (RT-PCR), and Western blot analysis. Thus, a genomic understanding of pathogenesis, mechanism, pathway, or regulation is achieved. For diseases, the information acquired will pave the way for improvement in preventive strategies, diagnosis, and therapeutic treatment.

Basically, specific full-length mRNA libraries from as few as 20 single microdissected cells from biopsied sections, cytology specimens with known pathological changes can be generated by the RNA-PCR method. Next, the quality of the mRNA libraries so generated can be assessed. Subsequently, global multisite gene expression (disease/condition-specific biomarkers) can be identified in different disease/condition-specific mRNA libraries by microarray analysis. The identified differentially expressed genes can be confirmed and correlated with clinical/experimental conditions. At that stage, functional significance of the identified genes can be determined.

1.2. Prostate Cancer as an Example

A novel approach to generating mRNA libraries from single cells obtained from prostate cancer histological slides from different cancer stages has been reported (*4*). Several potential applications of this method exist in studies of the conversion of neoplasia to carcinoma. One of the approaches is to couple the RNA-PCR-derived poly(A$^+$) RNA libraries with microarray technologies to identify specific differentially expressed genes from prostate cancer cells at various stages in the conversion from neoplasia to carcinoma. The results should permit the development of a useful diagnostic tool for early detection of genetic alterations associated with the conversion of neoplasia to carcinoma.

For the differentially expressed genes identified by RNA-PCR and microarray technologies, Northern and Western blot analyses, *in situ* hybridization, and immunocytochemical techniques are used to correlate the identified differentially expressed genes and their proteins with the specific pathological conditions. Subsequently, a functional significance study of the identified genes is performed to ascertain the clinical implication. Finally, the gene and its protein are examined to pave the way for potential gene therapy studies in the future.

1.2.1. Laser Capture Microdissection

Obtaining homogeneous single cells as the starting material for generating mRNA/cDNA libraries and subsequent molecular analyses has been a difficult challenge for many years. Laser capture microdissection has filled the gap by providing a rapid, efficient, and precise method for obtaining homogeneous single cells from complex tissue for molecular and cellular studies. Single cells of a tissue specimen can be obtained from histological tissue sections that are routinely formalin-fixed and paraffin-embedded *(5)* and this approach has been recently reviewed *(6)*.

There is still no effective cure for advanced prostate cancer. To understand the initiation and progression of the disease at the molecular level in single cells, laser capture microdissection (LCM) is a useful tool. To elucidate the etiologies of prostate cancer at the molecular level, pathological samples of benign, low-grade prostatic intraepithelial neoplasia (LGPIN), high-grade intraepithelial neoplasia (HGPIN), carcinoma, and metastasis can be obtained by laser-assisted preparation *(4)*. These samples may be used for generating full-length poly(A⁺)RNA libraries by the RNA-PCR method.

Briefly, the prepared tissue is shielded with a transparent film, and stained cells are identified and microdissected with a laser microbeam. In this way, a clear-cut gap is formed around the selected area. Because the dissected cells are adhesive to the film; the specimen may be directly delivered to a microfuge tube containing the extraction buffer. Recently, new laser microdissection systems have been developed to avoid heat generated by mechanical contact.

1.2.2. mRNA Amplification

This approach has been described in Chapter 12 in detail. During the first reverse transcription of intracellular mRNA, an oligo-(dT)-promoter primer is introduced as a recognition site for subsequent transcription of newly reverse-transcribed complementary DNAs (cDNAs). These cDNAs are further tailed with polynucleotide; the polynucleotide and the promoter primer of these cDNAs form binding templates for specific PCR amplification. After one round of reverse transcription, transcription, and PCR, a single copy of mRNA can be multiplied 2×10^9-fold. Coupling this method with a cell fixation and permeabilization step, the complete full-length cDNA library can be directly generated from a few single cells, avoiding mRNA degradation. Thus, cell-specific full-length cDNA libraries are prepared.

1.2.3. Identification of Differentially Expressed Genes by Microarray

Most current microarray techniques in prostate cancer research are capable of generating large gene expression datasets with the potential to provide novel insights into fundamental prostate cancer biology at the molecular level. However, there are limitations in currently available methods. For example, there are concerns that messenger RNA (mRNA) integrity may be damaged by fixatives, the large number of cells required to prepare cDNA/mRNA libraries may not be genetically homogeneous, what candidate genes should be probed for different stages of progression of the disease, how to determine the genes or pathways that mediate the formation or progression of prostate tumors, and whether unknown transcript bias is introduced. The RNA-PCR method coupled with microarray techniques partially solve these problems. We have demonstrated that full-length cDNA/mRNA libraries can be generated from less than 20 cells from a human prostate cancer histological slide, indicating that, with appropriate precautions and the high resolution of amplification, it is feasible to use routinely fixed tissue slides as the starting material. In this way, mRNA libraries could be built based on homogenous cell populations. Furthermore, with the control of cell type and stages of the progression of prostate cancer, bias introduced by unknown transcripts is eliminated. Moreover, with microarray analysis between two different stages in the same cell type, differentially expressed genes can be identified. This provides a tool for the study of the genes or pathways that lead to the conversion from neoplasia to invasive carcinoma.

1.2.4. Identification of Differentially Expressed Genes by Subtractive Hybridization

Several improved subtractive hybridization methods have been reported to provide a fast, simple, and reliable isolation of sequences not common to two compared DNA libraries (7), particularly novel genes. Coupled with the RNA-PCR method, subtractive hybridization methods will identify and isolate known and novel differentially expressed genes in cell-specific stage-specific full-length cDNA libraries from single cells isolated from human prostate cancer specimens (8).

1.2.5. Confirmation of the Identified Genes by Various Methods

To ascertain that the identified differentially expressed genes are associated with various stages of prostate cancer, Northern and Western blot analyses, *in situ* hybridization, and immunocytochemistry can be used to confirm the microarray data and the subtractive hybridization results. Consequently, identified genes and their proteins are shown to be correlated with the pathological conditions as well as localized in the the same types of cell microdissected as the starting materials. Because the *in situ* hybridization technique potentially localizes the gene expression qualitatively and quantatively in a pathological section, a quantitative confirmation of the identified genes can be conducted concurrently.

1.2.6. Examination of the Functional Significance of the Identified Genes

To complete the cycle of investigation, the identified genes are tested in vitro and in vivo for their functional significance. Numerous overexpression and gene knockout methods are available for the examination of functional significance of the identified genes. Any commercially available overexpressed system with the regulator of functions can be used to examine the effects of overexpression of a gene on cancer cell growth in vitro and in vivo. For genes identified as overexpressed, the elimination of the function of the gene can be accomplished by gene knockout, ribozyme, antisense, or RNA-mediated interference technologies (RNAi). Recently, an improved method has been developed to eliminate the function of a specific gene, termed mRNA–cDNA interference (D-RNAi) *(9)*. We were able to silence gene expression by a relatively long-term interference with specific gene expression by introducing a mRNA–cDNA hybrid into human prostate cancer cells. To the best of our knowledge, D-RNAi is the first instance of a long-term, sequence-specific cosuppression of homologous transcripts in human cells, using a hybrid of about 500 bp, with the RNA interference lasting for a longer time at much lower doses than those of RNAi. Incidentally, duplexes of 21-nucleotide RNAs produced RNA intereference in cultured mammalian cells, but nucleotides over 30 bp induced nonspecific RNA degradation *(10)*.

1.2.7. Specific Targeted Studies

The above-listed approaches can also be used for specific targeted studies. Several factors such as transforming growth fctor (TGF)-β, activin, tumor necrosis factor (TNF)-α, Fas, and radiation trigger apoptotic processes by initiating distinct and specific signal transduction in prostate cancer cells. Subsequently, cytoplasmic degradation that is associated with intracellular receptors such as Siva-like molecules takes place *(11–13)*. Consequently, death signals are initiated, resulting in increased levels of nucleases that lead to genomic DNA fragmentation. Several molecules, including protein kinase C (PKC) *(14)*, bcl-2, p53, bax, caspases, Ki-57 *(15–18)*, cyclooxygenase-2 (COX-2) *(19)*, and sulfated glycoprotein-2 (SPG-2) *(20)* may be associated with this postulated pathway. Any one of these genes can be used for specific targeted studies by either overexpression or knocking out the specific gene. Subsequently, mRNA/cDNAs are prepared that are analyzed by microarray technologies or subtractive hybridization. The resulting differentially expressed genes are the specific targeted genes for study.

1.2.8. Gene-Environmental Interaction

For complex diseases such as prostate cancer that are associated with heterogeneous loci and multiple factors and that are under a substantial degree of genetic control, the gene-envirnoment interaction or the genetic epidemiological etiology can be better studied with a global genomic approach, as outlined earlier. For example, consumption of soy-based food is about 50 times greater in Asian men than Caucasian men *(21)*; thus, it is pertinent to monitor the protective or preventive genes mediated by chronic exposure to soy-bean components.

For the following reasons, prostate cancer is used to illustrate this approach. First, the incidence rates and deaths from prostate cancer in Asian men are lower than in American men; therefore, the high consumption of soy-based foods in Asian men is thought to have chemoprotective or chemopreventive effects. Second, genistein, one of the components of soy, has been found to play a role in prostate cancer cell growth as a result of altering gene expression via inhibitions of tyrosine kinases. The role of these genes remains to be completely elucidated. Third, the advent of microarray technologies can allow monitoring of alteration of all genomic genes in prostate cancer cells concurrently. Finally, because frequently there are heterogenous loci in prostate cancer, it is desirable to make mRNA libraries from homogeneous single cells.

The possible goals of genetic epidemiological etiology are to use RNA-PCR-generated poly(A$^+$) RNA libraries and RNA microarray technologies to conduct a systematic characterization of gene expression in prostate cancer cells exposed to potential therapeutic agents such as genistein or to compare prostate cancer cells obtained from pathological slides of American and Asian men. mRNA libraries from these cells are prepared and hybridized with RNA microarrays. The changes in gene expression between high-soy-intake Asian men versus low-soy-intake American men can be analyzed as described earlier. By the same token, subtractive hybridization can also be used. Similar gene–environment interactions can be examined using the same approaches.

1.3. Research in Various Fields

1.3.1. Breast Cancer

The methodologies described can be adapted for breast cancer research. The pathogenesis of breast cancer involves ductal hyperproliferation, progression *in situ*, and invasion of carcinomal cells. Studies to elucidate differentially expressed genes among various stages (normal, DCIS [ductal carcinoma *in situ*], IDC [intraductal carcinoma], and metastatic carcinoma) of breast cancer could result in global targeted genes potentially useful for early detection and prevention strategies. Similarly, the differentially expressed genes between the ductal epitheial cells and their adjacent stromal cells of any specific stage could shed light on the epithelial–stromal interaction during the progression of breast cancer.

Another area of interest is angiogenesis in breast tumor, because the degree of vascularity is associated with prognosis of the patients exhibiting early-stage invasive breast cancer *(22)*. Indeed, there is a linear relationship between the probability of metastasis and the degree of vascularization of the primary breast tumor. Tumor growth and metastasis are inhibited by neutralizing angiogenic factors or by systemic therapy with endogenous antiangiogenic factors or agents. A number of growth factors and cytokines are involved in angiogenesis, including VEGF (vascular endothelial growth factor) *(23)*, PD-ECGF (platelet-derived endothelial cell growth factor, also known as TP [thymidine phosphorylase]) *(24)*, FGFs (fibroblast growth factors) *(25)*, interleukin-12 *(26)*, and seven thrombospondins *(27)*. Generally, these factors are capable of stimulating tumor growth by paracrine mechanisms, resulting in the formation of new blood vessels from the pre-existing vascular network of the tissue. With the above-described methodologies, endothelial cells collected at the site of initiation of new blood vessels, the formation of basement membrane, the extracellular matrix degradation of basement membrane, and the formation of the vascular lumen can be investigated for identification of differentially expressed genes as angiogenic biomarkers during the process of angiogenesis.

1.3.2. Neuronal and Glial Cell Studies

Our understanding of the structure and function of neuronal and glial cells has grown rapidly and new discoveries have been reported frequently during the past decade. To illustrate the research opportunities in this area using the methodologies described earlier, two potential approaches are described briefly.

One is the identification of genes that are dynamically regulated during neural differentiation as well as neural activity with the above-described methodologies. Genes involved in various stages of neural differentiation can be identified to shed light on the functional evolution of specific neurons. In addition, various populations of neurons that respond exclusively to one environmental factor vs neurons that responded to two or more environmental factors can be examined and the environment-specific genes can be identified.

Another potential area is to determine the molecular basis for the mechanisms underlying neurodegenration, including Alzheimer's disease. There has been dramatic progress in this area recently and several Alzheimer's disease-related genes have been identified. However, the above-described methodologies can provide the identification of global genes that are differentially expressed in normal cells versus diseased cells.

2. Diagnostic Utilization

Although the pathologic stage is currently the main prognostic indicator for patients with many diseases including cancer, increasing evidence suggests that, in its current form, this stage is insufficient to predict clinical outcome. Studies of the genetic etiology of diseases, particularly chronic and complex ones, offer a way of improving our understanding of pathogenesis, subsequently improving preventive strategies, diagnostic tools, and therapies. Considerable effort and expense have been expended in attempts to detect specific alteration of genes contributing to disease susceptibility. Up until now, mutation loci and genetic polymorphisms have been widely used for direct detection of the disease and its relation to disease phenotype. With the advent of the microarray era, innovative diagnosis of a particular disease and consequent risk profiling will facilitate early detection of the disease.

Numerous diseases have been routinely monitored to determine the appropriate time for initiation of treatment and prevention of disease sequelae. For example, through DNA analysis, mutations of genes for an enzyme that are responsible for inherited disorders such as congenital adrenal hyperplasia resulting in genital abnormalities can be diagnosed prenatally *(28)*. This type of prenatal diagnosis for a disease can be accomplished via DNA analysis; subsequently, the gene defects can be prevented or treated. Early detection of retinoblastoma mutations allow laser removal of eye tumors within the first few weeks of birth, substantially increasing the chances of saving an affected child's eyes *(29)*. The use of polymorphisms derived directly from DNA sequences is frequent, reflecting a high number of DNA sequence variants across the human genome. Indeed, molecular genetic testing is currently available for diseases such as cystic fibrosis *(30)*, sickle cell anemia *(31)*, and Tay–Sachs disease *(32)*, but the initial diagnosis of these disorders usually is established by other pathological methods.

In some cases, the low incidence of mutation requires techniques of high sensitivity; other cases may be associated with numerous potential obstacles to analysis of mutations. For example, K-*ras* mutation is an early event in the development of certain types of cancer and can be considered a biomarker in early detection (diagnosis) and in susceptibility assessment *(33)*. This may be more complicated than a single gene mutation because the mutant forms of *ras* family genes frequently result in constitutive activation of signaling cascades, including PI3k (phosphatidylinositol-3'-kinase), PKB/Akt (protein kinase-B [also known as Akt]), mitogen-activated protein kinase (MAPK), and stress kinases. Furthermore, the ratio of neoplastic or proneoplastic to normal cells may be extremely low and may vary in different target organs as well as in individuals, making it extremely difficult to pinpoint the so-called "hot spots" by the PCR technique in a histopathological examination. Although highly sensitive analyses of body fluids for the presence of mutant *ras* alleles have been developed, positive mutant *ras* alleles have been identified in heavy smokers at stages at which no pathological evidence for lung tumors can be detected. Another way to improve a diagnostic method is to enrich molecularly the target cells from crude clinical samples by microdissection and immunobinding sorting so that homogeneous cells can be used. Even with the optimization of a diagnostic method, it has been recognized that the presence of mutant *ras* alleles does not always correlate with the extent of tumor progression.

On the other hand, particular genes used as biomarkers for tumor detection have been found in patients who carry them in their non-neoplastic samples. Sometimes, levels of these genes in non-neoplastic cells could be higher than those in tumors. Given the biological importance and value of gene detection, there are still the issues of the gene *per se* vs the cooperation of the gene with other oncogenes and tumor suppressor genes to consider. Thus, it is equally important to analyze the targeted genes in isolation and in combination with other biomarkers known to be associated with the development of the tumor.

2.1. Prostate Cancer as an Example

2.1.1. Novel Approaches

The generation of cDNA/mRNA libraries from selected homogeneous cells by the RNA-PCR method and the RNA microarray allows gene expression measurements of thousands of genes in parallel, providing a powerful tool for pathologists seeking new genetic biomarkers for diagnosis. As a result, novel diagnostic approaches based on specific gene alterations are likely. This approach not only can provide biomarkers for known categories of neoplasia but will also lead to recognition of new diagnostic categories. However, the huge volumes of data so generated require thoughtful interpretation to produce information beneficial to patients. This section briefly outlines potential future contributions of a hypothetical DNA sequence variant or differential gene expression to disease diagnosis. To guide prospective users through the myriad decisions that must be made in the design and execution of a successful diagnosis, prostate cancer is given as an example. Similar approaches to identify cell- and stage-specific differentially expressed genes will undoubtedly be developed.

The functional significance of a specific gene is reflected by its protein gene product. For example, prostate-specific antigen has been successfully used as one of the diagnostics for the detection of prostate cancer. Given that understanding of the basis of a complex disease is different because of multiple genetic alterations, defined specific genetic changes are required to detect the progression of the disease. The elucidation of global gene expression would maximize the chances of finding data patterns indicative of the etiology of such a complex disease.

On the other hand, identification of differentially expressed genes at critical times, including initiation, development, and progression of a disease, would provide the means for deciphering the molecular etiology and, consequently, contribute to the goals of cancer detection and prevention. After the identification of a differentially expressed gene that is closely associated with a particular stage of prostate cancer, it might be possible to use this information to develop diagnostic tests.

2.1.2. Early Detections

The ability to determine the risk of prostate cancer before the onset of symptoms (i.e., neoplasia) would be potentially of great benefit in the prevention of this disease. The successful use of such diagnostic testing will depend on the nature and extent of the heterogeneity of the disease, the frequency of the differentially expressed genes at a particular stage that is closely associated with the progression of the disease, the functional significance of the identified genes in terms of disease progression, and its blockade of disease development after overexpression or knockout of the gene, as well as the interactions between the identified genes and factors such as age and gender. Further, there are operational problems such as false-negative responses and ethical and psychosocial concerns that must be taken into consideration. Nevertheless, using information obtained from the generation of cDNA libraries to develop an easy, rapid, and consistent diagnostic tool for early disease detection is the current trend. In this way, a gene diagnostic testing may be developed in the future.

To develop a promising new strategy for ultrafast, reliable, and low-cost detection of gene expression associated with the early stages of prostate cancer, a system involving laser capture microdissection, amplification of mRNA *in situ* by the RNA-PCR method, identification of differentially expressed genes by microarray technologies, confirmation of the correlation between clinical progression of the disease and expression of the identified gene and its proteins, and determination of the functional significance of the gene are all essential. To facilitate this procedure, homogeneous prostatic epithelial cells can be collected as the starting material, guaranteeing that the genes be identified will be pertinent. The most promising starting material for the gene diagnostic test is paraffin-embedded pathological tissue. We have demonstrated that the paraffin-embedded pathological section is suitable for LCM and mRNA amplification by the RNA-PCR method. With the microarray technique and D-RNAi method, the overexpressed genes identified from any particular stage of a disease can be ascertained as potential genes for early detection of the disease in a diagnostic test. Cell- and/or stage-specific differentially expressed genes in as many specimens as possible are required to assure that population-based screening will detect the earliest stages of prostate cancer.

Once a cluster of early-stage disease-specific genes is identified, a group of "housekeeping genes" will be used for the assessment of the quality of the mRNA prepared; this is called an internal control, which assures that these genes are consistently expressed in various stages of the disease, whereas the identified cluster of genes is only specifically expressed in an early stages, but not in the later stages of the disease. Then, the issue of false-negative readings will be explored. This is important because a negative test result may not determine conclusively that the patient is unaffected. After the establishment of different clusters of cell- and/or stage specific genes in a disease, a representative microarray plate can be designed to specifically identify these various clusters of genes as well as the cluster of genes associated with cell proliferation, apoptosis, adhesive proteins, signal transduction, and oncogene/suppressor gene-associated genes. With the same method described earlier (LCM and the RNA-PCR method), a small quantity of sample obtained from a patient can be used to generate a complete mRNA library. Subsequently, the patient's mRNA can be hybridized with the disease-specific microarray plate to detect the gene of interest for early detection of the disease. To validate this new strategy, this approach should be compared with currently commercially available methods using the same prostate cancer specimen.

3. Drug Development

Generation of cDNA is only the first step in achieving an understanding of normal and pathological states. However, the information so gathered in terms of mutations or differentially expressed genes can be used to provide a sufficient rationale for drug development and clinical application, resulting in an effective intervention in patients *(34)*.

Recent cDNA generation methodologies provide a cellular, biochemical, and molecular basis for drug development on the basis of interference with the cell cycle and the induction of apoptosis as a result of interaction with functional gene activities. Knowledge of the cell cycle phase specificity of a particular stage of a disease such as prostate cancer can be used to develop anti-prostate-cancer drugs that target specific genes. Similarly, apoptosis plays a critical role in the development and regulation of cell homeostasis and in the elimination of abnormal cells. Excessive or insufficient programmed cell death contributes to human disease, including neurodegenerative disorders and autoimmune and neoplastic diseases. Moreover, several anticancer drugs induce apoptosis of cancer cells, and apoptosis frequently correlates with cytotoxicity. Targeted gene manipulation during apoptosis can evolve into specific blocking agents for specific cancers at specific stages on the basis of the global differential gene expression. The biochemical and molecular approaches to develop targeted-gene-derived drugs is wide open. Already, several oncogenes, suppressor genes, and numerous growth factors appears to be excellent targets. Likewise, our understanding of numerous effective chemotherapeutic drugs can be further improved by a knowledge of the underlying molecular mechanism of action *(35)*.

To this end, knowledge in the following areas will help realize our vision of drug development: molecular biology, including transcription factors, antisense, protein, and cDNA characterizations, cell cycle and cell signaling networks, molecular-structure-based search and rational design, as well as chemical and biological combinatorial libraries. Thus, it becomes more important to search and validate target genes that control specific diseases. To achieve this goal, it is often critical to obtain a homogeneous sample or isolate individual cells. In addition, the integration of mRNA amplified is essential. The RNA-PCR method described earlier fulfills these requirements by LCM, and preservation of mRNA that is subsequently amplified. In this way, a rapid, high-standard, simple manipulation can be developed in conjunction with further miniaturization and automation for target identification, and target validation leading to better models for human disease.

The trial-and-error days of drug development have given place to an era in which researchers attempt to do drug design logically, based on the underlying biology. This includes transcription factors, antisense, protein, and cDNA characterizations, cell cycle and cell signaling networks, and so forth. In addition, molecular-structure-based search, design, and combinatorial libraries will facilitate synthesis of novel drugs for specific diseases. This approach aids researchers in developing a full understanding of the complex biological mechanisms that are the molecular causes for natural and disease states, enabling them to select the most effective drugs that work most specifically with the fewest side effects.

In an attempt to discover new drug targets for cancer, we have used a combination of methods, including identification of genetic biomarkers as differentially expressed genes by the RNA-PCR method, subtractive hybridization, microarray, and functional significance assay technologies such as D-RNAi *(36)*. D-RNAi is a novel posttranscriptional procedure for silencing gene expression by transfection of mRNA–aDNA (antisense DNA) hybrids. This phenomenon has been observed in the effects of bcl-2 on phorbol ester-induced apoptosis in human prostate cancer LNCaP cells, in chicken embryos, and in the human CD4$^+$ cell line H9. In the former, the in vivo transduction of β-catenin D-RNAi was shown to knock out more than 99% of endogenous β-catenin gene expression. In the latter, a human immunodeficiency virus (HIV)-1 D-RNAi homolog suppressed viral gene replication completely. This approach provides long-term gene knockout effects that may lead to the selection of differentially expressed genes as potential targets in drug development.

4. Clinical Therapy

The methods of generation of cDNA libraries and other techniques described in this volume will also provide crucial information for developing future causal therapeutic intervention as a strategic program designed to investigate gene function at the molecular, cellular, and organism levels. In this way, methods used to decipher the function of single genes can be applied to entire genomes. Thus, understanding gene function provides insight into disease pathogenesis. This information can be linked to understanding the function of human genes by exploiting the respective features of model systems. Frequently, the gene alterations identified in the tumor cell at specific sites, at specific stages, and in terms of specific gene perturbation is most effective.

Exploring gene functions and/or gene alterations for the development of causal therapies (i.e., gene therapies) is not new. The global approach of gathering information on targeted genes in various stages of a disease certainly is novel and will pave the way for more therapeutic applications. The ideal original gene therapy is a therapeutic intervention introducing a gene (transgene) that is efficiently delivered, durably transferred, and stably expressed to correct a defect underlying an inherited disease. For example, cystic fibrosis is the result of a genetic defect in the cystic fibrosis transmembrane conductance regulator (CFTR). The protein of this gene is a chloride transporter, regulating transmembrane voltage, notably in the airway epithelium. Subsequent to adenovirus-mediated gene therapy for this disorder, adenoviral transfection can be assessed by quantitation of Ad-CFTR vector plaque-forming units. No adverse effects resulting from the adenovirus were observed, CFTR RNA was expressed in vivo, and transepithelial voltage normalization was also noted *(37)*. However, the present technology does not permit sustained expression of the CFTR gene for the long-term treatment of cystic fibrosis.

A second example is the introduction of VEGF121 cDNA to individuals with clinically significant severe coronary artery disease by direct myocardial injection into the area of reversible ischemia for the treatment of vascular insufficiency *(38)*. Other areas of gene therapy are neurological diseases associated with motoneuron cell death *(39)*, autoimmune disorders such as rheumatoid arthritis *(40)*, and severe combined immune deficiency (SCID) *(41)*. However, the chief challenge facing clinical gene therapy strategies is the lack of efficient gene delivery systems.

In another approach, known as ex vivo gene therapy, cells, including peripheral blood lymphocytes *(42)*, skin fibroblasts *(43)*, and stem cells *(42)*, collected and genetically modified ex vivo, are used. Using these approaches, genes capable of controlling cell proliferation, the pathway for transmission of positive and negative signals from cell surface to nucleus, angiogenesis, apoptosis, DNA repair, mutation, and differential expression, certainly offer potential targets. In addition to delivering "corrective" nucleic acid sequences into the target cells to compensate for the hereditary defect in a specific gene, the generation of cDNA libraries and other methods described in this volume may provide insight into disease pathogenesis in terms of functional genomics. In this way, better understanding of disease mechanisms at the molecular level may lead to potential targets that might indicate novel entry points for pharmaceuticals and consequent clinical application. One such approach is gene-driven knockout, which generally requires prior knowledge of genomic information such as overexpressed genes. To test the novel predicted overexpressed genes by gene knockout, an RNA-mediated interference (RNAi) technology has been recently reported.

RNA-mediated interference results in silencing of specific gene expression in the presence of double-stranded RNA (dsRNA) homologous to the silenced gene *(44)*. The dsRNAs are thought to be associated with an RNA endonuclease and produce RNA fragments of 21–23 nucleotides. Once recognized, the mRNAs are cleaved at sites 21–23 nucleotides apart *(45)*, therefore silencing the gene activity. This is known as posttranscriptional gene silencing (PTGS). The introduction of a transgene or dsRNA probably evokes an intracellular sequence-specific RNA degradation that affects all highly homologous transcripts. This method has practical implications for functional genomics and provides a rapid method to test the function of genes *(46)*.

Recently, an improved RNAi phenomenon known as D-RNAi was developed in human prostate cancer cells. This method silences gene expression via relatively long-term interference with specific gene expression *(36)*. This method introduces an mRNA–cDNA hybrid into human cells. To the best of our knowledge, D-RNAi is a long-term, sequence-specific cosuppression of homologous transcripts in human cells. Although this type of gene therapy has been explored previously, the idea of using D-RNAi for several upregulated genes associated with apoptosis in human prostate cancer cells is a new one. Incidentally, duplexes of 21-nucleotide RNAs produce RNA intereference in cultured mammalian cells, but nucleotides over 30 bp induce nonspecific RNA degradation *(10)*. However, the D-RNAi method uses a hybrid of about 500 bp, with the RNA interference lasting for a longer time.

The propects of gene therapies have increased tremendously as a result of recent advances in our understanding of gene alteration at molecular, cellular, and organism levels. However, outstanding issues remain to be addressed before an effective, durable, and stable delivery system is possible. In vivo cellular and body responses to the treatment can reduce the therapeutic benefits, resulting in variance in the pharmacological actions among various patients. It is hoped that a simple, effective method of introducing genes for expression or overexpression to compensate for the activity of the defective endogenous genes will be developed. Although much work needs to be done to assess the safety, risks, and efficiency of RNAi and/or D-RNAi as well as alternative approaches, the excitement and prospects for a better gene therapy based on novel understanding of differentially expressed genes have grown substantially and will be increased exponentially among scientists and clinicians in the future. Eventually, the promise of better treatments of various disease by gene manipulation is realistically within reach, as we enter the era of gene medicine.

5. Future Challenge

The generation of cDNA/mRNA and high-throughput technologies has revolutionized expression of genes that may be a seminal development for simulating and predicting biomarkers for diagnostic utilization, drug development, and clinical therapy. However, current limitations include the following difficulties.

First, there is the issue of standardizing procedures to provide reliable and reproducible results and to assure that comparable data are generated from different laboratories. Another hurdle is the lack of a validated, robust, automatic preparation of single cells and the generation of full-length mRNA/cDNAs libraries containing both rare and abundant genes. Using incomplete libraries as the starting material results in incomplete gene analysis. Third, improved software is needed for data analysis to allow a healthy assimilation and interpretation of information flowing from various high-throughput programs. Fourth, for the differentially expressed genes, there is a risk that transcription levels of the identified genes do not necessarily correlate with those of proteins, particularly for rarely expressed genes. Thus, studies of peptide libraries, gene–protein interactions, and protein–protein interactions are urgently needed. Some problems in the original model of gene therapy are stilled unsolved, including efficient delivery, durable transfer, and stable expression of transgenes to correct a gene defect underlying an inherited disease. Fifth, the novel gene knockout methods for functional significance studies such as D-RNAi raises our hopes for adopting personalized drug development; however, there is a void of technology for efficient gene overexpression to allow the functional significance of identified genes to be determined. Finally, there is a need in gene therapy for an efficient delivery system, particularly in vivo, to generate a durable, stable expression or knockout to correct the differentially expressed genes underlying an inherited disease. Therefore, the progress of gene medicine from laboratory results to diagnosis, drug development, and clinical therapy will depend on how effectively we can provide the answers to these questions.

In sum, limited effectiveness, insufficient gene transduction, nonviral gene transfer, nuclear import, DNA condensing agents, other vaccines, and antiangiogenesis will be future challenges for clinical gene therapy. In addition, resistance of living cells to invasion by foreign materials and interference with cellular function need special attention. Novel approaches for delivery systems including D-RNAi, synthetic peptide-containing nuclear localization signals bound to DNA *(47)*, and inhibitors of telomerase *(48)* will provide the future perspectives in gene therapy.

References

1. Lakhani, S. R. and Ashworth, A. (2001) Microarray and histopathological analysis of tumours: the future and the past? *Nat. Rev. Cancer* **1,** 151–157.
2. Nuttall, M. E. (2001) Drug discovery and target validation. *Cells Tissues Organs* **169,** 265–271.
3. Truong, A. H. and Ben-David, Y. (2000) The role of Fli-1 in normal cell function and malignant transformation. *Oncogene* **19,** 6482–6489.
4. Lin, S. L., Chuong, C. M., Widelitz, R. B., and Ying, S. Y. (1999) In vivo analysis of cancerous gene expression by RNA-polymerase chain reaction. *Nucleic Acids Res.* **27,** 4585–4589.
5. Becker, I., Becker, K. F., Rohrl, M. H., and Hofler, H. (1997) Leser-assisted preparation of single cells from stained histological slides fro gene analysis. *Histochem. Cell. Biol.* **108,** 447–451.
6. Rubin, M. A. (2001) Use of laser capture microdissection, cDNA microarrays, and tissue microarrys in advancing our understanding of prostae cancer. *J. Pathol.* **195,** 80–86.
7. Embleton, M. J., Gorochov, G., Jones, P. T., and Winter, G. (1992) In-cell PCR from mRNA: amplifying and linking the rearranged immunoglobulin heavy and light chain V-genes within single cells. *Nucleic Acids Res.* **20,** 3831–3837.
8. Ying, S. Y. and Lin, S. L. (1999) High performance subtractive hybridization of cDNAs by covalent boding between specific complementary nucleotides. *BioTechniques* **26,** 966–979.
9. Lin, S. L., Chuong, C. M., and Ying, S. Y. (2001) A novel mRNA–cDNA interference phenomenon for silencing bcl-2 expression in human prostate cancer LNCaP cells. *Biochem. Biophy. Res. Commun.* **281,** 639–644.

10. Elbashir, S. M., Harbath, J., Lendeckel, W., Yalcin, A., Weber, K., and Tuschl, T. (2001) Duplexes of 21-nucleotide RNAs mdiate RNA interference in cultured mammalian cells. *Nature* **41,** 494–498.

11. Prasad, K. V., Ao, Z., Yoon, Y., Wu, M. X., Rizk, M., Jacquot, S., et al. (1997) Protein CD27, a member of the tumor necrosis factor receptor family, induces apoptosis and binds to Siva, a proapoptotic protein. *Proc. Natl. Acad. Sci. USA* **94,** 6346–6351.

12. Lin, S. L. and Ying, S. Y. (1999) Differentially expressed genes in activin-induced apoptotic LNCaP cells. *Biochem. Biophys. Res. Commun.* **257,** 187–192.

13. Henke, A., Launhardt, H., Klement, K., Stelzner, A., Zell, R., and Munder, T. (2000) Apoptosis in coxsackievirus B3-caused diseases: interaction between the capsid protein VP2 and the proapoptotic protein siva. *J. Virol.* **74,** 4284–4290.

14. O'Brian, C. A. (1998) Protein kinase C-alpha: a novel target for the therapy of androgen-independent prostate cancer? [Review–hypothesis]. *Oncol. Rep.* **5,** 305–309.

15. Johnson, M. I. and Hamdy, F. C. (1998) Apoptosis regulating genes in prostate cancer. *Oncol. Rep.* **5,** 553–557.

16. Moul, J. W. (1999) Angiogenesis, p53, bcl-2 and Ki-67 in the progression of prostate cancer after radical prostatectomy. *Eur. Urol.* **35,** 399–407.

17. Howell, S. B. (2000) Resistance to apoptosis in prostate cancer cells. *Mol. Urol.* **4,** 225–229.

18. Kyprianou, N., Chon, J., and Benning, C. M. (2000) Effects of alpha(1)-adrenoceptor (alpha(1)-AR) antagonists on cell proliferation and apoptosis in the prostate: therapeutic implications in prostatic disease. *Prostate* **9(Suppl.),** 42–46.

19. Fosslien, E. (2000) Biochemistry of cyclooxygenase (COX)-2 inhibitors and molecular pathology of COX-2 in neoplasia. *Crit. Rev. Clin. Lab. Sci.* **37,** 431–502.

20. Lee, C., Janulis, L., Ilio, K., Shah, A., Park, I., Kim, S., et al. (2000) In vitro models of prostate apoptosis: clusterin as an antiapoptotic mediator. *Prostate* **9(Suppl.),** 21–24.

21. Messina, M. and Bennink, M. (1998) Soyfoods, isoflavones and risk of colonic cancer: a review of the in vitro and in vivo data. *Baillieres Clin. Endocrinol. Metab.* **12,** 707–728.

22. Weidner, N., Semple, J. P., Welch, W. R., and Folkman, J. (1991) Tumor angiogenesis and metastasis correction in invasive breast carcinoma. *N. Engl. J. Med.* **324,** 1–8.

23. Verheul, H. M. and Pinedo, H. M. (2000) The role of vascular endothelial growth factor (VEGF) in tumor angiogenesis and early clinical development of VEGF-receptor kinase inhibitors. *Clin. Breast Cancer* **1(Suppl.),** S80–S84.

24. Peters, G. J., De Bruin, M., Fukushima, M., Van Triest, B., Hoekman, K., Pinedo, H. M., et al. (2000) Thymidine phosphorylase in angiogenesis and drug resistance. Homology with platelet-derived endothelial cell growth factor. *Adv. Exp. Med. Biol.* **486,** 291–294.

25. Cross, M. J. and Claesson-Welsh, L. (2001) FGF and VEGF function in angiogenesis: signalling pathways, biological responses and therapeutic inhibition. *Trends Pharmacol. Sci.* **22,** 201–207.

26. Hiscox, S. and Jiang, W. G. (1997) Interleukin-12, an emerging anti-tumour cytokine. *In Vivo* **11,** 125–132.

27. de Fraipont, F., Nicholson, A. C., Feige, J. J., and Van Meir, E. G. (2001) Thrombospondins and tumor angiogenesis. *Trends Mol. Med.* **7,** 401–407.

28. Lo, J. C. and Grumbach, M. M. (2001) Pregnancy outcomes in women with congenital virilizing adrenal hyperplasia. *Endocrinol. Metab. Clin. North. Am.* **30,** 207–229.

29. Smith, B. J. and O'Brien, J. M. (1996) The genetics of retinoblastoma and current diagnostic testing. *J. Pediatr. Ophthalmol. Strabismus* **33,** 120–123.

30. Rowley, P. T., Loader, S., and Levenkron, J. C. (1997) Cystic fibrosis carrier population screening: a review. *Genet. Test.* **1,** 53–59.

31. Verlinsky, Y., Rechitsky, S., Verlinsky, O., Ivachnenko, V., Lifchez, A., Kaplan, B., et al. (1999) Prepregnancy testing for single-gene disorders by polar analysis. *Genet. Test.* **3,** 185–190.

32. Kaplan, F. (1998) Tay–Sachs disease carrier screening: a model for prevention of genetic disease. *Genet. Test.* **2,** 271–292.

33. Kimura, W., Zhao, B., Futakawa, N., Muto, T., and Makuuchi, M. (1999) Significance of K-ras codon 12 point mutation in pancreatic juice in the diagnosis of carcinoma of the pancreas. *Hepatogastroenterology* **46,** 532–539.

34. Debouck, C. and Goodfellow, P. N. (1999) DNA microarrays in drug discovery and development. *Nat. Genet.* **21(Suppl.),** 33–37.

35. Marton, M., DeRisi, J., Bennett, H., Iyer, V., Stoughton, R., Burckard, J., et al. (1998) Drug target validation and identification of secondary drug target effects using DNA microarray. *Nat. Med.* **4,** 1293–1301.

36. Lin, S. L , Suksaweang, S., Chuong, C. M., and Ying, S. Y. (2001) D-RNAi (messenger RNA–antisense DNA interference) as a novel defense system against cancer and viral infections. *Current Cancer Drug Targets* **1,** 241–247.

37. Perricone, M. A., Morris, J. E., Pavelka, K., Plog, M. S., O'Sullivan, B. P., Joseph, P. M., et al. (2001) Aerosol and lobar administration of a recombinant adenovirus to individuals with cystic fibrosis. II. Transfection efficiency in airway epithelium. *Hum. Gene Ther.* **12,** 1383–1394.

38. Rosengart, T. K., Lee, L. Y., Patel, S. R., Kligfield, P. D., Okin, P. M., Hackett, N. R., et al. (1999) Six-month assessment of a phase I trial of angiogenic gene therapy for the treatment of coronary artery disease using direct intramyocardial administration of an adenovirus vector expressing the VEGF121 cDNA. *Ann. Surg.* **230,** 466–470.

39. Keir, S. D., Xiao, X., Li, J., and Kennedy, P. G. (2001) Adeno-associated virus-mediated delivery of glial cell line-derived neurotrophic factor protects motor neuron-like cells from apoptosis. *J. Neurovirol.* **7,** 437–446.

40. Gouze, J. N., Ghivizzani, S. C., Gouze, E., Palmer, G. D., Betz, O. B., Robbins, P. D., et al. (2001) Gene therapy for rheumatoid arthritis. *Hand Surg.* **6,** 211–219.

41. Rosen, F. S. (2002) Successful gene therapy for severe combined immuno-deficiency. *N. Engl. J. Med.* **346,** 1241–1243.

42. Ng, Y. Y, Bloem, A. C., van Kessel, B., Lokhorst, H., Logtenberg, T., and Staal, F. J. (2002) Selective in vitro expansion and efficient retroviral transduction of human CD34+ CD38- haematopoietic stem cells. *Br. J. Haematol.* **117,** 226–237.

43. Veelken, H., Jesuiter, H., Mackensen, A., Kulmburg, P., Schultze, J., Rosenthal, F., et al. (1994) Primary fibroblasts from human adults as target cells for ex vivo transfection and gene therapy. *Hum. Gene Ther.* **5,** 1203–1210.

44. Fire, A., Xu, S., Montgomery, M. K., Kostas, S. A., Driver, S. E., and Mello, C. C. (1998) Potent and specific genetic interference by double-stranded RNA in *Caenorhabditis elegans. Nature* **391,** 806–811.

45. Carthew, R. W. (2001) Gene silencing by double-stranded RNA. *Curr. Opin. Cell. Biol.* **13,** 244–248.

46. Barstead, R. (2001) Genome-wide RNAi. *Curr. Opin. Chem. Biol.* **5,** 63–66.

47. Aronsohn, A. I. and Hughes, J. A. (1998) Nuclear localization signal peptides enhance cationic liposome-mediated gene therapy. *J. Drug Target.* **5,** 163–169.

48. Hamilton, S. E., Simmons, C. G., Kathiriya, I. S., and Corey, D.R. (1999) Cellular delivery of peptide nucleic acids and inhibition of human telomerase. *Chem. Biol.* **6,** 343–351.

Index

A

Adaptor, 13,
    Capfinder, 14, 19
    *Eco*RI-*Not*I-*Bam*HI, 35, 39,
    ligation, 41, 62, 63, 110, 117, 175,
        186, 191, 276
    phosphorylation, 276
    pseudo-double-stranded, 47
Affymetrix U95A2 gene chip, 133
AFLP analysis, 25
Agarose gel electrophoresis, 155–158
AGPC method, 88
Alkali denaturation, 56
Amplification,
aRNA, 94, 95, 119, 133,
    cDNA, *see* Library
    GLGI, 215, 216,
    RNA, 130, 291, 313
AMV, *see* Reverse transcriptase
Annealing temperature, 20
Antisense RNA, 93, 98, 99, 117, 124
aRNA, *see* Antisense RNA

B

Bacterial alkaline phosphatase, 76
BacterioMatch two-hybrid system,
    295, 303,
    bait, 295, 298, 302
    target, 295, 298, 303
BAP, *see* Bacterial alkaline
    phosphatase
Biotin-streptavidin interaction, 14, 241
BLAST, 84, 86, 151, 153, 204,
    219, 250

C

CapFinding, 13
cDNA library, 1, 5–10, 33,
    5'-end enriched, 76
    full-length, 7, 8, 23, 74, 87, 88, 95,
        117, 130, 223–230
    lambda vector library, 277–282, 285
    single-cell, 117
    size fractionated, 60, 63
    solid phase, 25
cDNA synthesis,
    classical, 104, 107, 109
    double-stranded, 62, 99, 123, 199,
        248, 249
    first strand, 28, 54, 81, 97, 109, 122,
        137, 212
    full-length, 23, 95
    second-strand, 29, 55, 97–99, 109,
        110, 124, 212, 213
    SMART, 105–107, 190
CDS, *see* Continuous protein coding
    region
Cloning,
    ligation-assisted, 60, 65, 67
    recombination, 60, 64, 66
CMV, 279
Competent cells, 198
Continuous protein coding region, 73
Cre recombinase, 191
cRNA, 164, 165
Crosslinking, 175
Cytomegalovirus, *see* CMV

D

Database,

EST, 204, 219
Eukaryotic Promoter Database, 87
GenBank, 204
Human Genome Sequence, 204
NR, 219
RefSeq, 86, 204
UniGene, 219
DEPC, 27, 34, 88, 95, 138, 173, 199,
    246, 254, 290
Diethyl pyrocarbonate, *see* DEPC
D-looped plasmid, 224
DNA,
    artificial, 70, 262
    ligase,
        E. coli ligase, 199
        T4, 64, 106
    polymerase,
        E. coli DNA polymerase I, 95
        Klenow fragment, 18, 98, 175
        Pfu, 186, 274
        T4, 26, 53, 55, 96, 106
        Taq, 19, 27, 53, 209, 246,
            249, 261
DNase I, 35, 81
D-RNAi, *see* Gene knockout
Dynabeads, *see* Paramagnetic beads

E

E. coli,
    BM25, 180, 191, 192
    DH5α, 262, 266
DH10B, 61,
    SOLR, 180, 188, 189
    TOP10, 85
    XL1-Blue, 180, 188, 189, 192,
        295, 302
    XLOLR, 283
Enrichment probes, 288
ESTs, 33, 104, 197
Ethidium bromide, 53, 156
Expressed sequence tags, *see* ESTs

F

Fluorescent interacting dye, 156

G

Gel electrophoresis, 155, *see also*
        Agrose gel electrophoresis
GenBank, *see* Database
Genchip, 93; *see also* Microarray
Gene diagnostic test, 318–320
Gene knockout, 315,
    RNAi, 315, 324
    D-RNAi, 315, 322–324
Gene profiling, 103
Genetic epidemiological etiology, 316
Genetic polymorphisms, 318
GLGI, 207, 209, 215, 220; *see also*
        Amplification

H

HAP absorption, 202, 204
Hexaminecobalt chloride, 54
HPLC, 37
Hybridization, 2, 37, 38, 170, 176, 228,
        239, 240, 249,
    in situ, 170,
    subtractive, *see* Subtractive
        hybridization

I

In vivo excision, 188, 191, 283, 284
In vitro transcription, 97, 121, 137,
        138, 142, 258
In vitro translation, 291, 292
IPCR, *see* PCR, inverse
IPTG, 136, 192, 193, 203, 272, 301
Isopropyl-thio-β-galactoside, *see*
        IPTG
IVT, *see* In vitro transcription

L

Labeling, *see* Probe
Laser capture microdissection, 135, 313, 320, 321
LCM, *see* Laser capture microdissection
Library,
    cDNA, *see also* cDNA library,
        amplification, 16, 17, 41, 81–83, 122, 123
        packaging, 188, 191, 280, efficiency, 286
        screening, 230, 284, 285, 306, blue/white, 148
            functional assay, 285
            PCR, 148, 149
            panning assay, 284
        titering, 146, 188, 191, 282,
    mRNA, 155, 169, 312, 314
    peptide, 289
Ligation, 186, 191, 214, *see also* Adaptor
Linker, 13, 208, 214, 255

M

MCS, *see* Multiple cloning site
Microarray, 93, 245, 311, 314, 316
MMLV, *see* Reverse transcriptase
mRNA, 74,
    isolation, 28, 34–36, 79, 104, 108, 181, 182, 211
    oligo-capping, 74–76
mRNA-cDNA interference, *see* D-RNAi
Multiple cloning site, 182, 298

N

National Cancer Institute, 133
NCBI, 150
Normalization, 33, 38, 39, 197, 202, 203, efficiency, 39

Northern blot, 130, 169, 248, 312, 313
*Not*I, *see* Restriction sites

O

Oligo (dT), 25, 36, 74, 76, 97, 122, 180, 255
Oligonucleotide, 23, 25, 42, 208; *see also* Primer

P

Packaging, *see* Library
Paramagnetic beads, 25, 27, 28
pBluescript, 182, 188, 193, 266, 277
pBT, 296, 302, 303
PCR, 6, 13, 26, 41, 76, 261, 313, 319,
    anchor, 51, 267
    cloning, 266
    inverse, 51–57
    long-distance, 149
    nested, 42
    RNA, 10, 130–140, 312, 320
    step-out, 42
PCR-suppression effect, 42
pCR4-TOPO, 209, 211, 217
pGEM-T, 46
pGEM5ZF, 198, 201
Phagemid, 182, 188, 189, 277, 285
Phage titering, 146
pME18s-FL3, 84, 90
Polymerase chain reaction, *see* PCR
Primer, 46, 74, 87, 93, 136, 204, 256, 267,
    adaptor-specific, 47
    annealing, 176, 256, 257
    colony PCR, 78
    extension, 87
    fragment-specific, 45
    gene-specific, 14, 17, 42, 47, 51
    random, 7, 96, 99
    SAGE, 208
    sequencing, 78

Probe,
    biotinylated, 224, 226, 227
    denaturation, 177
    double-stranded DNA, 283
    labeling, 248
    radioactive, 175, 176
Pseudo-double-stranded adaptor, 47
pSP, 66
pTRG, 298, 303, 307, 308

R

RACE, 13–18, 48,
    first stage, 45, 46
    second stage, 46, 47
    3'-RACE, 14, 43
    5'-RACE, 13, 19, 20, 25, 43, 87
RecA, 223, 224, 227, 228, 235
Reporter gene, *see also* Two hybrid
        system,
    β-galactosidase (lacZ), 295, 300, 307
    β-lactamase (Ampr), 295, 301, 307
Representational difference analysis,
        243
Restriction sites,
    *Eco*RI, 18, 39, 182, 266, 272, 277
    *Kpn*I, 38
    *Nla*III, 199, 209, 210
    *Not*I, 29, 201, 202
    *Sal*I, 29
    Sfi, 190
    *Sph*I, 201
    *Xho*I, 38, 274, 279
Reverse transcriptase,
    AMV, 68, 96, 120, 121, 135, 137, 246
    MMLV, 13, 53, 95, 105, 212
    SuperScript, 63, 77, 107
RNA,
    degradation, 112, 113
    ligase, 55, 76
    poly(A), 16, 36, 37, 62, 74
    polymerase,
        T3, 39, 161

    T4, 117
    T7, 23, 96, 121, 132, 137, 280
RNAi, *see* Gene knockout
RNA-PCR, *see* PCR
RNase H, 54, 84
RNasein, 34, 122

S

SAGE, 207, 208, 245,
    tags, 207, 220
Screening, *see* Library
Sense RNA, 139
Sequence analysis, *see* BLAST
Sequence replacement reaction, 93, 97
Sequencing, 218, 219,
    large-scale, 87, 88
    single-passing, 33
Serial analysis of gene expression,
        see SAGE
Shine–Dalagarno translational signal, 301
Size fractionation, 64, 67, 83, 84, 186,
        190, 277
SMART, *see* cDNA synthesis
Southern blot, 21
SPGI, 197,
Subtractive hybridization, 239–241,
        243, 244, 257, 314,
    driver, 239, 243, 249
    tester, 239, 240, 243
SV40, 279, 285
SYBR Green, 61, 65, 66

T

T7 bacteriophage RNA promoter
        element, 97, 122
Tailing, 29, 118, 120, 122, 123, 141, 257
TAP, *see* Tobacco acid
        pyrophosphatase
Taq polymerase, *see* DNA, polymerase
Template-switching, 41, 95, 110
Terminal transferase, 117, 118, 123

Titering, *see* Library
Tobacco acid pyrophosphatase, 76, 77
Transcription start sites, 88
Transformation, 69, 805, 304
Trizol, 34, 104, 162, 165, 181,
 198, 208
Two-hybrid system, *see* BacterioMatch
 two-hybrid system

U

Udist primer, 46,
U-DNA, *see* Uracil-DNA
UNG, *see* Uracil-DNA, glycosylase
UniGene, *see* Database
Uracil-DNA, 241, 249,
 glycosylase, 248, 249
 subtraction assay, 242, 250
USA, *see* Uracil-DNA

V

VecScan, 150

Vector, 198, 121, 271; *see also* cDNA
 Library,
 cloning, 117
 *Dra*III, 85
 *Eco*RI, 18
 *Hin*dIII, 18
 λgt10, 39, 148
 λgt11, 148
 λZAPII, 39
 TripEx2 system, 182
 Uni-ZAP XR system, 182, 188, 193,
 271, 279

W

Western blot, 193, 312, 314

X

X-Gal, 192, 203, 272, 296

Y

Yeast-two-hybrid system, 295